Electron Cyclotron Emission and Electron Cyclotron Resonance Heating (EC-16)

Proceedings of the 16th Joint Workshop

Electron Cyclotron Emission and Electron Cyclotron Resonance Heating (EC-16)

Proceedings of the 16th Joint Workshop

Sanya, China, 12 – 15 April 2010

editor

Ronald Prater

General Atomics, USA

NEW JERSEY • LONDON • SINGAPORE • BEIJING • SHANGHAI • HONG KONG • TAIPEI • CHENNAI

Published by

World Scientific Publishing Co. Pte. Ltd.

5 Toh Tuck Link, Singapore 596224

USA office: 27 Warren Street, Suite 401-402, Hackensack, NJ 07601

UK office: 57 Shelton Street, Covent Garden, London WC2H 9HE

British Library Cataloguing-in-Publication Data
A catalogue record for this book is available from the British Library.

ELECTRON CYCLOTRON EMISSION AND ELECTRON CYCLOTRON RESONANCE HEATING (EC-16)
Proceedings of the 16th Joint Workshop

ISBN-13 978-981-4340-26-7
ISBN-10 981-4340-26-X

Printed in Singapore by World Scientific Printers.

PREFACE

The Sixteenth Joint Workshop on Electron Cyclotron Emission and Electron Cyclotron Resonance Heating was held in Sanya, China, on April 12-15, 2010. The workshop was hosted by the Institute of Plasma Physics Chinese Academy of Sciences (ASIPP).

The members of the Local Organizing Committee were:

Jiangang Li (chair)
Baonian Wan
Shaohua Dong
Yuan Wang.

The International Program Committee was responsible for selecting the invited and overview papers and for organizing the meeting sessions. Its members comprised:

Ron Prater, General Atomics, USA, chair
Young Soon Bae, KSTAR, Korea
Daniela Farina, IFP, Italy
Bob Harvey, CompX, USA
Shin Kubo, NIFS, Japan
Mark Henderson, ITER, France
Y.-R. Lin-Liu, National Central University, Taipei
John Lohr, General Atomics, USA
Gary Taylor, PPPL, USA
George Vayakis, ITER, France

The topics of the meeting were organized in four broad areas: Electron cyclotron theory, electron cyclotron emission, electron cyclotron heating and current drive, and electron cyclotron technology. The presentations in these areas were summarized in the closing session by G. Giruzzi, G. Taylor, R. Prater, and W. Kasparek, respectively, and their written summaries are printed in this volume.

The EC-16 Workshop was attended by 60 participants from 12 countries. There were 11 invited talks, 28 oral presentations, and 34 posters. The success of this Workshop is due to the diligence of the participants in preparing presentations and papers of high quality and interest. The discussions following each paper and during the poster sessions and coffee breaks were lively and insightful and frequently entertaining. The generosity of the host institution ASIPP and its support for this meeting, the hard work of the members of the Local Organizing Committee and the International Program Committee, and the contributions of the session chairs and the summary speakers are warmly acknowledged.

The papers in this volume have been reproduced from the authors' pdf files as supplied, without changes. The figures in some papers are better viewed in color, so an accompanying CD, in which the use of color is supported, is attached. The files are also available on the web at http://EC16.ipp.ac.cn.

The 16 biannual EC workshops have been held over 31 years in many of the countries that carry out research on or with millimeter waves oriented toward magnetic fusion. This list shows the location and organizers of the workshops:

EC-1	College Park, Maryland, USA, 1979 (D.A. Boyd)
EC-2	Oxford, UK, 1980 (A.E. Costley and A.C. Riviere)
EC-3	Madison, Wisconsin, USA, 1982 (D.A. Boyd)
EC-4	Frascati, Italy, 1984 (P. Buratti and M. Bornatici)
EC-5	San Diego, California, USA (R. Prater and J. Lohr)
EC-6	Oxford, UK, 1987 (A.E. Costley and A.C. Riviere)
EC-7	Hefei, Anhui, China, 1992 (Wan Yan Xi and A.E. Costley)
EC-8	Gut Ising, Germany, 1992 (H.J. Hartfuss and V. Erckmann)
EC-9	Borrego Springs, California, USA, 1995 (J. Lohr and T. Luce)
EC-10	Ameland, The Netherlands, 1997 (T. Donne and T. Verhoeven)
EC-11	Oh-arai, Japan, 1999 (T. Imai and K. Sakamoto)
EC-12	Aix-en-Provence, France, 2002 (G. Giruzzi)
EC-13	Nizhny Novgorod, Russia, 2004 (A. Litvak)
EC-14	Santorini Island, Greece, 2006 (A. Lazaros)
EC-15	Yosemite National Park, California, USA, 2008 (J. Lohr)
EC-16	Sanya, China (R. Prater and J. Li)

The next workshop in the series, EC-17, will be hosted by Eindhoven Technical University in the Netherlands in 2012. The Local Organizing

Committee will be chaired by P. Nuij and the International Program Committee will be chaired by E. Westerhof.

I would personally like to thank Prof. Li and his extremely able staff at ASIPP for their hard work in creating such a productive environment for the workshop. I would also like to give warm thanks to the members of the International Program Committee for their success in producing a program of balance and interest. Most of all I would like to thank the workshop participants for their outstanding presentations and collegial discussions.

Ron Prater
San Diego, August 2010

CONTENTS

III. Electron Bernstein Waves

IV. Electron Cyclotron Theory

V. Gyrotrons and EC Technology

I. Electron Cyclotron Heating

SUMMARY OF PAPERS ON ECH AND ECCD

R. PRATER

General Atomics, San Diego, CA 92138 USA

The presentations focusing on the physics and applications of heating and current drive in the electron cyclotron range of frequencies are summarized.

Electron Cyclotron Heating and Current Drive

An overview of the objectives of ECH and ECCD in toroidal high performance confinement systems for steady-state power generation was presented by T. Luce. This paper stressed the sound theoretical basis for understanding and predicting the effects of electron cyclotron waves and their flexibility in supporting steady-state operation in both tokamaks and stellarators. Three primary roles are envisioned for EC effects in steady-state systems: access to the burning plasma state, support of the toroidal current in tokamaks, and control of the operating point. Access to the operating conditions includes preionization and burn-through and plasma heating to fusion temperatures, a role that ECH is expected to largely fill on ITER. For noninductive current support, all auxiliary current drive systems are too inefficient energetically to support all the plasma current of a tokamak reactor, but with large bootstrap current fractions this can be done by ECCD while preserving adequate energy multiplication. ECCD, because of its localizability, can be used to optimize the current profile to provide improved confinement and stability. Active control of the reactor operating point may be necessary if small perturbations tend to grow. Also, ECCD has been demonstrated to be effective at suppressing neoclassical tearing modes and to a lesser degree sawteeth instabilities, and this is important for maintaining high performance operation.

Preionization and startup

The first stage of operation of any tokamak is the preionization and startup phase. Until recently, preionization has usually not required ECH, but with the advent of superconducting tokamaks, ECH startup assist has proven necessary. Fully superconducting tokamaks have severe limits on the loop voltage, but it was reported at EC16 that second harmonic ECH was successful in KSTAR and EAST tokamaks in improving the startup capabilities, following up on previous work on the copper machines DIII-D and JT-60U and others. This is a positive results for ITER, should it have initial operations at half its normal toroidal field so that the 170 GHz ECH system would interact at the second harmonic.

Work on KSTAR at 110 GHz addressed the pre-fill gas pressure, the mode—X-mode or O-mode—of the injected power, and the vertical magnetic field. It was found, as expected, that the X-mode power was more effective at startup than O-mode power. Breakdown at the 3rd harmonic of the X-mode was unsuccessful. Injecting the EC beam at 10 deg counter to the plasma current direction was most beneficial to the startup, but 20 deg injection was also successful. Work on ASDEX Upgrade showed that fundamental O-mode was more effective than second harmonic X-mode, but not greatly so, suggesting that a dedicated fundamental breakdown system was not necessary for ITER.

Electron Cyclotron Heating

There was little experimental work reported at EC16 on the physics of ECH or ECCD. This may be attributed to the mature state of understanding of the physics and modeling capabilities for ECH. A point of contention remains the role of the upper hybrid layer in cases where the ray tracing models neglect any interaction there. Work with a full wave code suggests that under some circumstances this interaction can be unexpectedly important, and experiments are needed to resolve the physics.

The effect of ECH on transport can be important, however. Work on DIII-D showed examples of discharges in which the 'density pumpout' effect caused a strong decrease in global confinement, while in other discharges the confinement remained constant when the ECH was added. In ASDEX Upgrade, which has tungsten walls, it was reported that central ECH is needed in nearly all cases to reduce the neoclassical accumulation of metallic impurities to manageable levels, and that the ECH power had to be deposited within the central 20% of the minor radius. This effect was attributed to an outward pinch in low density discharges where trapped electron modes dominate transport. At higher densities where the ion temperature gradient modes dominate, the pumpout effect is not seen. Similar effects were seen with central ion cyclotron heating.

Some applications of ECH that were presented include a study of turbulent transport, in which the ECH power was applied at two nearby minor radii. By square-wave modulating the powers with the radial locations out of phase, the electron temperature gradient was modulated with constant total power. This modulated the nature of the turbulence for purposes of validating the turbulence models. Also, third harmonic X-mode was studied on ASDEX Upgrade to allow heating at lower toroidal field and second harmonic O-mode to provide central heating at densities well above the limit for X-mode. This work used wave reflectors on the centerpost that provided the optimum polarization of the reflected wave. Work on DIII-D showed that H-mode threshold power is lower for ECH than for neutral injection in deuterium discharges, but this effect is much smaller in hydrogen and helium plasmas. Experiments on ASDEX Upgrade, however, showed similar power levels required for the H-transition in ECH and NBI cases. An interesting observation from the HL-2A tokamak is that an internal transport barrier was found to form when the far off-axis ECH was switched *off*; no explanation for this effect was known. Interesting results from ASDEX Upgrade showed suppression of high beta disruptions when ECH was applied close to or outside the q=3/2 surface. Experiments with EC heating on T-10 using diagnostics with improved spatial and temporal resolution showed nonuniform rotation and fixed frequency ratios for different mode numbers. Very high highly peaked central electron temperatures, to 15 keV, have been obtained in low density discharges in the LHD device using 1 MW at 77 GHz.

Several initiatives for major new ECH installations were described at EC16. A major project for a 10 MW 170 GHz ECH system for JET was described in a series of papers. The main objective of this system is support of the JET operational scenarios, but the concept of such a system on the world's largest tokamak was viewed by many participants as a long-overdue demonstration of the readiness of ECH for use as a primary heating system on ITER. If funded, this system is designed to be operational in 2015. The twelve 1-MW gyrotrons would be obtained in partnership with the Russian Federation. Extensive modeling was presented, showing that the system would have sufficient power and flexibility to meet its objectives. Another major installation of EC power will take place on the EAST tokamak using 140 GHz 1 MW gyrotrons for a total power of 4 to 5 MW. Also, a 3 MW ECH system using six 0.5 MW gyrotrons at 68 GHz is being developed for HL-2A.

Electron Cyclotron Current Drive

A major advance in the technology of control of neoclassical tearing modes (NTMs) was made in the TEXTOR tokamak. In this work, the mode was detected using electron cyclotron emission that was collected along by the same antennas that launched the EC power in response. This is impressive because the forward EC power is of order 1 MW while the emitted EC radiation is less than a microwatt. Because the ECE collection followed the same path as the EC waves and used several frequencies surrounding the applied frequency, the location and phase of the instability relative to the local current drive location could be unambiguously determined. This information was used to control the

EC steering angles and phase to optimize the control efficiency. The NTM was simulated by an island created by fields from the dynamic ergodic divertor system. Fully automated control was obtained.

Tearing mode control was also demonstrated on other devices. On HL-2A, improvement in the confinement was obtained when heating near the q=2 surface was applied to cause the mode amplitude to decrease and disappear. Use of steerable mirrors to control the NTM was shown on ASDEX Upgrade and DIII-D. Heating was also used to control the sawtooth period in HL-2A, thereby allowing the density to increase.

ECCD experiments on the Heliotron J device showed that very energetic electrons heated by EC waves drive current consistent with the Fisch-Boozer effect. The dependence of the driven current with the magnetic well depth was as expected from theory.

Electron Bernstein Waves

Angular scanning of the EBW emission has been used in the MAST spherical tokamak to estimate the pitch angle of the magnetic field at the mode conversion location. The radial location of the mode conversion layer is found using data from Thomson scattering. EBW at a range of frequencies can determine the pitch angle profile, providing an alternative to the more usual determination using the motional Stark effect.

EBW heating at 28 GHz was reported on the WEGA stellarator. A very prompt transition into the O-X-B heating condition was obtained as soon as the density threshold was reached, in confirmation of the theory.

Current drive experiments were done using a new phased array antenna system on the QUEST low aspect ratio tokamak. O-mode power at the 30 kW power level generated a strong nonthermal electron distribution and the plasma current of 10 kA was fully supported by the RF for 0.8 sec. Systematic simulations of heating and current drive in spherical tokamaks more generally show that EBWs are a viable option for NSTX, NHTX, and MAST.

ROLE OF ECH AND ECCD IN HIGH-PERFORMANCE STEADY-STATE SCENARIOS

T.C. LUCE
General Atomics, PO Box 85608, San Diego, California 92186-5608, USA

The ability to control the location and localization of energy and current deposition in fusion plasmas with electron cyclotron waves is unmatched by any other auxiliary heating and current drive method. The establishment of the physics basis of the wave propagation and absorption allows accurate prediction of the deposition given the launcher optics and the profiles of magnetic field, electron density, electron temperature, and impurity density. This, along with the development of the technology of high-power millimeter-wave sources and the means to transmit them to the plasma, has facilitated experiments in present-day fusion devices to demonstrate the utility of heating and current drive with electron cyclotron waves (ECH and ECCD, respectively). Three primary roles are envisioned for ECH or ECCD in burning plasmas. First, all burning plasma scenarios require some access conditions, whether it is merely heating the plasma to fusion-relevant conditions or generating specific profiles of the magnetic field necessary to access steady-state operation in a tokamak. Second, tokamak-based steady-state scenarios require efficient auxiliary current drive to sustain the magnetic configuration. Finally, control of the operating point of a burning plasma is essential. This role includes applications such as response to off-normal events and active control of plasma instabilities. Each of these roles will be discussed using examples from present-day tokamak and stellarator experiments. Prospects for application in ITER and fusion power plants will also be presented.

1. Introduction

The use of fusion energy to satisfy the world's fundamental energy needs is an attractive long-term vision. Within the range of possible fusion energy systems, steady-state operation of magnetic confinement systems is an attractive option [1,2]. The schemes for realizing steady-state magnetic confinement fusion that are closest to achieving the conditions necessary for fusion energy production are the stellarator and the tokamak [3].

Both schemes start with a strong toroidal magnetic field. But such a field alone cannot confine plasma due to charge separation [1]. In both cases, a helical field is applied to prevent charge separation and provide plasma confinement. The two schemes apply the helical field in very different ways. In the stellarator, the magnetic configuration is generated dominantly (in many cases, solely) by external coils that are arranged in a three-dimensional configuration. With proper design, this results in closed flux surfaces and plasma confinement. As long as the coils are energized, the magnetic configuration can be sustained and steady-state operation is possible. In the tokamak, the helical field is generated by a toroidal current flowing in the plasma. Typically, this is achieved by means of induction, so the magnetic configuration cannot be maintained in steady-state. However, noninductive means of generating currents in tokamaks are well-established [4] and fully noninductive operation has been achieved on several tokamaks [1].

Three general issues arise for steady-state magnetic fusion energy systems — accessing the steady-state fusion energy regime, sustaining that regime, and control of the operating point in normal and off-normal conditions. The theme of this paper is that advances in the technology and physics of electron cyclotron heating (ECH) and current drive (ECCD) establish the basis, both theoretically and experimentally, to address these issues. The focus here will be on the physics aspects; the technology aspects will be reviewed in other contributions to these proceedings [5].

The outline for the remainder of this paper is as follows. The fundamental aspects of ECH and ECCD — wave propagation and absorption, and the plasma response to the absorption — will be reviewed briefly with emphasis on experimental verification. Then the role of ECH and ECCD in addressing the three general steady-state issues for stellarators and tokamaks will be discussed in turn. Finally, a brief discussion of the prospects for ECH and ECCD playing a role in the steady-state operation of the next generation of magnetic confinement fusion devices will be given, along with some conclusions.

2. Validation of Basic EC Wave Physics in Plasmas

Propagation of electromagnetic waves with frequency near the electron cyclotron frequency is described well by the dispersion relations derived from cold plasma theory [6,7]. (Electron Bernstein waves, electrostatic waves in this same frequency range, will not be considered here.) Two wave polarizations are supported by the plasma, with distinct dispersion relations. In the limit of propagation perpendicular to the plasma magnetic field, the two polarizations correspond to waves with the wave electric field parallel (O mode) and perpendicular (X mode) to the plasma magnetic field. In the limit of propagation parallel to the plasma magnetic field, these become left-handed and right-handed circularly polarized, respectively. They are commonly referred to by O mode and X mode independent of the direction of propagation. The O mode propagates when the wave frequency is above the electron plasma frequency

$\omega_p \equiv \left(ne^2/\varepsilon_0 m\right)^{1/2}$, while the X mode propagates above the right-hand cutoff, which is a combination of the electron cyclotron frequency and the electron plasma frequency

$$\omega_{rh} \equiv (\Omega/2)\left[1+\sqrt{1+4\left(\omega_p^2/\Omega^2\right)}\right] ,$$

where $\Omega \equiv eB/m$ is the gyrofrequency. With these dispersion relations, the propagation of the waves is predicted accurately by ray tracing using WKB theory [8]. The physics of absorption by the electron cyclotron resonance is long developed [9] and has been validated extensively. An example of experimental validation of the propagation and absorption physics from the DIII-D tokamak is shown in Fig. 1 [10]. Waves at 110 GHz with mixed O and X mode polarization are launched at an oblique angle with respect to the major radius from side where the magnetic field is lower, as shown in the left figure. The lines in the figure are projections of the predicted paths of propagation projected back on the poloidal cross-section of the launch point. The density and magnetic field are such that both polarizations experience significant refraction away from the trajectory predicted in vacuum. The right figure shows the amplitude response of the electron cyclotron emission measurement of the electron temperature when the power is modulated with a square wave of 50% duty cycle. The peaks are broadened by heat transport in the plasma, but it is clear that the peaks correspond to the locations predicted by the ray tracing code, including a hint of the second pass through the resonance for the X mode and the strong absorption on the second pass of the O mode. This correspondence requires both the propagation and absorption to be correctly predicted. This is merely one example of a extensive set of experimental validations of the basic theory of electron cyclotron wave propagation and absorption [11].

Perhaps the most stringent test of the theory is in the prediction of the plasma response in cases where the electron cyclotron waves are predicted to drive substantial current in the plasma. As mentioned above, the ability to drive current noninductively is essential to steady-state operation of a tokamak. The current arises because the absorption can occur in a very localized region in both real space and velocity space due to the resonance condition $\omega = \ell\Omega/\gamma + k_{||}v_{||}$,

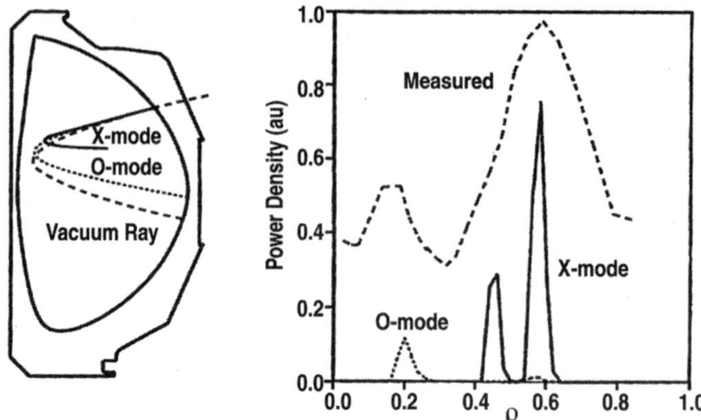

Fig. 1. (a) Projection of vacuum (dashed), X mode (solid), and O mode (dotted) ray trajectories onto a poloidal cross section of the DIII-D tokamak. The outer line shows the location of the plasma-facing components and the inner line shows the last closed flux surface of the plasma. (b) Comparison of the EC power deposition profile for X mode (solid) and O mode (dotted) with the response of the electron temperature measured by ECE to the modulation of the EC power (dashed). Reprinted courtesy of Ref. [10].

where

$$\gamma \equiv 1/\sqrt{1-(\mathrm{v}/c)^2}$$

[7]. Even without putting momentum into the plasma, the waves can drive current due to the asymmetry [12]. Precise prediction of the plasma response, in this case the total current or the local current density, requires an accurate theory of the propagation, absorption, and the dynamics of the electrons in the plasma. The generation of current through ECCD has been demonstrated on several devices. Figure 2 shows a particularly clear example from the TCV tokamak where the ECCD is responsible for current greater than the total plasma current [13]. The total current is under feedback control, using the inductive drive as the actuator. The overdrive from ECCD results in the inductive current reversing sign to oppose the increase in current that would result if the feedback were not applied. This is evidenced by the negative voltage at the plasma surface and the change in sign of the charge in current in the primary of the inductive transformer.

Quantitative validation of the ECCD theory is shown by examples from the DIII-D tokamak in Fig. 3. The left figure compares the total current driven by ECCD to that predicted by the most accurate method — a relativistic Fokker-Planck calculation of the distribution function, including a momentum-conserving collision operator and the remaining parallel electric field [14]. The agreement between theory and experiment is generally within the experimental uncertainties. The right figure compares the current density profile measured during ECCD with the theoretical prediction of the Fokker-Planck calculation [15]. Again, given the experimental accuracy possible, the measurement agrees well with the theoretical prediction. Including examples from other experiments and extensive benchmarking of codes [16], there is a high degree of confidence in the predictive theory of EC wave physics.

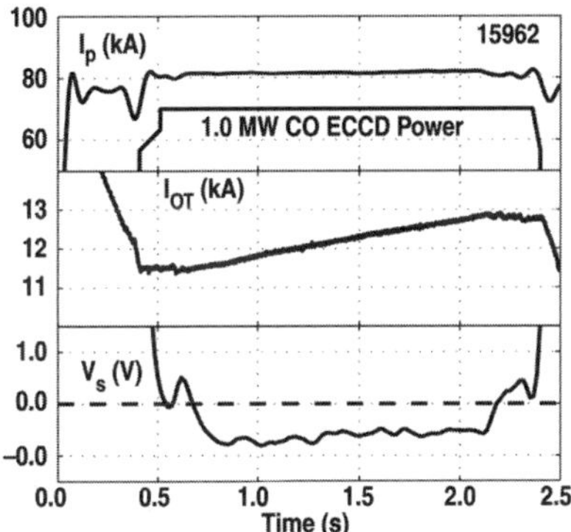

Fig. 2. Plasma with ECCD greater than the preset feedback value for plasma current from the TCV tokamak. Shown are time histories of the plasma current and EC power (upper box), current in the solenoid (middle box), and voltage at the plasma boundary (bottom box). Note the suppressed zero in the upper two boxes. Reprinted courtesy of Ref. [13].

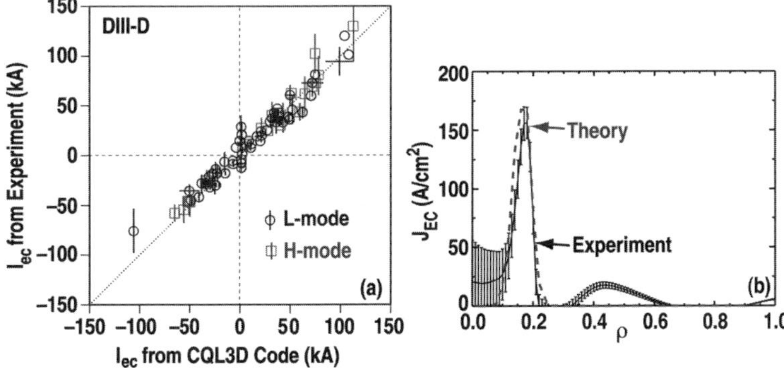

Fig. 3. (a) Comparison of measured current driven by ECCD to that predicted by a quasi-linear Fokker-Planck calculation with the experimental density, temperature, and impurity profiles including the effects of inductive parallel electric field. Adapted from Fig. 8 of Ref. [14]. (b) Comparison of measured current density profile from ECCD to that predicted by a quasi-linear Fokker-Planck calculation. Adapted from Fig. 2(b) of Ref. [15].

3. Access to Steady-State Operating Conditions

The first general issue for steady-state operation of magnetic confinement fusion devices is accessing the relevant operating conditions. Initiation of the plasma and heating it to the necessary pressure for fusion energy are generic requirements for stellarators and tokamaks. For the tokamak, achieving the necessary pressure while maintaining high-energy gain is expected to require access to the high-confinement mode or "H mode" regime. This regime has certain access conditions. Stellarators have also observed H mode, but it is not clear that this is necessary or even beneficial for stellarator operation. Finally, for the tokamak, only certain classes of current profiles will support steady-state operation at high gain, and these profiles must be formed prior to going to the high temperature conditions associated with fusion energy because redistribution of current in plasmas of power plant size and temperatures has a time scale of 100s of seconds and might require capacity much greater than needed to sustain the steady-state. These topics will be examined here in turn.

Stellarators have relied on EC waves to initiate the plasma for many years [11]. Since the vacuum magnetic configuration usually is able to confine plasma, it is necessary simply to have a means to breakdown the neutral gas and accelerate the electrons sufficiently to build more plasma [17]. At the fundamental resonance ($f = f_{ce}$), EC waves can easily do this. At second harmonic ($f = 2f_{ce}$), it is perhaps surprising that EC waves can be sufficiently absorbed because absorption at higher harmonics is attributed to finite Larmor radius effects [9]. But many devices have shown that startup with second harmonic EC waves is effective, even with oblique launch and the corresponding reduced absorption [17–21]. Beyond the generation of the plasma, sufficient power must be available to overcome the radiation due to low charge states of impurities in the device (typically carbon and oxygen). The radiation is strongest for electron temperatures below 100 eV since higher temperatures lead to higher charges states and eventually fully ionized atoms that no longer radiate through line radiation. This phenomenon is known as "burnthrough" and is greatly facilitated by the EC wave power absorbed in the formation phase. This is especially critical for the next generation of tokamaks that will have superconducting coils. These new machines will have much lower inductive electric fields than present-day tokamaks due to limits on the voltages that can be applied to the coils and the significant shielding they require from the neutrons generated through the fusion reactions. Since plasma current is necessary for confinement, the tokamak must achieve burnthrough promptly or the plasma current will decay away. The demonstrated ability to achieve burnthrough at low electric field on present-day experiments using EC assist [22,23] gives assurance that future tokamaks will be able to start up reliably at low electric field.

Bringing the plasma to sufficient pressure that the self-heating from fusion products (α particles in the case of deuterium-tritium fusion) becomes significant is essential to fusion energy production at high gain. The stellarator community envisions an operational point without auxiliary heating (ignited operation) [2], while steady-state tokamaks will require some level of auxiliary heating and current drive to maintain the magnetic configuration. In either case, auxiliary heating must be applied to start the fusion burn. The physics basis for predicting the amount of power required for tokamaks is based largely on experiments using injection of neutral beams (NBI) at energies higher than the thermal plasma temperature [24]. This form of heating has some similarity to heating by the α particles (transfer of energy through collisions), but also some significant differences (correlated particle source, possibility of correlated torque applied, energy transfer can be dominantly to ions in present experiments). The prediction of the power requirement is largely done using 0-D scalings derived from regression analysis of data from multiple experiments. Heating with EC waves does not have correlated particle or torque sources and the time scale for collisional coupling is two orders of magnitude shorter than for NBI, so the possibility of different power requirements in the EC case must be investigated. Since EC startup has been used extensively on stellarators, the standard scalings derived for stellarators have a significant fraction of the data from ECH [25,26]. No clear difference in the power requirements for different heating schemes is seen in these stellarator databases; however, the stellarator configuration strongly damps rotation that would arise from any applied torque. For tokamaks, the standard scalings [24,27] are heavily dominated by data from NBI. Figure 4 shows confinement data from EC heating of the FT-U tokamak plotted against the multi-tokamak database [28]. Within the accuracy of a log-log plot, the ECH data appears to agree well with the other data. However, the plasma in this dataset are not in the H mode confinement regime needed for fusion energy from steady-state tokamaks. As will be discussed below, there is no problem with access to H mode with ECH, but the lack of experimental data with EC systems on a scale comparable to the neutral beam heating systems on present-day tokamaks is of concern. The limited H mode data with EC agree well with the standard H mode scaling; it is of great interest to see whether this agreement continues into the regime where strong NBI implies significant core fueling and applied torque.

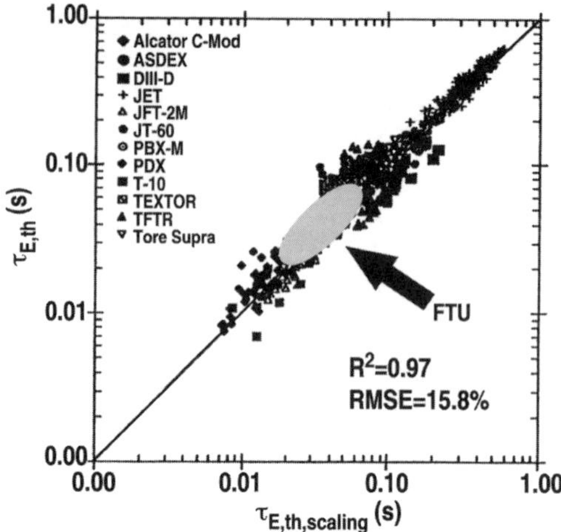

Fig. 4. Thermal confinement time from ECH plasmas in the FT-U tokamak plotted against the ITER-97L scaling [29]. Reprinted courtesy of Ref. [28].

In tokamaks, access to the H-mode confinement regime is characterized by crossing a threshold level of power flowing through the plasma boundary [30]. ECH is effective at accessing the H-mode regime [31] and recent experiments confirm that the access conditions are similar for EC and NBI [32]. Figure 5 shows data from the AUG tokamak indicating that the power required to access H mode is the same within the uncertainties of the experimental data. It also shows one advantage of ECH — the absence of NBI data at low density is due to restrictions on NBI operation at low density. It is possible to use ECH at lower densities where the transparency of the plasma to the neutral beam excludes its application. In this case, there is little advantage to the lower density operation, because the power threshold increases again. The scaling of the minimum density is not known, so there may be a clear advantage at low density for ECH in future tokamaks. H mode has been observed in stellarators first with ECH [33]. Figure 6 shows time histories from the WVII-AS stellarator where H mode is obtained with ECH as the sole heating source. There appears to be no power threshold for H mode access in the WVII-AS stellarator; instead, there is a requirement on the configuration. It is not clear whether H mode is an advantage to stellarators — present conceptual designs of power plants do not make use of it, despite the need for improvement of stellarator confinement — but the capability for H mode operation exists and was discovered using ECH.

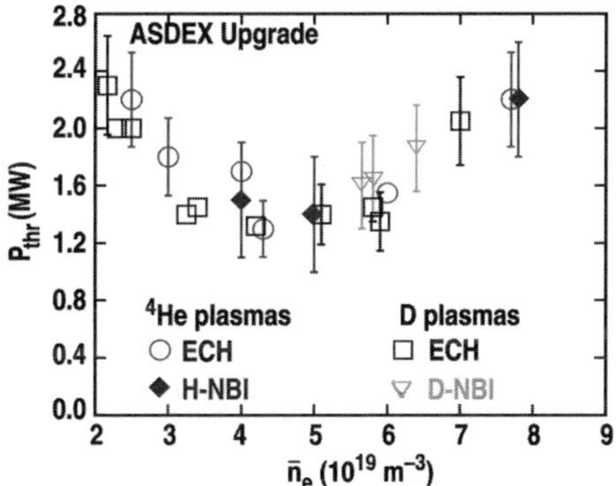

Fig. 5. Threshold power for entering H mode in the AUG tokamak using ECH and NBI in helium and deuterium plasmas as a function of line-averaged density. Adapted from Fig. 4 of Ref. [32].

The last topic relevant to accessing steady-state operation discussed here is formation of the desired current profile. This issue is specific to the steady-state tokamak. At constant total plasma current, the profile of the inductive current density in the tokamak will diffuse toward a profile consistent with a constant electric potential in radius and the conductivity profile, which is normally very peaked ($\propto T_e^{3/2}$). The total current density profile will vary during this time in proportion to the local fraction of inductive current. A broad current profile is favored for steady-state tokamak operation [1], both because it implies a substantial bootstrap current at the edge (large noninductive fraction) and it is favorable for stability in the wall-stabilized regime. The large difference between the "natural" profile toward which inductive current density evolves and the desired profile for steady-state tokamak operation motivates using the current rise phase of the discharge to "steer" the current profile toward the desired value [34]. The current rise phase has several clear advantages for this task. First, the inductive drive supplies current from the edge, so the profiles are naturally broad during this time. The power from the external heating dominates the power balance during this time, whereas at current flattop the fusion self-heating can be large. The lower temperatures during the current rise phase allow modifications of the profile on a shorter time scale. Finally, keeping the plasma close to the desired stable current profile avoids the need to make large changes in the current profile, which might lead to ideal or resistive instabilities during the transition.

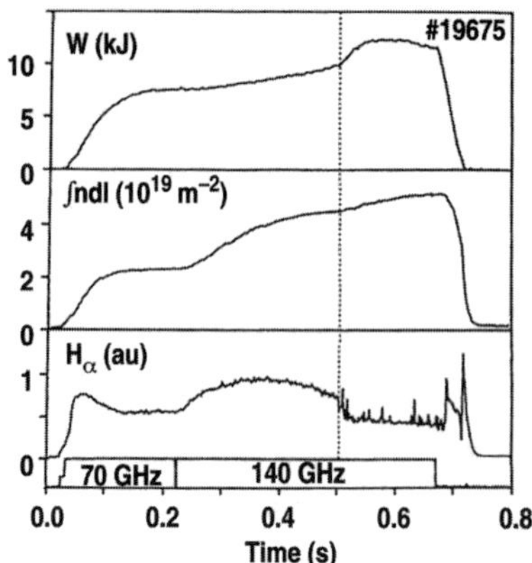

Fig. 6. Time histories of the stored energy (top), line-integrated density (middle), and edge D_α emission (bottom) from a plasma with a transition to H mode using ECH in the WVII-AS stellarator. The bottom figure shows the timing of the application of the two different ECH systems. The dotted vertical line indicates the time of the transition to H mode. Reprinted courtesy of Ref. [33].

ECH is an ideal tool for profile modification, especially under feedback control, since it can be localized in both space and time. Two examples of feedback control are given here. On the TCV tokamak, the power applied to two fixed radial locations has allowed feedback control of the electron temperature profile [35]. Figure 7 shows how the peak value and the profile width, defined as the fraction of the simple profile average to the peak value, are controlled to match target waveform values by varying the power at normalized radii of 0.2 and 0.5. The feedback algorithm has been specifically designed for this application. Modeling of the plasma dynamics allows the system to follow rapid changes in the target values without overshoot or undershoot as long as the required power is in the range available to the system. On the DIII-D tokamak, the EC power at a single radial location near the half radius has been used to modify the electric conductivity during the current ramp to feedback control the value of the central safety factor $q(0)$ [36]. A real-time estimate of $q(0)$ is provided to the feedback system through inclusion of the motional Stark effect (MSE) measurements of the internal vertical magnetic field in the real-time equilibrium reconstructions. A proportional-integral gain controller is used to determine the ECH power using the simple model that more power (hotter plasma) slows the reduction in $q(0)$. Therefore, if the measured value falls below the target value more power is applied to slow the downward evolution (and vice versa). Again, this controller does a reasonable job of providing a $q(0)$ evolution that follows the target trajectory until the requested power encounters the system limits. Note that in the context of tailoring the current profile during the current rise, both the TCV and DIII-D schemes act on the conductivity — the ECCD is much less than the inductive current and is not effective at modifying the current profile. More sophisticated feedback schemes are being tested that take into account the expected plasma response and schedule the future actuator trajectories [34].

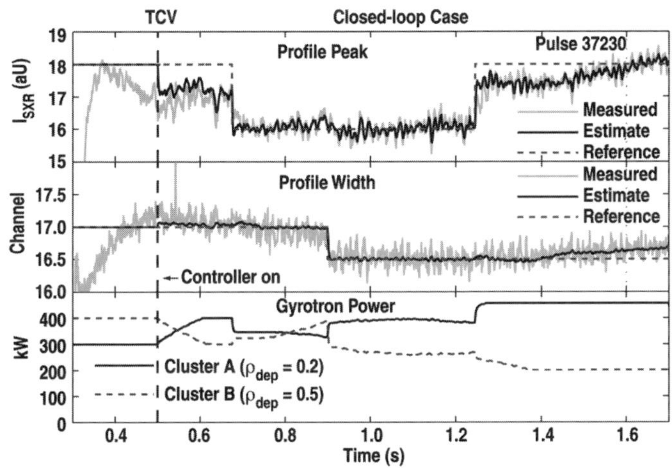

Fig. 7. Time histories of peak soft x-ray intensity (top), the normalized average soft x-ray intensity (middle), and the ECH power in two separate systems aimed at a normalized radius of 0.2 (solid) and 0.5 (dashed), respectively (bottom). The dashed lines in the top boxes indicate the reference value input to the controller. Adapted from Fig. 2 of Ref. [35].

4. Sustaining the Steady-State Regime

As discussed earlier, the present vision for stellarator power plants is to operate in the ignited regime, where the self-heating sustains the pressure in a magnetic configuration determined largely by the external coil set. Given this view, there is no need for ECH or ECCD in routine operation. In the next section, a possible role for fine tuning the operating point of a stellarator with ECCD will be discussed. For the steady-state tokamak, however, ECCD is envisioned to play a critical role in maintaining the current profile noninductively in a stable configuration.

As shown above, experiments have demonstrated full sustainment of the current profile in a tokamak by ECCD. However, sustaining the total current by external sources at the expected parameters for power plant operation is not practical. The current drive efficiencies are too low by about a factor of 2.5 to allow economic operation, due to the large fraction of the electrical output of the plant that would need to be recirculated to power the current drive systems. (See Ref. 1 for a more complete discussion of this point.) This necessitates that the self-generated "bootstrap" current [37] must dominate the current generation. But this current is not likely to yield full current drive nor perhaps a stable current profile on its own. Here is where the flexibility of the ECCD to supply current at various radii and magnitudes is essential to steady-state operation. The ECCD can be deployed to make up the difference locally between the desired total current density profile that is stable to ideal and resistive instabilities at the desired operating pressure and the bootstrap current profile that is largely dependent on the shape of the density and temperature profiles.

An example of ECCD used in this role from the DIII-D tokamak is shown in Fig. 8 [38]. The fraction of current generated noninductively is sustained at 100% for nearly 1 s [Fig. 8(a)], which is still less than a global resistive relaxation time for this plasma. The duration is limited by the technical limitations of the NBI system and not plasma effects in this case. The fact that the current profile is not yet in equilibrium is clear from the fact that the remaining inductive current still has some spatial structure [Fig. 8(b)]. The calculated noninductive currents [Fig. 8(c)] show that the dominant noninductive current is the bootstrap current at all radii. The edge current is exclusively bootstrap current, while the NBCD and ECCD contribute significantly in the plasma interior. The ECCD is not positioned to maximize the driven current, but to optimize stability to tearing modes and drive current to make up the difference between the sum of the bootstrap current and the NBCD. In the case shown, it appears that the combination of current sources is nearly optimal for sustaining the current noninductively when the inductive current equilibrates.

The optimization of the ECCD distribution has been done empirically, as there is no equivalent to the ideal MHD codes for surveying large numbers of equilibria for tearing stability. An example of this optimization is shown in Fig. 9. Identical target plasmas are made with only the three variations in the ECCD profile shown [Fig. 9(c)] — two profiles with very narrow deposition at two locations and one profile with broad deposition spanning the region of the two narrow cases. At β_N=3.2, the plasma is stable to n=1 tearing modes for the duration of the high performance phase. At β_N=3.4, however, the two narrow deposition cases are unstable, while the broad deposition case remains stable. In general, broad ECCD deposition results in stability to n=1 tearing modes at higher β_N in DIII-D.

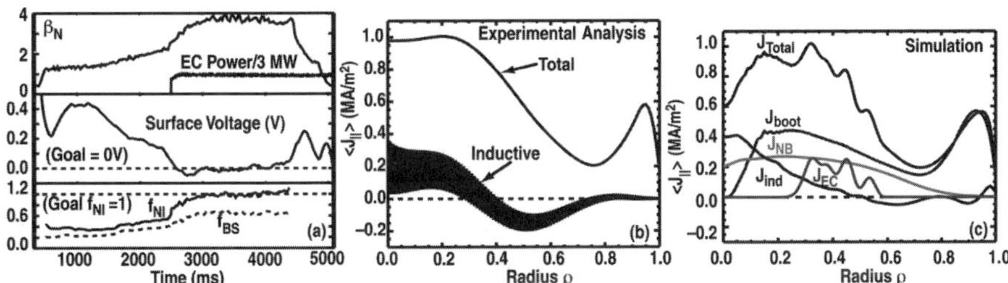

Fig. 8. (a) Time histories of normalized pressure (β_N) and EC power (top), surface inductive voltage (middle), and noninductive current fraction (solid) and bootstrap current fraction (dashed) (bottom). (b) Radial profiles of the flux-surface average current density and the inductive current density inferred from experimental equilibrium reconstructions. (c) Radial profiles of components of the current density calculated from modeling. Reprinted courtesy of Ref. [38].

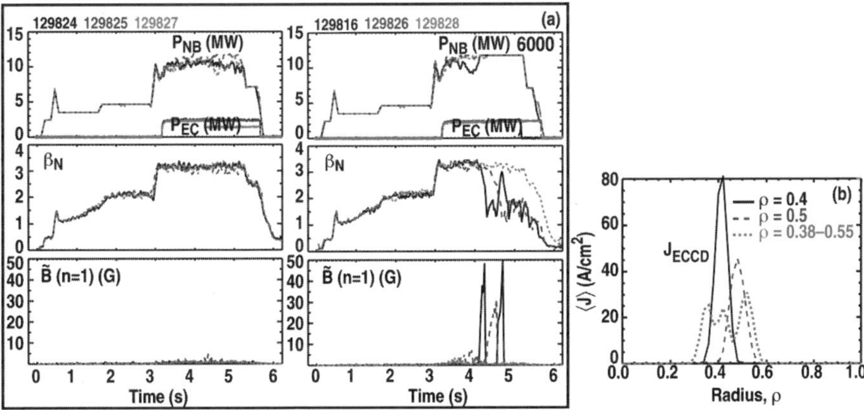

Fig. 9. (a) Time histories of NBI and EC power (top), normalized pressure (β_N) (middle), and n=1 magnetic fluctuation amplitude at the vacuum vessel (bottom) for three steady-state scenario plasmas at β_N = 3.2 with various ECCD spatial distributions. (b) Same as (a), but with β_N = 3.4. (c) Radial profiles of the ECCD current density showing the three different spatial distributions.

5. Control of the Operating Point

From the point of view of power plant operation, a steady-state scenario that has an operating point that is passively stable, i.e., it returns to the original parameters after small perturbations without external intervention, is an optimal solution. Solutions like this seem reasonable for stellarators, since there is little free energy in the magnetic configuration and stellarator power plants are being designed to lie on the stable part of the fusion power curve to perturbations in temperature. They will need to be continuously fueled, however, and may require some feedback control of the heating due to the refueling process, depending on the method used. Depending on the type of stellarator and the sensitivity of the plasma transport to details of the internal magnetic configuration, some tuning of both the pressure and the magnetic configuration may be needed to remain at the desired operating point. ECH and ECCD are unique tools to allow local tuning, and due to the ability to act locally, they should be able to carry out this control with minimum impact on the energy gain.

As an example, Fig. 10 shows the sensitivity of the W7-X divertor design to the details of the magnetic configuration [39]. The divertor is designed to handle the heat and particle fluxes in the 3-D magnetic geometry. However, it is fixed in space and optimized for a given configuration, taking advantage of the proximity of a low-order rational surface to increase the "wetted" area in the divertor to reduce the heat flux. The simulation shown compares the magnetic configuration in vacuum to that with full operating pressure. A relatively small amount of current (43 kA) due to the bootstrap current changes the magnetic configuration in the divertor region significantly [left side of Fig. 10]. The same EC system that is used to increase the pressure to the operating point can be used to compensate temporarily the bootstrap current and restore the proper magnetic configuration for the divertor until the coil currents can be changed to do this in steady state. (The deposition of the ECCD is not optimized in radius in the case shown, leading to magnetic shear inside the plasma that is probably undesirable. It is expected that this can be further optimized.)

While this simulation was done in the context of accessing the desired operating point in W7-X, the same general issues can be anticipated at the operating point of a stellarator power plant with similar design philosophy. If the divertor has been optimized for a particular magnetic configuration, then perturbations away from this optimized configuration must be rapidly dealt with in order to preserve the integrity of the divertor in the presence of full-performance heat fluxes in this non-optimized state. The large inductances and low voltage limits on the superconducting coils expected for steady-state power plants imply a more rapid response is needed. A rapid compensation system with ECCD to restore temporarily the magnetic configuration seems much more desirable that a full unscheduled shutdown and could maintain the fusion power output at a reasonable level with only a temporary reduction of gain. Experiments on the WVII-AS stellarator showed the ability of ECCD to compensate the bootstrap current globally as part of a larger experiment to validate the theoretical model of ECCD in that configuration [40].

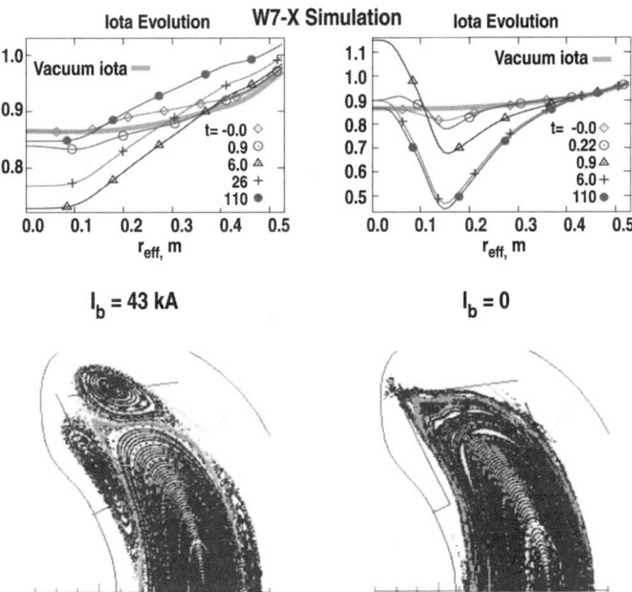

Fig. 10. The upper figures show the vacuum ι profile for the case without (left) and with (right) correction for the net bootstrap current with ECCD. The lower figures are Poincaré plots at a single poloidal cross-section showing the effect of each case on the magnetic configuration in the divertor region.

For tokamaks, a passively stable operating point may be possible, but it is much less likely to exist against all perturbations, due to the strong coupling of the pressure and current profiles through the bootstrap current and the large free energy in the plasma current required to sustain the magnetic configuration. The most critical control task is to stabilize the operating point following a perturbation to allow an orderly shutdown and restart of the plasma, rather than activation of the protection systems required for rapid shutdown. Again ECCD has a proven ability to control instabilities in present-day tokamaks that appears to be transferable in both physics and technology to future uses.

Control of MHD instabilities by precise localization of ECCD under feedback control has been demonstrated in several experiments. Tearing modes are a significant concern for steady-state tokamaks, as they can alter the magnetic configuration and the energy confinement significantly and, if not dealt with, can lead to a complete loss of thermal and magnetic stored energy (disruption). Several experiments have demonstrated suppression of tearing modes by ECCD deposited precisely in the magnetic islands generated by the tearing modes [41]. Experiments on JT-60U [42] have demonstrated automated detection of tearing modes using an electron cyclotron emission (ECE) diagnostic and subsequent suppression by feedback control using an EC launcher with a steerable mirror (Fig. 11). This technology seems feasible for use in a power plant. Other experiments have used plasma motion or change in the toroidal magnetic field to demonstrate the feasibility of feedback control, but these methods are not amenable to the power plant environment. Pre-emptive stabilization has been demonstrated as a possible means of enhanced operation by active feedback control [43]. Experiments on Tore Supra demonstrated control of the sawtooth instability by means similar to the JT-60U tearing mode experiment [44]. While the sawtooth instability is unlikely to be encountered in steady-state tokamak operation, the proof of principle of ECCD for feedback control of MHD instabilities bolsters the physics and technology basis needed for future tokamaks.

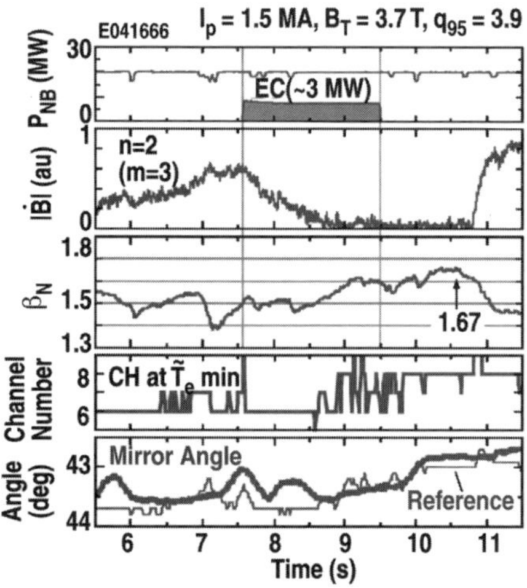

Fig. 11. Time histories of (a) NB and EC power, (b) n=2 magnetic fluctuations at the vacuum vessel, (c) normalized pressure (β_N), (d) ECE channel number for mode location, and (e) aiming mirror command and response for a plasma from the tearing mode suppression by ECCD experiment on the JT-60U tokamak. Adapted from Fig. 11 of Ref. [42].

Fusion power plants are likely to have plasma-facing components made from high-Z materials, such as tungsten. If these metals get into the core of the plasma, they can radiate copious quantities of energy, cooling the plasma and spoiling the fusion output. In extreme cases, a radiative collapse could lead to disruption. The optimal solution is a divertor design that keeps these high-Z impurities out of the plasma. However, it is necessary to anticipate off-normal events that may bring these metals into the plasma. The AUG tokamak has nearly complete coverage of the plasma-facing components with tungsten. Accumulation of tungsten at the plasma center can be mitigated by localized ECH [45]. By scanning the location of the ECH from plasma center to off-axis, the sensitivity to the heating location is clear (Fig. 12). The concentration of tungsten rises exponentially once the heating moves to a relatively small radius off-axis. Similar dependence on ECH deposition location is seen in scans of the toroidal magnetic field, which move the resonance in major radius. A validated physical model for this behavior is not yet available.

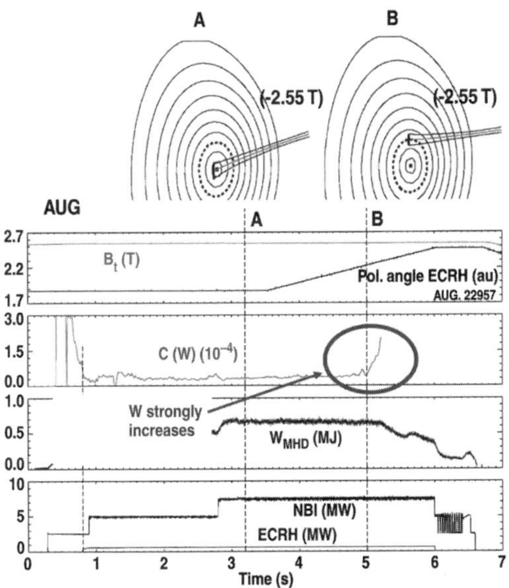

Fig. 12. Time histories of (a) toroidal field magnitude and poloidal angle of the EC aiming mirror, (b) tungsten concentration, (c) stored energy, and (d) NB and EC powers for a scan of the EC heating deposition location. The cross-sections above labeled A and B show the heating location of the EC at the corresponding vertical lines in the time histories. Reprinted courtesy of Ref. [45].

6. Prospects for Use of ECH and ECCD in Fusion Energy Devices

The energy gain in magnetic fusion devices scales strongly with magnetic field [46]. This motivates power plant designers to operate at the maximum field allowed by the superconducting materials from which the magnets are likely to be made. Present superconducting materials limit the central vacuum field in fusion devices to ≤7 T, which implies a fundamental cyclotron resonance frequency of <200 GHz. This frequency is only slightly above that of the presently developed high-power cw gyrotron sources (≤170 GHz). The electron plasma frequency, which sets the wave cutoff for the O mode, is ~$5x10^{20}$ m^{-3}. A simple estimate using $\langle\beta\rangle = 5\%$ as a typical value for the operational point of a fusion power plant implies that the operating density will be below the O mode cutoff, independent of B, if $\langle T\rangle > 6$ keV, which is likely due to the increase in fusion power faster than the square of the temperature in this range (trading density for temperature at fixed β increases fusion power). This argument indicates ECH and ECCD are viable tools for fusion power plants.

A secondary advantage for ECH and ECCD systems is the high power density possible in the transmission lines. Power densities up to 1 GW/m^2 have been demonstrated in evacuated transmission line [47]. This high power density allows substantial power to be applied without a corresponding decrease in the breeding blanket and shielding area of the fusion device. The ability to guide the power with transmission lines also has a very limited impact on the coil design, which is especially critical for stellarator systems that can have complex 3-D coil sets. Other heating and current drive schemes under consideration for fusion power plants have more significant impact on the design of the physical plant.

These considerations, in addition to the proven abilities of ECH and ECCD to deliver functions essential to steady-state operation, have led to the inclusion of EC systems in the design of the next-step fusion experiments and consideration in the design for power plant demonstration facilities. The next large stellarator, W7-X, has EC as a primary heating system and allows for the possibility of ECCD [48]. An EC system is already in use on LHD [49], the largest operating stellarator in the world. The JT-60SA tokamak, which aims at demonstration of conditions necessary for steady-state operation, employs an EC system for heating and current drive [50]. The ITER tokamak, envisioned to be the first demonstration of burning plasmas in steady-state conditions, also has an EC system as one of its primary auxiliary heating and current drive systems [46]. Two potential steady-state scenarios that use ECCD as a essential element are shown in Fig. 13. The upper case is a simulation, based on the DIII-D scenario shown in Fig. 8, that approaches the ITER physics objective of steady-state operation with fusion energy gain Q of 5 [1]. The role of ECCD in this case is to supply off-axis current drive to obtain fully noninductive operation with $\min(q) > 1$. The lower case takes advantage of the local ECCD to induce a transport barrier (due to the form of the energy transport model used) to generate significant bootstrap current in addition to the off-axis ECCD. (This simulation is similar to those published results that include lower hybrid current drive [51]). Both types of scenarios are under study in present-day experiments. These simulations indicate that EC systems may play a critical role in meeting the ITER physics objective of steady-state operation with Q=5.

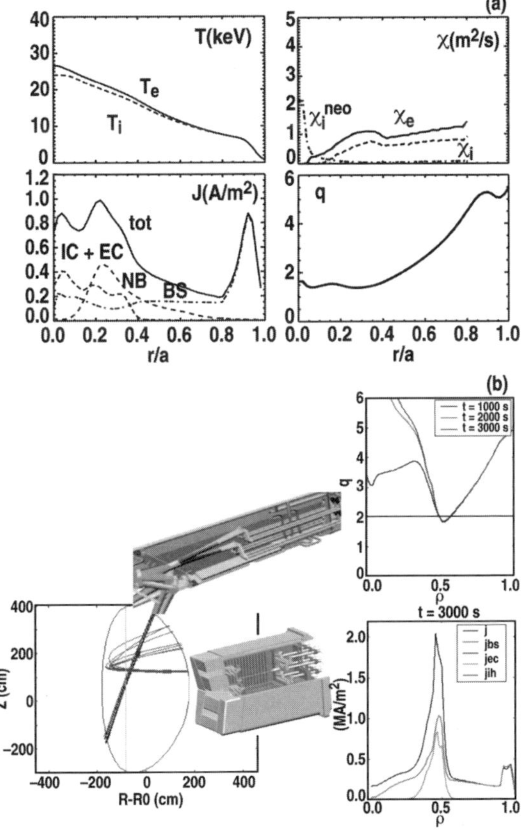

Fig. 13. Upper boxes show the temperature, diffusivity, current density, and q profiles from a simulation of an ITER steady-state scenario based on a DIII-D plasma. Reprinted from [1] by permission. Lower boxes show the q and current density profiles from a simulation of ITER steady-state scenario based on a model allowing formation of a transport barrier.

7. Conclusions

The evidence presented here indicates that ECH and ECCD can perform vital roles for steady-state magnetic fusion energy systems, from accessing the desired operating conditions to sustaining those conditions and controlling the operating point. In some cases such as flexible off-axis current drive and instability control, these roles are uniquely possible for EC systems — no other means are known. These unique capabilities, along with the technological advantages of EC systems and the proven technology developments over the past decades, enhance the prospects for steady-state fusion energy systems.

Acknowledgments

This work was supported by the U.S. Department of Energy under DE-FC02-04ER54698.

References

[1] T.C. Luce, "Realizing Steady-State Tokamak Operation for Fusion Energy," submitted to Phys. Plasmas (2010).
[2] F. Najmabadi, et al., *Fusion Sci. Technol.* **54**, 655 (2008).
[3] J.P. Freidberg, *Rev. Mod. Phys.* **54**, 801 (1982).
[4] N.J. Fisch, *Rev. Mod. Phys.* **59**, 175 (1987).
[5] K. Sakamoto, these proceedings (2010).
[6] T.H. Stix, *Waves in Plasmas*, 2nd ed., AIP Press, New York (1992).
[7] R. Prater, *Phys. Plasmas* **11**, 2349 (2004).
[8] S. Weinberg, *Phys. Rev.* **126**, 1899 (1962).
[9] M. Bornatici, et al., *Nucl. Fusion* **23**, 1153 (1983).
[10] C.C. Petty, et al., Radio Frequency Power in Plasmas, *AIP Conf. Proc.* **485**, AIP, Melville, NY (1999) p. 245.
[11] V. Erckmann and U. Gasparino, *Plasma Phys. Control. Fusion* **36**, 1869 (1994).
[12] N.J. Fisch and A.H. Boozer, *Phys. Rev. Lett.* **45**, 720 (1980).
[13] O. Sauter, et al., *Phys. Rev. Lett.* **84**, 3322 (2000).
[14] C.C. Petty, et al., *Nucl. Fusion* **42**, 1366 (2002).
[15] L.L. Lao, et al., in Radio Frequency Power in Plasmas, *AIP Conf. Series* **595**, 310 (2001).
[16] R. Prater, et al., *Nucl. Fusion* **48**, 035006 (2008).
[17] G.L. Jackson, et al., *Nucl. Fusion* **47**, 257 (2007).
[18] B. Lloyd, et al., Proc. 13th European Conf. (Schliersee) vol 10C part II (European Physical Society) 266 (1986).
[19] Zhan R. et al., *Int. J. Infrared Millimeter Waves* **11,** 765 (1990).
[20] K. Kajiwara, et al., *Nucl. Fusion* **45,** 694 (2005).
[21] Y.S. Bae, et al., *Nucl. Fusion* **49**, 022001 (2009).
[22] G.L. Jackson, et al., *Nucl. Fusion* **48**, 125002 (2008).
[23] A.C.C. Sips, et al., *Nucl. Fusion* **49**, 085015 (2009).
[24] ITER Physics Basis Editors, et al., *Nucl. Fusion* **39**, 2175 (1999).
[25] U. Stroth, et al., *Nucl. Fusion* **36**, 1063 (1996).
[26] H. Yamada, et al., *Nucl. Fusion* **45**, 1684 (2005).
[27] P.N. Yushmanov, et al., *Nucl. Fusion* **30**, 1999 (1990).
[28] S. Cirant, et al., *Plasma Phys. Control. Fusion* **41**, B351 (1999).

[29] S.M. Kaye, et al., *Nucl. Fusion* **37**, 1303 (1997).
[30] J.A. Snipes, et al., *Nucl. Fusion* **36**, 1217 (1996).
[31] John Lohr, et al., *Phys. Rev. Lett.* **60**, 2630 (1988).
[32] F. Ryter, et al., *Nucl. Fusion* **49**, 062003 (2009).
[33] V. Erckmann, et al., *Phys. Rev. Lett.* **70**, 2086 (1993).
[34] Y. Ou, et al., *Plasma Phys. Control. Fusion* **50**, 115001 (2008).
[35] J.I. Paley, et al., *Plasma Phys. Control. Fusion* **51**, 124041 (2009).
[36] J.R. Ferron, et al., *Nucl. Fusion* **46**, L13 (2006).
[37] J. Wesson, D.J. Campbell, *Tokamaks*, 3rd ed., Oxford University Press, Oxford, UK, 2004.
[38] C.T. Holcomb, et al., *Phys. Plasmas* **16**, 056116 (2009).
[39] H. Maaßberg, private communication (2010).
[40] V. Erckmann et al., *Fusion Eng. Design* **53**, 365–375 (2001).
[41] R.J. La Haye, *Phys. Plasmas* **13**, 055501 (2006).
[42] A. Isayama, et al., *Nucl. Fusion* **43**, 1272 (2003).
[43] R. Prater, et al., *Nucl. Fusion* **47**, 371 (2007).
[44] M. Lennholm, et al., *Fusion Sci. Technol.* **55**, 45 (2009).
[45] J. Stober, et al., these proceedings (2010).
[46] C. Gormezano, et al., *Nucl. Fusion* **47**, S285 (2007).
[47] T.C. Luce, *IEEE Trans. Plasma Sci.* **30**, 734 (2002).
[48] T. Klinger, "The Construction of the W7-X Stellarator," Proc. 22nd IAEA Fusion Energy Conf., Geneva, http://www-naweb.iaea.org/napc/physics/FEC/FEC2008/papers/ft_1-4.pdf.
[49] S. Kubo, et al., these proceedings (2010).
[50] T. Fujita, et al., *Nucl. Fusion* **47**, 1512 (2007).
[51] J. Garcia, et al., *Phys. Rev. Lett.* **100**, 255004 (2008).

ECRH ON ASDEX UPGRADE - SYSTEM EXTENSION, NEW MODES OF OPERATION, PLASMA PHYSICS RESULTS -

J. STOBER[1,*], D. WAGNER[1], L. GIANNONE[1], F. LEUTERER[1], M. MARASCHECK[1], A. MLYNEK[1], F. MONACO[1], M. MÜNICH[1], E. POLI[1], M. REICH[1], D. SCHMID-LORCH[1], H. SCHÜTZ[1], J. SCHWEINZER[1], W. TREUTTERER[1], H. ZOHM[1], A. MEIER[2], TH. SCHERER[2], J. FLAMM[2], M. THUMM[2], H. HÖHNLE[3], W. KASPAREK[3], U. STROTH[3], A.V. CHIRKOV[4], G.G. DENISOV[4], A. LITVAK[4], S.A. MALYGIN[5], V.E. MYASNIKOV[5], V.O. NICHIPORENKO[5], L.G. POPOV[5], E.A. SOLUYANOVA[5], E.M. TAI[5]

[1]*Max-Planck Institut für Plasmaphysik, EURATOM-Association, Garching, Germany*
[2]*Karlsruhe Institute of Technology, EURATOM-Association, Karlsruhe, Germany*
[3]*Institut für Plasmaforschung, Universität Stuttgart, Stuttgart, Germany*
[4]*Institut of Applied Physics, Russian Academy of Science, Nizhny Novgorod, Russia*
[5]*GYCOM Ltd, Nizhny Novgorod, Russia*
[*]*E-mail: Joerg.Stober@ipp.mpg.de*

The ECRH system at ASDEX Upgrade is currently extended from 1.6 MW to 5 MW. The extension so far consists of 2-frequency units, which use single diamond-disk vacuum-windows to transmit power at the natural resonances of these disks (105 & 140 GHz). For the last unit of this extension two additional intermediate non-resonant frequencies are foreseen, requiring new window concepts. For the torus a polarisation-independent double-disk window has been developed. For the gyrotron a grooved diamond disk is actually favoured, for which the grooved surfaces act as anti-reflective coating. Since ASDEX Upgrade operates with completely W-covered plasma facing components, central ECRH is often applied to suppresses W-accumulation in the plasma center. In order to extend the operational range for central ECRH, X3- and O2-heating schemes were developed. Both are characterized by incomplete single-path absorption. For X3 heating, the X2 resonance at the pedestal on the high field side is used as a 'beam-dump', for the O2 scheme a specific reflector tile on the inner heat shield enforces a second path through the plasma center. The geometry for NTM control had to be modified to allow simultaneous central heating. In real-time the ECRH position can be determined either by ray-tracing based on real-time equilibria and density profiles or from ECE for modulated ECRH power. Fast real-time ECE also allows to determine the NTM position. Further major physics applications of the system are summarized.

Keywords: Multi-frequency gyrotron, Tokamak, Tungsten, Impurity control, ECRH, NTM stabilisation

1. Introduction

ECRH is operated on ASDEX Upgrade since 1995. In 2000 the first system was completed consisting of 4 units of 500 kW and 2 s pulse length. This system is described in detail in [1].

The ECRH on ASDEX Upgrade is currently extended by a completely new system with an additional power of 4 MW for 10 sec (4 units of 1 MW each). The new launchers are steerable during the discharge to allow feedback-control of the power deposition with respect to the location of MHD modes, which shall be controlled using ECCD. The new system is a multi-frequency system for the frequency range from 105-140 GHz. This means that not only the gyrotrons must be able to operate at several frequencies, but also matching optics unit (MOU) and transmission line must be sufficiently broad-band. Last but not least the high power windows at the gyrotron and the torus must not reflect more than 1% of the power at all desired frequencies to ensure proper operation. In [2] it is reported how broadband operation of MOU and transmission line is achieved. In [3] gyrotron operation at 9 different frequencies at the MW level has been reported for up to 100ms using a BN window, which limited the pulse length. The remaining crucial components are the high-power long-pulse multi-frequency windows. The long pulse requirement can only be fullfilled with synthetic diamond disks, which should be thick enough to withstand air-pressure, but otherwise as thin and small as possible to reduce costs. A relatively simple solution for 2-frequency operation is possible if the gyrotron has suitable modes with a frequency ratio of small numbers, in our case 140/105=4/3. If the disk thickness is a multiple of $\lambda/2$, the disk acts as a fabry-perot resonator, strongly reducing reflection. For the chosen frequencies a disk thickness of 1.8 mm equals $4 \times \lambda_{140}/2 = 3 \times \lambda_{105}/2$. Based on this concept, 3 dual-frequency gyrotrons have been operated successfully on ASDEX Upgrade [4]. For intermediate frequencies a new window concept is necessary as will be discussed in section 3.

Although the new system was primarily designed for MHD control, the dominant request for ECRH on ASDEX Upgrade is 'simple' central heating. This is related to the complete coverage of the plasma facing components with tungsten in 2006. Especially for H-modes at high plasma current, low gas puff and heating power moderately above the H/L threshold central accumulation of W is observed, which leads to excessive core radiation and a degradation of the discharge. Empirically it is found that this situation can be overcome by central ECRH. For a given frequency, central ECRH is only possible for a fixed toroidal field. Additionally its applicability is limited to densities below the cut-off-density. Since the density increases empirically with the plasma current there is an effective limit on the plasma current. For standard ECRH operation with X2 heating at 2.5T the current was limited to ≈1MA. The corresponding value of q_{95} is > 4.5, in comparison to 3.0 foreseen for the ITER Q=10 scenario. Operation at higher triangularity as also foreseen for ITER makes the situation even worse, since the density for a given current is usually found to be higher. For these reasons X3- and O2-heating schemes were considered which allow central heating at lower toroidal field or higher density, but with incomplete single pass absorption. Means had to be found to handle and control the power not absorbed in the first path. This is described in section 4

As mentioned above the original motivation to construct the new system was the control of MHD modes by ECCD using the fast movable launchers to drive current at the appropriate flux surface. With respect to the stabilisation of neoclassical tearing modes (NTMs) the original concept had to be modified to allow for simultaneous central heating. This is described in section 5 together with the tools developed for feedback controlled NTM stabilisation.

The paper closes with a summary on other major physics results with a focus on particle transport.

2. Status of the new system

The design and construction progress of the new multi-frequency ECRH system have been described in a number of papers [4,5]. The first 2-f gyrotron was taken into operation on ASDEX Upgrade in 2006, confirming the broadband design of MOU and transmission line. According to the original planning, this 2-f gyrotron should be followed by three 4-f gyrotrons. Due to reasons related to the multi-frequency gyrotron window, which will be discussed in the next section, two more gyrotrons were manufactured as 2-f gyrotrons and are actually being installed at ASDEX Upgrade. This window concept using two resonances works very reliably. Only the last gyrotron is still foreseen as 4-f system, if a viable window concept can be found in due time.

3. Multi-frequency windows

High power vacuum windows are necessary on the gyrotron side and on the torus side, but the required properties are not completely identical. The gyrotron emits linearly polarized radiation and the direction of this polarization does not depend on frequency, such that the reflectivity needs only to be small for that polarization. The polarisation has to be modified to an elliptical one appropriate to couple to the desired plasma mode (X or O) in the direction defined by the launcher settings. Therefore the polarisation can be arbitrarily modified in the MOU box located between gyroton and torus window. Hence the reflectivity of the latter must be polarisation independent. On the other hand the window at the gyrotron has to be sufficiently broadband to cover the full frequency chirp ($\geq$ 200 MHz) of the gyrotron, such that the start-up sequence of modes is not disturbed. This condition can be relaxed for the torus window, since an inclination of 2 degrees is sufficient to decouple the reflected power from gyrotron operation [6]. This power will add to the stray radiation in the mirror box close to the torus and in the MOU and may potentially lead to arcing there, but it could be tolerable for some milliseconds at the beginning of the pulse, when the frequency variation is strongest. Two different approaches to a multi-frequency window have been followed, either based on Fabry-Perot (destructive interference) or Brewster-angle principles. For the torus a double-disk diamond window has been constructed from two 1.8 mm disks and tested at low power in the whole frequency band between 105 GHz and 140 GHz. Interference of reflections from both disks can be tuned by the adjustable distance between the disks. In the whole frequency range it is possible to reduce the reflectivity below 1% [7]. As described in [4,8] the test results were close to the theoretical expectations, especially the bandwidth at intermediate frequencies is $\geq$ 300 MHz. For the gyrotron window, a Brewster-angle type window has been constructed, which makes use of the well defined polarisation and is intrinsically broadband. For one polarization and one incident angle (Brewster-angle) the reflection is much smaller than 1% and the frequency band is not limited by the window, but only by the quality of the incident beam (angle, polarisation, focussing), which may vary somewhat with frequency. For the interface vacuum-diamond the Brewster angle is 67.2 degrees with respect to the surface normal. Since the diameter of diamond disks is limited by the production process and also by air pressure the beam has to be strongly focused to pass a window under this angle, raising some concern with respect to arcing, especially since MOU and transmission line are operated in normal air. After initial high power tests with a BN window demonstrated successful operation for nine frequencies [3], a long pulse Brewster angle window was fabricated using diamond. Unfortunately the diamond disk turned out to be less forgiving with respect to arcing during the conditioning phase and the diamond window broke. Analysis revealed that for a disk with full aperture of 106 mm and a thickness of 1.8 mm, the force due to the air pressure is approximately half of the breaking force, leaving not enough margin for shock waves due to arcs in air. At this point, it was decided to equip two of the planed three 4-f gyrotrons with a 2-f window to prevent further delays in delivery. For the last gyrotron alternative window concepts were studied. For the Brewster-window a thicker disk would give more stability, but costs would be significant, especially since the brazing process is not fully settled. Alternatively a double-disk window for the gyrotron was discussed, but the narrow bandwidth is a concern. The bandwidth is limited by the minimum distance between the two disks. The volume between the disks has to be evacuated to prevent arcing and a minimum distance is given by pumping considerations. This problem can be overcome if additional reflective layers can be created within one disk. In practice, both sides of a disk resonant at 105 and 140 GHz can be grooved to create anti-reflective coatings. The minimum of reflectivity is broad in frequency. If the grooved layers are designed for a minimum at 122.5 GHz, the reflectivity is below 1% in the whole frequency band from 105 GHz to 140 GHz. See [9] for details. Such surface structures have been produced on test samples and are currently under investigation.

4. X3- and O2-heating

As discussed in the introduction, a major application of ECRH in fully W-coated ASDEX Upgrade is to provide central electron heating to counteract central accumulation of W. To fullfill this requirement for a given frequency the toroidal magnetic field is almost fixed since the deposition must be within $\rho_{tor} < 0.2$ [4]. For H-mode operation at higher currents (≥ 0.8 MA) the density is typically above the X-mode cutoff for 105 GHz, so that only 140 GHz can be used for X2-heating, fixing $B_t \approx 2.5$ T and $q_{95} > 4.5$ for $I_p \leq 1.0$ MA. Higher currents can hardly be accessed because the central density tends to exceed the X-mode cut-off for 140 GHz. The operational range can be extended by the use of O2- or X3-heating which both have incomplete single-pass absorption. The non-absorbed power is a potential danger for the machine, especially for absorbing insulators. For the case of central X3 heating, there is additional X2 absorption at the pedestal on the high field side serving as a beam dump. The O2 absorption is most efficient for toroidal angles of $\approx$15 degrees of toroidal inclination, i.e. the beam geometry is rather inflexible. This means on the other hand that the location where the non-absorbed power hits the inner heat-shield is well defined. In

ASDEX Upgrade special holographic reflector plates have been installed at these locations which redirect and refocus the reflected beam in the proper polarisation toward the plasma center. Details are described in a separate contribution to these conference proceedings [10] including a demonstration of a $q_{95} = 3.0$ discharge at 1 MA using X3-heating. Both, X3- and O2-heating work well if the central plasma temperature is above 2-3 keV, for lower T_e the level of stray radiation in the vessel increases. In order to protect the vessel from excessive stray radiation, three sniffer probes [11,12] have been installed on ASDEX Upgrade close to the launching systems. Their signals are used to generate an additional interlock for the ECRH system based on peak values and 30 ms averages. Figure 1 illustrates a case of X3-heating used to access a high triangular H-mode at moderate q_{95}. In the discharge on the left the ECRH partly tripped (due to gyrotron interrupt) and the ECR power was too low to suppress W accumulation. As W accumulates radiation increases leading to a loss of H-mode and a substantial reduction of the temperature. At this point the sniffer probe signal increases, exceeding occasionally the 30ms average threshold but not

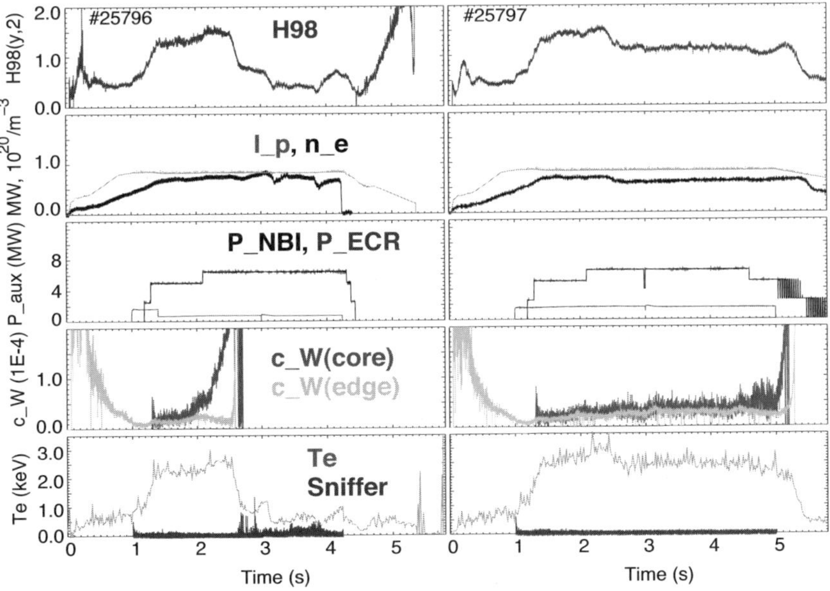

Figure 1: X3 heating (140 GHz), δ=0.34, B_t=1.8T, q_{95}=4, $\beta_{N,max}$=3.5, control parameters are identical except ECRH power. Sniffer refers to a stray radiation antenna as described in [11]. The red dashed line marks the ECRH switch-off threshold for a sliding 30 ms average of the sniffer signal. The switch-off threshold for the raw sniffer signal is twice as high.

the peak threshold, such that the ECRH is not switched off. It is obviously debatable if the threshold should be reduced somewhat for this way of operation. This will evolve as we establish this scenario as a standard operation. The left part of the figure shows that sufficient ECRH is well able to prevent accumulation. This plasma has a very high H-factor and will be addressed by several proposals in the coming campaign.

5. NTM control

As mentioned above, the new ECRH system was designed to control MHD instabilities by ECCD. The fast movable launcher can move with a speed of 10 degrees in 100 ms. Taking acceleration phases into account it takes for example ≈30 ms to move by 2 degrees (figure 3d) which is small with respect to the NTM growth rate of ≈ 100 ms. Originally NTMs have been stabilized at the high field side of the resonant flux surface (figure 2 left and middle). There, the ECCD efficiency is highest since no trapped electrons are present. Also the driven current density is highest, because the resonance is parallel to the flux surface such that a finite beam width does not broaden the width of the driven current density if plotted as a function of a flux surface label such as ρ_{pol}, which is the square root of the normalised poloidal magnetic flux. Independent of geometry the driven current increases as the toroidal injection angle increases but in this geometry the current density has a maximum at rather small angles of 5 degrees. At higher angles the Maxwell distribution of the parallel velocity leads to a significant broadening of the driven current along the beam and in this geometry also along ρ_{pol}. Most efficient NTM stabilisation has been found for highest current densities. Current profiles wider than the island are less effective or, if too wide, even destabilizing [13]. The draw back of this geometry is that no central heating is possible with the same frequency. Since at the required densities central heating has to be applied with 140 GHz, simultaneous NTM stabilisation at the high field side is impossible with the given frequencies. Using 140 GHz for both applications NTMs can only be stabilized above or below the magnetic axis as shown in figure 2 (left and right). In this configuration the beam is parallel to the flux surface. Here, the beam width determines the width of the current profile which is only 3 cm at ASDEX Upgrade and a widening of the current profile along the beam due to toroidal angle variations does not influence the current profile in zeroth order. All these effects can be well seen quantitatively in figure 2, which is based on a high β discharge and the TORBEAM code [14]. In terms of feedback control the new configuration is even favourable for several reasons. The operational point of the high-field side scheme cannot be at the midplane, because it is impossible to decrease ρ by changing the launching angle. Fig. 2, middle shows that the benefits in terms of the narrowness of the current profile decrease significantly as the beam moves upwards or downwards. For a reasonable radial margin, there is no difference to the top/bottom scheme. For the latter the variation of the current profile shape with angle is much smaller, which should even facilitate NTM control with the new scheme. Fig. 3 shows the first feed forward tests of the new scheme and proves that also with this scheme NTMs can be stabilized [15]. Additionally, the stabilizing beam does not have to pass through the plasma center, where the density is highest. Therefore NTM stabilisation with X2 should be possible even if O2 is needed to heat the plasma center, or with X2 at 105 GHz when the plasma center is heated with X3 at 140 GHz. The latter would for example allow to address the NTM present in the high beta discharge shown in fig. 1.

For feedback control of MHD, the ECCD profile has to be located

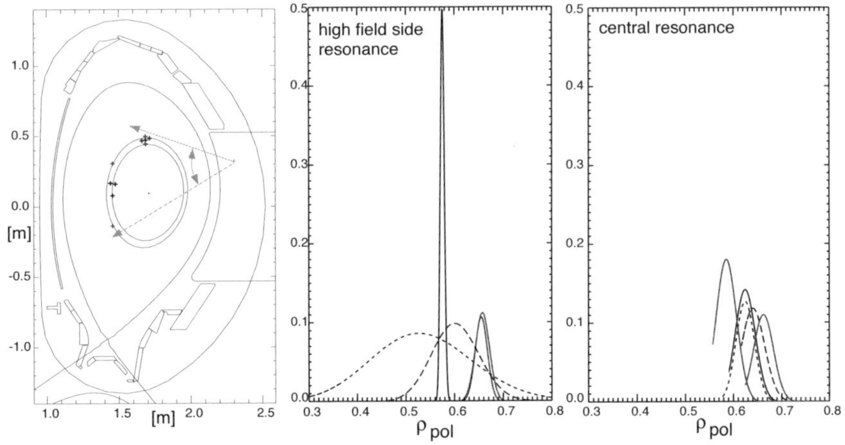

Figure 2: Beam tracing results for different values of toroidal field B_t and launching angles. Kinetic and (scaled) magnetic data from discharge 24061 at 3.5 s. Left: Position of the maxima of j_{ECCD} for the various cases, projected on a poloidal cross section of the torus, together with a sketch of the launching geometry. The radial position of the maxima depends on the value of B_t. Middle: $j_{ECCD}(\rho)$ ($MA/(m^2 MW)$) for the resonance located on the high field side (B_t = 2.1 - 2.2 T). Solid curves correspond to 5° toroidal angle at 2.1 T, dashed curves correspond to 15° and 20° at 2.2 T. Right: $j_{ECCD}(\rho)$ for the resonance through the plasma center (B_t = 2.5 T). Solid curves correspond to 15° toroidal angle, dashed curves correspond to 10° and 20°.

accurately relative to the position of the respective MHD mode. The basic concept is to determine both radii, ρ_{ECCD} and ρ_{MHD} with sufficient accuracy and steer the launchers iteratively to the coincidence position. If the accuracy is not high enough, additional feedback on the mode amplitude may be necessary. Even in the latter case it would be beneficial to have a high relative accuracy of ρ_{ECCD} for the four gyrotrons with different launcher geometries. ρ_{ECCD} and ρ_{MHD} shall both be determined from the upgraded fast ECE system [16] and, independently, from magnetics. For the latter a real-time equilibrium had to be produced, so far based on function parameterization [17] (forward solution of the Grad-Shafranov equation is actually implemented). The radius of the resonant flux surface can be obtained from the equilibrium, but its accuracy depends on the available experimental information on the current profile. Since the diagnostic of the Motional Stark Effect did have hardware problems recently, there are no reliable data from this branch yet. Significant progress has been made on the determination of the ECRH deposition

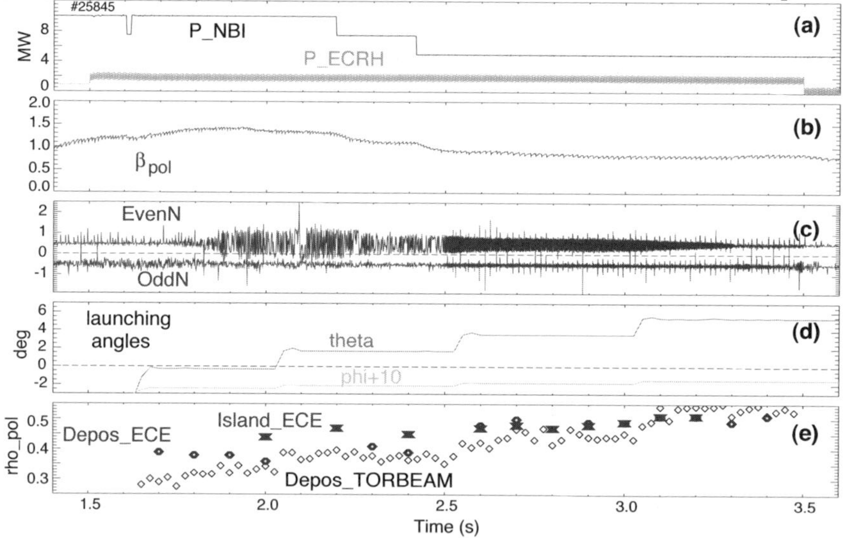

Figure 3: Feed-forward NTM stabilisation combining central ECRH heating (cw) and off-axis ECCD (here modulated, 250Hz). The upper curve in (c) corresponds to the amplitude of the (3,2)-NTM. After its stabilisation at 3.3 s, β_{pol} rises slightly. In (d) the launching angles are shown, a coupled variation rotating the mirror around one axis. (e) shows the real-time estimate for the location of the ECRH deposition using TORBEAM and the offline evaluations for the location of the ECRH deposition and the NTM-island using ECE.

using the real-time equilibrium data. First these data are used to determine a real-time density profile fitting a 4th order polynomial to 5 interferometer channels [18]. Comparing the results to our best off-line analysis [19] including edge-data from the Lithium-beam diagnostic, significant differences are found for ρ >0.8, but further inside they are less than 10%. For H-modes we estimate a fixed shape temperature profile on the basis of $n_e(0.5)$ and β_{pol}. The latter is already available in the discharge control system (DCS), as are the poloidal and toroidal angles of the ECRH launchers. With this information we use a simplified version of TORBEAM (essentially going back from beam-tracing to ray-tracing) to follow the central ray and to determine the maximum of power absoption which we identify with ρ_{ECRH}. A second calculation with slightly different poloidal launching angle allows to invert the problem locally and to estimate the necessary change in angle to get the required change in ρ. Figure 3(e) shows the results of this real-time calculation in comparison to the offline results using modulated ECRH and ECE correlation [20]. There are discrepancies which correspond to half of the width of the deposition profile (at half maximum) as shown in the left part of fig. 2. These first results are now analysed and we expect to be ready for operation with closed feedback loops in the next campaign. As described above, quite complex and CPU-time consuming algorithms have to be run in real-time. These requirements were met by the implementation of an intermediate real-time data-analysis structure separated from the core of the discharge control system (DCS). This intermediate layer does the required analysis and delivers the results to the DCS or other analysis programs using shared memory architectures and standardized interfaces [21].

6. Other major physics applications

In ASEDX Upgrade ECRH and ECCD are used for a large variety of physics studies. In the first part of this section the status of understanding of the ECRH effect on particle and impurity transport is summarised, which gained additional interest in the fully W-coated device. In the second part of this section references to selected other results are given which were highlighted during the oral presentation.

6.1. *Particle transport*

Since the early days of ECRH the effect of 'density pump-out' during application of ECRH has been observed. This flattening of the core density profile is found in low density L-mode plasmas. In [22] it could be shown that this is related to an outward pinch driven by the electron temperature gradient if trapped electron modes (TEM) dominate transport. At higher densities, when transport is dominated by ion temperature gradient modes (ITG), such flattening is not observed, again in agreement with theory, which predicts that ITGs do not drive an outward pinch under these conditions. Studies in H-mode are technically more difficult since ECRH usually is not the dominant heating method. The first systematic study on ASDEX Upgrade is described in [23], which analyses a hot-ion case at low density ($\bar{n}_e \approx 4.0 \cdot 10^{19}$ m^{-3}, $I_p = 1.0$ MA) and case with $T_e \approx T_i$ at $\bar{n}_e \approx 5.5 \cdot 10^{19}$ m^{-3}. In the first case the density peaking is reduced with central ECRH inside $\rho_{tor} < 0.2$, but in the latter case there was no difference with and without ECRH. Again for the low density case TEMs were found to be dominant and for the higher density case ITGs were dominant.

The first study of the effect of ECRH on impurity transport in ASDEX Upgrade is [24] with Si as tracer ions. The inward pinch v_{neo} predicted by Neoclassical theory is sufficiently large to explain the observed inward pinch v_{exp} of Si. In the case without central wave heating, i.e. with NBI heating only (which is largely off-axis), neoclassical particle diffusivity D_{neo} even fits the observed ratio v/D in the plasma center. With central ECRH D_{exp} exceeds the neoclassical value significantly, leading to a reducion of the central Si concentration. In [25] it has been noticed that ECRH is more efficient in suppressing central W accumulation as compared to ICRH, indicating that central electron heating may play a crucial role. Indeed one often finds that central ECRH peaks the T_e-profile inside $\rho_{tor} \approx 0.3$. To gain more insight into the underlying mechanisms, the most unstable micro-instabilities have been analysed for a certain range of T- and q-profiles. The analysis [26] revealed that a mode due to non-adiabatic passing electrons is dominant in the plasma center in cases of low central shear and $1/L_{Ti} << 1/L_{Te}$, which gives rise to an outward particle pinch . Although the experiments confirm qualitatively the low shear and $1/L_{Ti} << 1/L_{Te}$ conditions for cases in which central ECRH suppresses W accumulation, a quantitative comparison to theory is difficult due to saw teeth (and (1,1)-modes) and poor central T_i data in high density H-modes. By using Argon ions as tracer, it was recently possible to determine experimentally the presence of and outward impurity convection in the plasma center under these conditions [27]. Also with SXR it is possible to observe under these conditions hollow radiation profiles in the very center, which also indicate the presence of an outward impurity pinch [28]. The latter results also indicate that the structure of the MHD activity at the $q = 1$ is related to the observation of these hollow radiation profiles.

For H-mode plasmas with high q_{95} (lower I_p), the W-concentration is low even without central ECRH and the effect of central ECRH on the central W-concentration is weak. A theoretical analysis of the most instable modes in these cases is ongoing. In these plasmas the electron heating by ECR leads to a strong increase of the transport in the ion channel (energy and toroidal momentum), similar to the observations in [23]. These new observations and their interpretation especially with respect to the significant changes in plasma rotation are described in [29].

6.2. *Other applications*

This paper so far may give the impression that the usage of the ECRH system on ASDEX Upgrade is restricted to tasks and questions related to operation with the W-coating. This is actually not the case. The following three cases are referenced for illustration, but are not aiming for complete coverage of the whole program:

Studies on the L-H threshold in Deuterium and He are described in [30]. Especially for He plasmas, ECRH allows cleaner experiments than NBI, since the latter cannot use He, but has to inject H or D atoms modifying the ion composition of the plasma. Additionally ECRH allows to address lower plasma densities. For D and He the L/H transition is found at similar power levels for similar densities.

High-beta disruptions can be mitigated in ASDEX Upgrade, if the ECRH is applied at the q=1.5 surface or further outside (but with reducing efficiency). The ECRH is triggered by the locking of a (2,1)-mode, which does lead to a severe disruption in the reference cases without ECRH [31].

In a common effort of several machines, ECRH pre-ionization of the gas prior to plasma breakdown by application of loop-voltage has been studied. The background being the limited available loop voltage in machines with superconducting solenoids such as ITER but also EAST and KSTAR. The successful contribution from ASDEX Upgrade is described in [32].

References

1. LEUTERER, F. et al., Fusion Sci. Technol. **55** (2009) 31.
2. WAGNER, D. and LEUTERER, F., Int. J. Infrared Millimeter Waves **26** (2005) 165.
3. DENISOV, G. et al., Multi-frequency gyrotron with BN Brewster window, in *Conference Digest of the Joint 31th International Conference on Infrared and Millimeter Waves and 14th International Conference on Terahertz Electronics, Shanghai, China (2006)*, page 75.

4. WAGNER, D. H. et al., IEEE Trans. Plasma Sci. **37** (2009) 395.
5. LEUTERER, F. et al., Fusion Eng. Design **74** (2005) 199.
6. DAMMERTZ, G. et al., IEEE Trans. Plasma Sci. **28** (2000) 561.
7. HEIDINGER, R. et al., Tunable double disk window for ECH&CD system of ASDEX Upgrade, in *Conference Digest of the Joint 29th Int. Conf. on Infrared and Millimeter Waves and 12th Int. Conf. on Terahertz Electronics, Karlsruhe, Germany (2004)*.
8. STOBER, J. et al., On the way to a multi-frequency ECRH system for ASDEX Upgrade, in *Proceedings of thw 15th Joint Workshop on Electron Cyclotron Emission and Electron Cyclotron Resonance Heating, EC-15, Yosemite National Park, California, USA, 10-13 March 2008*, edited by LOHR, J., page 433, Singapore, 2009, World Scientific Publishing Co.
9. BELOUSOV, V. et al., Broad band matched windows for gyrotrons, in *Conference Digest of the Joint 34th Int. Conf. on Infrared and Millimeter and Terahertz Waves, Busan, Korea (2009)*.
10. HÖHNLE, H. et al., O2- and X3-heating experiments at ASDEX Upgrade, These conference proceedings.
11. GANDINI, F. et al., Fusion Eng. Design **56-57** (2001) 975.
12. WAGNER, D. et al., Fusion Sci. Technol. (2010), accepted for publication.
13. MARASCHEK, M. et al., Phys. Rev. Lett. **98** (2007) 025005.
14. POLI, E. et al., Fusion Eng. Design **53** (2001) 9.
15. REICH, M. et al., Fusion Sci. Technol. (2010), accepted for publication.
16. HICKS, N. et al., Fusion Sci. Technol. **57** (2010) 1.
17. GIANNONE, L. et al., Fusion Eng. Design **84** (2009) 825.
18. MLYNEK, A., *Real-time control of the plasma density profile on ASDEX Upgrade*, PhD thesis, Universität München, Munich, Germany, 2010, submitted.
19. FISCHER, R. et al., Integrated density profile analysis in ASDEX Upgrade H-modes, in *Europhysics Conference Abstracts (CD-ROM, Proc. of the 35th EPS Conference on Plasma Physics, Hersonissos, Crete, 2008)*, edited by LALOUSIS, P. and MOUSTAIZIS, S., Vol. 32D, p. P–4.010, Geneva, 2008, EPS.
20. REICH, M. et al., ECCD-based NTM control using the ASDEX Upgrade real-time system, in *Europhysics Conference Abstracts (CD-ROM, Proc. of the 37th EPS Conference on Plasma Physics, Dublin, Ireland, 2010)*, p. P2.189, EPS, 2010.
21. TREUTTERER, W. et al., Fusion Sci. Technol. (2010), accepted for publication.
22. ANGIONI, C. et al., Phys. Plasmas **12** (2005) 040701.
23. MANINI, A. et al., Plasma Phys. Controlled Fusion **46** (2004) 1723.
24. DUX, R. et al., J. Nucl. Mater. **313–316** (2003) 1150.
25. DUX, R. et al., Plasma Phys. Controlled Fusion **45** (2003) 1815.
26. ANGIONI, C. et al., Plasma Phys. Controlled Fusion **49** (2007) 2027.
27. SERTOLI, M., *Local effects of ECRH on Argon Transport at ASDEX Upgrade*, PhD thesis, Universität München, Munich, Germany, 2010, submitted.
28. GUDE, A. et al., Hollow central radiation profiles and inverse sawtooth-like crashes in ASDEX Upgrade plasmas with central wave heating, in *Euro-*

physics Conference Abstracts (CD-ROM, Proc. of the 37th EPS Conference on Plasma Physics, Dublin, Ireland, 2010), p. P4.124, EPS, 2010.

29. MCDERMOTT, R. et al., Core toroidal rotation changes observed with ECRH power in NBI heated H-modes on ADSEX Upgrade, in *Europhysics Conference Abstracts (CD-ROM, Proc. of the 37th EPS Conference on Plasma Physics, Dublin, Ireland, 2010)*, p. P1.1062, EPS, 2010.
30. RYTER, F. et al., Nucl. Fusion **49** (2009) 062003.
31. ESPOSITO, B. et al., Avoidance of disruptions at high betan in ASDEX Upgrade with off-axis ECRH, in *Proc. of the 23rd IAEA Conference Fusion Energy (CD-Rom), Daejon, Korea, October 2010*, p. EXW/10–2Ra, Vienna, IAEA.
32. SIPS, A. et al., Nucl. Fusion **49** (2009) 085015.

FEEDBACK CONTROL OF TEARING MODES THROUGH ECRH WITH LAUNCHER MIRROR STEERING AND POWER MODULATION USING A LINE-OF-SIGHT ECE DIAGNOSTIC

B.A. HENNEN[†], E. WESTERHOF, M.R. DE BAAR, W.A. BONGERS, D.J. THOEN[††]

FOM-Institute for Plasma Physics Rijnhuizen, Association EURATOM-FOM, Trilateral Euregio Cluster, PO Box 1207, NL-3430 BE Nieuwegein, The Netherlands

P.W.J.M. NUIJ, M. STEINBUCH

Control Systems Technology Group, Eindhoven University of Technology PO Box 513, NL-5600 MB Eindhoven, The Netherlands

A. BÜRGER, J.W. OOSTERBEEK[‡] AND THE TEXTOR-TEAM

Institut für Energieforschung - Plasmaphysik, Forschungszentrum Jülich, Association EURATOM-FZJ, Trilateral Euregio Cluster, 52425 Jülich, Germany

A demonstration of real-time feedback control for autonomous tracking and stabilization of $m/n = 2/1$ tearing modes in a tokamak using Electron Cyclotron Resonance Heating and Current Drive (ECRH/ECCD) is reported. The prototype system on TEXTOR combines in the same sight-line an Electron Cyclotron Emission (ECE) diagnostic for tearing mode sensing and a steerable ECRH/ECCD antenna. The mode location is retrieved from the ECE measurements and serves as input for a control loop, which aligns the ECRH/ECCD deposition with the tearing modes by steering of a launcher mirror. The alignment is achieved by matching the mode location in the sensor spectrum with the fixed ECRH/ECCD actuator frequency. The control response is dominated by the response of the mechanical launcher. Analysis of the launcher dynamics receives special emphasis in the control design. In addition, the ECRH/ECCD power is modulated in phase with the rotation frequency of the O-point of the tearing modes using a feedback loop, which extracts the mode's frequency and phase from the ECE data. The experimental results demonstrate the capabilities of the control system to track and suppress tearing modes in real-time. A relatively simple control design suffices to meet the performance requirements demanded for effective tearing mode suppression.

[†] Author is both affiliated to FOM Rijnhuizen and the CST group at the Eindhoven University of Technology, NL.

[††] Author is currently affiliated to the Kavli Institute of Nanoscience Delft, NL.

[‡] Author is currently affiliated to the FUSION group at the Eindhoven University of Technology, NL.

1. Introduction

On the mid-sized, circular tokamak TEXTOR (major radius $R = 1.75$ m, minor radius $a = 0.46$ m),[1] a powerful and flexible Electron Cyclotron Resonance Heating and Current Drive (ECRH/ECCD) system (140 GHz, 1 MW, 10 s)[2] is operated in combination with a line-of-sight Electron Cyclotron Emission diagnostic[3,4] and the Dynamic Ergodic Divertor (DED) coils to study the suppression of tearing modes. The DED is routinely used at TEXTOR to perturb the magnetic field topology. Previous studies demonstrated its merits in the creation of large $q = m/n = 2/1$ and 3/1 tearing modes.[5] The instrumentation provides TEXTOR with an ideal set of tools for studies on real-time control of tearing modes. Stabilization of tearing modes by localized heating and current drive, either in open loop or feedback controlled, has previously been demonstrated or is presently being developed on many machines.[6-12] This work presents the application of real-time feedback control for autonomous tracking and stabilization of $m/n = 2/1$ tearing modes using ECRH/ECCD with controlled launcher mirror steering and power modulation using a line-of-sight ECE diagnostic.

2. Feedback control of tearing modes and line-of-sight ECE

Suppression of tearing modes requires precise positioning of the localized ECRH/ECCD deposition with respect to the mode.[13] An alignment accuracy of less than 1-2 cm is required. The alignment between ECRH/ECCD deposition and the tearing mode can be obtained and maintained using feedback control schemes.[6,8,9,11,12] A high power gyrotron and a steerable ECRH/ECCD launching antenna are generally operated as actuators in these schemes, while diagnostics such as Mirnov coils, soft x-ray and electron cyclotron emission are applied as feedback sensor. Alternative actuations via rigid horizontal shifts of the plasma column or changes in the magnetic field have been reported.[6,8] Three important control tasks are distinguished: First, based on measurements of the ECRH/ECCD deposition location and the radial location of a tearing mode, the control system should be able to align the ECRH/ECCD deposition with respect to the centre of the tearing mode;[13] second, whenever the alignment of ECRH/ECCD deposition onto the mode is perturbed, the control system should compensate for this perturbation by offering tearing mode tracking capabilities in real space; third, to achieve the most efficient and adequate suppression, the ECRH/ECCD power must be modulated in phase with the rotation of the tearing modes in the poloidal cross-section.[10,14] A real-time feedback control system is required to perform these tasks. It must align the ECRH deposition fast and accurate with the mode whilst guaranteeing disturbance rejection and robustness of the alignment.

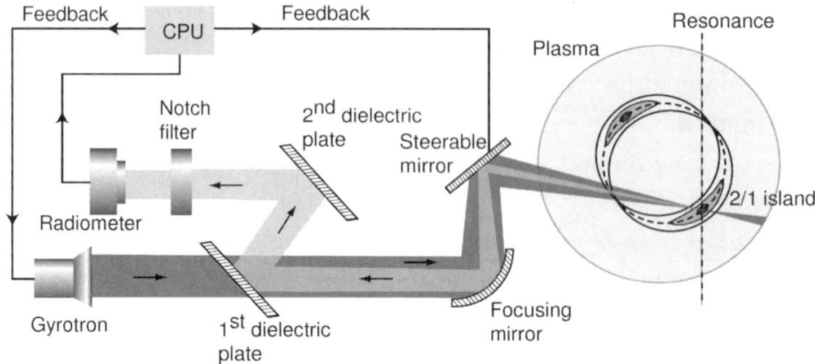

Figure 1: Outline of a real-time tearing mode control system using an identical line-of-sight for the ECRH/ECCD actuation and the ECE sensing. The ECE signals gathered near the ECRH/ECCD deposition location on the high-field side are fed back into a feedback loop for controlled suppression of tearing modes using ECRH/ECCD, by generating appropriate feedback action on a steerable mirror and the gyrotron power supply.

Here, a dedicated tearing mode control system (see Figure 1) is presented which was designed, implemented and operated at TEXTOR. A unique feature of this system is the use of a single transmission path for measurement of the ECE emitted by the plasma and transmission of ECRH/ECCD power into the plasma.[3] The steerable ECRH/ECCD launching antenna, which in the case of TEXTOR can be steered both in toroidal and poloidal directions, is then used to collect ECE besides transmitting ECRH/ECCD power. The 'line-of-sight ECE concept' thus merges the observation of relevant tearing mode properties with the application of ECRH/ECCD as the actuator for suppression of the modes.[3] The 'line-of-sight' concept has advantages over conventional tearing mode suppression schemes as it provides information on a mode's location and phase in the direct vicinity of the ECRH/ECCD-deposition. The use of a single transmission path guarantees identical refractive properties for the EC emission and the ECRH/ECCD beam, which is justified under the assumption of identical emission and absorption profiles. Hence, this system provides inherently accurate ECRH/ECCD beam steering and alignment of the ECRH/ECCD deposition with respect to a mode while avoiding the use of real-time equilibrium reconstruction and ray-tracing codes in the feedback control loop, as required by comparable systems which operate actuators and sensors in different coordinate frames with respect to each other. The line-of-sight concept also guarantees accurate alignment in the presence of disturbances.

Although the line-of-sight ECE approach simplifies feedback control of tearing modes from a conceptual point of view, its technical implementation is challenging: EC emission at nW power levels must be separated from the high power ECRH/ECCD beam. A prototype line-of-sight system[4] has been installed and operated on TEXTOR in a quasi-optical transmission line, where a resonant dielectric quartz plate, based on the Fabry-Perot interference principle, was employed to separate the low-power ECE signal components at selected ECE frequencies from the high power ECRH/ECCD component.[4] The plate is highly transparent (> 95%) at 140 GHz, i.e. the ECRH/ECCD frequency, but has maxima in reflection (at about 35%) at 132.5, 135.5, 138.5, 141.5, 144.5, and 147.5 GHz. Hence, ECE is measured at six radial locations distributed equally in space around the actuator frequency of 140 GHz. The radial spacing of the radiometer channels is 3 GHz ~ 3 cm. A second resonant dielectric plate in tandem with the first, followed by an 80 dB notch filter protects the ECE radiometer from 140 GHz stray radiation.[4]

3. Methods

3.1. *Retrieval of a tearing mode's radial location*

The line-of-sight ECE diagnostic has been incorporated into the real-time tearing mode control system. For control purposes, the radial location of a tearing mode must be determined from the line-of-sight ECE measurements. The radial location of a mode residing on the resonant surface r_s, is directly inferable from the telltale fluctuations of the electron temperature, caused by the rotation of the mode. These fluctuations show a 180° phase reversal when comparing ECE measurements on opposite sides of the resonant surface r_s. This fact is used to extract an estimate of the $m = 2$, $n = 1$ tearing mode location. The tearing mode location is derived from ECE measurements using correlation techniques. For this purpose, the six ECE channels are sampled at 100 kHz and fed into a field programmable gate array (FPGA). On the FPGA, the information contained in all channels is combined and processed by a correlation algorithm, which distinguishes the radial location where the 180° phase reversal occurs within the ECE spectrum.[15] The output of the algorithm is an estimate of the radial location of the $m/n = 2/1$ tearing mode r_s, specified as an EC frequency $f_{EC,\ tearing\ mode}(r_s)$ in GHz. The processing time of the FPGA implementation of the algorithm is 16 μs.

3.2. *Feedback controlled alignment of ECRH/ECCD with respect to a tearing mode*

The tearing mode controller[15] used throughout the experiments executes the following steps: First, the signals from all six ECE channels are fed into the tearing mode recognition and localization algorithm discussed in the previous section. When a mode is detected, the algorithm provides the EC frequency at which the mode is localized. In the second step, this frequency is compared to the reference 140 GHz ECRH/ECCD frequency. A feedback control error e = 140 – $f_{EC,\ tearing\ mode}$ GHz is defined and a standard PI controller is designed, which should minimize the error e to establish alignment. The controller provides a correction to the poloidal steering angle of the launching mirror which is added to the current steering angle and fed into a separate real-time controller for positioning of the launching mirror. When alignment is achieved, i.e. when the EC frequency at which the mode is detected matches the ECRH/ECCD frequency (and e approximately $\leq$ 0.5 GHz), the ECRH/ECCD power is switched on. Subsequently, the position control loop remains active to keep the ECRH/ECCD aligned with the mode. All steps are implemented on the field programmable gate array (FPGA).

Flexible, fast and accurate positioning of the ECRH/ECCD-beam is assured by a real-time position controller which has been derived through experimental analysis of the dynamics of the electro-mechanical ECRH/ECCD launcher using frequency response function measurement techniques. The frequency response measurements allow the extraction of transfer function estimates for the launcher, which in turn are used to design a dedicated feedback controller. The result is a standard proportional integral derivative (PID) controller, a lead/lag filter and a low-pass filter.[15] A feed-forward is added to compensate for the highly dominant stick-slip friction behavior of the launcher. In simulations, the stability robustness and performance of the feedback controlled system have been verified. The controller was then implemented in real-time control experiments on a mock-up of the actual launcher system. After verifying its performance, the controller was transferred to the real launcher system and the performance was further optimized. The poloidal injection angle can be swept from –30° to +30° within 100 ms. The maximum steady-state positioning error is 0.6° for this large sweep.

3.3. *Synchronous ECRH/ECCD power modulation*

A parallel control loop accounts for the modulation of the ECRH power synchronously in frequency and phase with the mode rotation. To this end, the signal of one ECE channel (typically channel 2 at 135.5 GHz) is fed into an analog phase locked loop (PLL) circuit. The sinusoidal output signal of the PLL is locked in phase and frequency to the first harmonic of the noisy ECE input signal, and is translated into a block-wave pulse train, which is used as input to the ECRH/ECCD power supply.[15] The output power is controlled between a constant lower level of 70 kW and a peak level up to 850 kW by varying the electron beam voltage from 53.5 to 71 kV. The power modulation is controllable in a frequency range from 300 Hz up till 5 kHz during 10 s for the particular PLL implementation considered here.

4. Experimental results

Tearing mode control has been studied in plasmas with a toroidal field $B_t = 2.25$ T, plasma current $I_p = 300$ kA, and central line average density $n_e = (2\text{-}2.2) \times 10^{19}\ \mathrm{m}^{-3}$. The presented experiments focus on the control of $m/n = 2/1$ tearing modes. During most of the discharge, co-tangential Neutral Beam Injection (NBI) is applied at a power level of 300 kW. In these plasmas a locked 2/1 tearing mode is created by ramping the current in the DED-coils to 2 kA in AC+ mode. This 2/1 tearing mode rotates at 1 kHz as it is locked in frequency and phase to the DED perturbation field. The toroidal injection angle of the launcher is fixed at –4° effecting co-ECCD throughout all experiments presented here.

Figure 2 shows an experimental result revealing the capability of the controller to align ECRH/ECCD with a tearing mode and suppress the mode in real-time. An $m = 2$, $n = 1$ mode is excited by the DED, and is kept locked in rotation with the DED perturbation field. At $t = 2$ s the real-time controller is switched on and the mirror is steered to align ECRH/ECCD with the mode location as reflected by the signals in the third and fourth panel of Figure 2. As soon as the achieved alignment is sufficiently accurate (i.e. ≤ 0.5 GHz) the gyrotron is triggered by the control system and ECRH/ECCD is applied to suppress the mode to a constant width. After switch off of the ECRH/ECCD the mode grows to a saturated width again. Note that a comparable experiment has been done, with a short DED pulse only in the beginning of the discharge. In that case, the $m = 2$, $n = 1$ mode is excited by the DED but evolves naturally for the remainder of the discharge. This naturally evolving mode is not locked in rotation to the DED and its growth is not driven by the DED perturbation field anymore. It was demonstrated that a complete suppression of a $m = 2$, $n = 1$ mode is feasible under full feedback control in such experiment with an ECRH/ECCD power of 200 kW.[15]

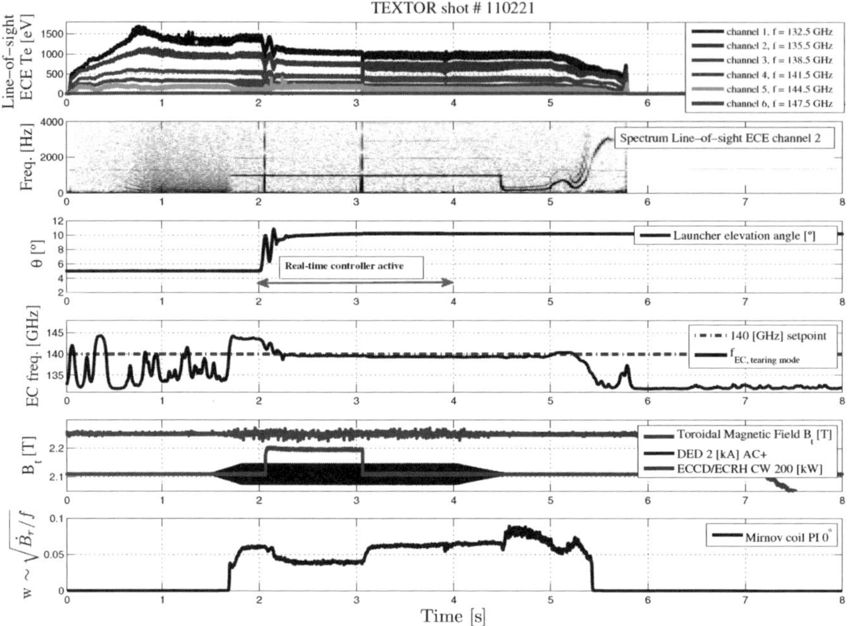

Figure 2: Real-time tearing mode suppression in TEXTOR discharge 110221. The top panel shows the six line-of-sight ECE time traces, the 2nd panel shows the time-frequency spectrum of one of the line-of-sight ECE signals revealing the rotation frequency of a DED-excited $m = 2$, $n = 1$ tearing mode. The 3rd panel shows the launcher poloidal injection angle, which is feedback controlled in the time interval $t = 2$-4 s. The 4th panel shows in black the EC frequency identifying the tearing mode location as time evolves and the 140 GHz actuation frequency serving as the control reference as a dashed line in red. The 5th panel displays the discharge scenario: in green the toroidal magnetic field B_t, in red the applied ECRH/ECCD pulse and in black the DED current, which is ramped in AC+ to trigger a $m = 2$, $n = 1$ mode. The DED is operated in flat-top to keep the mode locked to the DED perturbation field. The last panel shows the width of the $m = 2$, $n = 1$ mode as derived from a Mirnov coil placed poloidally on the low field side at the equatorial axis.

Figure 3 demonstrates the capability of the real-time control system to follow a changing mode location during application of 200 kW of ECRH/ECCD. An AC+ DED locked tearing mode is excited at $t = 1.7$ s. The real-time controller is switched on at $t = 2$ s. When alignment is achieved the controller switches on ECRH/ECCD. Again, the mode is not suppressed completely as the DED remains active to drive the mode. From $t = 2.5$ s a slow ramp down of the toroidal magnetic field B_t mimics a change in real space of the resonant surface at which the mode is located. The controller correctly tracks the 'changing position of the island' by compensation of the perturbed alignment through steering of the launcher poloidal injection angle.

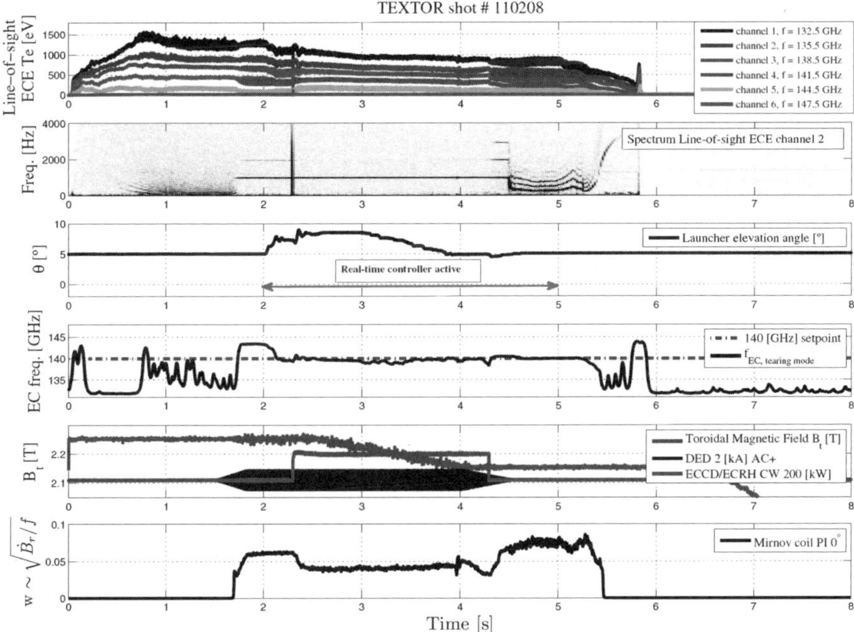

Figure 3: Real-time tracking of a partially suppressed tearing mode in TEXTOR discharge 110208. The subsequent panels show the same signals as in Figure 2. A tracking experiment is shown where the alignment between the ECRH/ECCD deposition location and the tearing mode centre is perturbed by a B_t ramp from 2.25 T to 2.15 T. Continuous ECRH/ECCD at 200 kW is applied in the time interval t = 2.3-4.3 s. The launcher poloidal injection angle is actively controlled to keep the ECRH/ECCD deposition aligned with the tearing mode centre as the alignment is perturbed by the changing magnetic field.

Figure 4 illustrates the capabilities of the PLL controller to follow rapid changes in the rotation frequency of the mode and to modulate the ECRH/ECCD power in phase with the passage of the island center in front of the beam. A natural tearing mode is generated by early neutral beam heating. After ECRH/ECCD switch-on the mode spins up in frequency as it is partly suppressed by ECRH/ECCD. In both the early low rotation frequency phase and the subsequent high rotation frequency phase the PLL controlled ECRH/ECCD power pulse train accurately tracks the mode rotation frequency and phase.

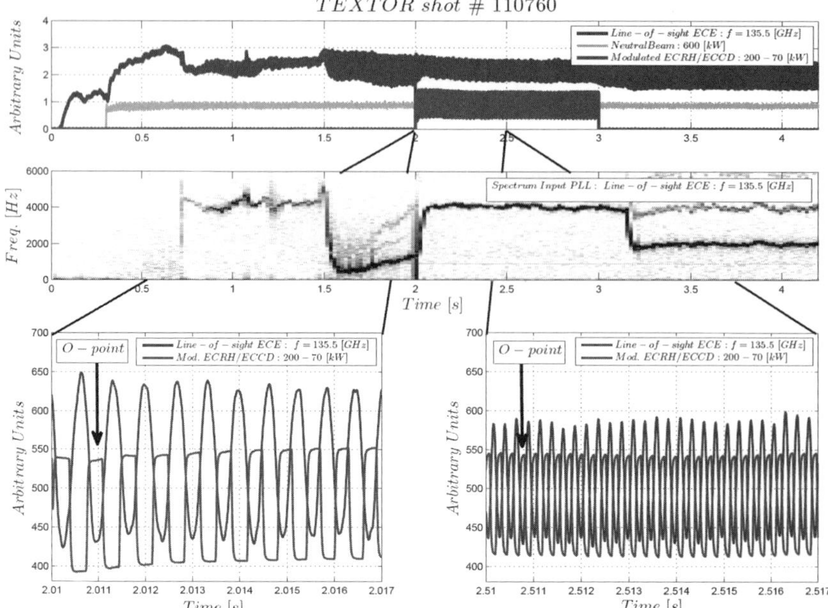

Figure 4: Real-time ECRH/ECCD power modulation synchronous to the rotation of a partially suppressed tearing mode in TEXTOR discharge 110760. The top panel shows the discharge evolution: In green the neutral beam power which is switched on early to excite a natural $m = 2$, $n = 1$ tearing mode. The line-of-sight ECE signal at 135.5 GHz serving as the input to the PLL circuit is indicated in blue. The modulated ECRH/ECCD (in red) is switched on from $t = 3\text{-}4$ s and modulated between 200 kW and 70 kW. The second panel shows the time-frequency spectrum of the 135.5 GHz ECE signal, revealing the evolution of the rotation frequency of the $m = 2$, $n = 1$ mode. The bottom two panels show a zoom of the 135.5 GHz ECE signal and the modulated ECRH/ECCD power during the early slow mode rotation frequency phase (i.e. 1 kHz) and the later fast rotation phase (i.e. 4 kHz), respectively. Passages of the island center (indicated as O-point) coincide with minima in the 135.5 GHz ECE signal.

5. Summary and Conclusions

This paper demonstrates real-time feedback control for autonomous tracking and stabilization of $m/n = 2/1$ tearing modes in a tokamak using ECRH/ECCD. The control system based around a line-of-sight ECE receiver suppresses the tearing modes through ECRH/ECCD with active launcher mirror steering and modulation of the ECRH/ECCD power under full closed loop feedback control. The results convincingly demonstrate the capabilities of a real-time tearing mode controller based on co-aligned ECRH/ECCD actuation and ECE sensing. ECRH/ECCD is aligned with the modes with sufficient accuracy, the system provides tracking capabilities in case the alignment is perturbed and is able to modulate the ECRH/ECCD power in synchronization with a changing mode rotation frequency. The simplicity and robustness of the applied techniques makes this system ideal for implementation in future tokamaks. Significant further developments, however, are needed to realize this. Most importantly, the system should be adjusted to offer CW capability. The presented same-sight-line ECE diagnostic is limited in pulse length: the high power mm-wave absorption in the dielectric plate used to separate the ECRH/ECCD and ECE radiation results in heating of the plate, which changes its resonant properties. Moreover, the present line-of-sight system uses quasi-optical transmission with wide beams. For implementation on other tokamaks, a waveguide compatible line-of-sight diagnostic is being developed.[16]

Acknowledgements

This work has been performed in the framework of the NWO-RFBR Centre of Excellence (grant 047.018.002) on Fusion Physics and Technology. The work was supported by NWO, ITER-NL and the European Communities under the contract of the Association EURATOM/FOM, and was carried out within the framework of the European Fusion Programme. The views and opinions expressed herein do not necessarily reflect those of the European Commission.

References

1. U. Samm, *Fusion Sci. Techn.* **47** (2005) 73.
2. E. Westerhof, et al., *Fusion Sci. Techn.* **47** (2005) 108.
3. E. Westerhof, et al., *Proceedings of the 13th Workshop on ECE & ECH* Nizhny Novgorod, Russia, 17-20 May (2005) 357.
4. J.W. Oosterbeek, et al., *Rev. Sci. Instrum.* **79** (2008) 093503.
5. H.R. Koslowski, et al., *Nucl. Fusion* **46** (2006) L1.
6. R.J. La Haye, et al., *Phys. Plasmas* **13** (2006) 055501.
7. C.C. Petty, et al., *Nucl. Fusion* **44** (2004) 243.
8. D.A. Humphreys, et al., *Phys. Plasmas* **13** (2006) 056113.
9. A. Isayama, et al., *Nucl. Fusion* **43** (2003) 1272.
10. M. Maraschek, et al., *Phys. Rev. Letters* **98** (2007) 025005.
11. A. Manini, et al., *Fusion Eng. Des.* **82** (2007) 995.
12. J. Berrino, et al., *Nucl. Fusion* **45** (2005) 1350.
13. R.J. La Haye, et al., *Nucl. Fusion* **48** (2008) 054004.
14. A. Isayama, et al., *Nucl. Fusion* **49** (2009) 055006.
15. B.A. Hennen, et al., *Plasma Phys. Control. Fusion* **52** (2010) accepted for publication.
16. W.A. Bongers, et al., *Fusion Sci. Technol.* **55** (2009) 188.

STATUS OF THE ECRH SYSTEM ON HL-2A*

JUN ZHOU, JUN RAO, BO LI, MEI HUANG, ZHIHONG LU, ZIHUA KANG, HE WANG KUN FENG, MINGWEI WANG, GANGYU CHEN, YINGNAN BU, CHAO WANG, BO LU

Southwestern Institute of Physics, Chengdu, 610041, China

3MW ECRH system with six 68GHz/500kW gyrotrons has been developed on HL-2A. H mode discharges have been realized with ECRH and NBI. NTM has been observed during ECRH heating. Transient ITB after ECRH switch-off has been achieved. Stabilization of tearing mode and sawtooth control with ECRH has been studied

1. Introduction

HL-2A tokamak with double null diverter has the major radius of 1.65 m and minor radius of 0.40m [1]. NBI system with four ion sources (power up to 1.5MW) and LHCD (1 MW/2.45 GHz) systems were improved or installed. In order to explore ECRH/ECCD experiments, such as transport study and MHD control, a 2MW /68GHz/1s ECRH system had been developed in HL-2A tokomak [2]. The developed ECRH system consists of four 500kW subsystems The output of the gyrotrons, manufactured by GYCOM has been demonstrated as 68GHz /500kW/1s, horizontal linear polarization, Gaussian beam and its purity is up to 98.4% after matching optical unit. An electronic efficiency of 50% has been determined with optimized parameters.

The transmission line includes a matching optical unit, 7m oversized corrugated circular waveguide, a miter bend with a plane mirror or a polarizer mirror, sliding waveguides and a DC break. A sinusoidal grooved polarizer has been developed. With the low power test results, the X-mode purity of the polarizer reaches 85% and almost 100% when the toroidal angle is 15^0, which is the best angle for current drive efficiency. The four transmission lines have been aligned carefully. The biggest tilt angle of the lines is 0.14^0. 500kW/500ms high power EC wave can be transmitted through the line with the polarizer in air and the transmission efficiency is over 90%.

The Gaussian beam is injected into plasma from the low field side by two launchers. The steering mirrors in the launcher can be rotated to choose the

* this work partially supported by Core-University Program on Plasma and Nuclear Fusion

angle of injection in the toroidal and poloidal direction between 0-30^0. It is possible to explore on and off-axis plasma heating over half of the plasma radius and the electron cyclotron current drive (ECCD).The location of the Gaussian beam waist is 580mm from the center of plasmas and the beam radius is 37mm in the center of HL-2A, which can fulfill the requirements of the localized heating experiments and small enough for local MHD control, compared to the minor plasma radius (40mm).

2. A new 1MW ECRH system

In order to do more powerful ECRH/ECCD experiments, especially H mode discharges triggered by ECRH, a new 1MW ECRH system has been developed. The new ECRH system consists with two 68GHz/0.5MW/1.5s subsystems. The new system is conditioning now. Six gyrotrons are going to operate together this year.

The size of the collectors of the two new gyrotrons was increased, which can sustain more energy burden. The main specified parameters of the two gyrotrons are given in table 1

Table 1.Main parameters of the 68GHz/0.5MW/1.5s gyrotron.

Output power	500kW after MOU	Frequency	68.2GHz
Power in MOU	22kW	Pulse duration	1.5s
Output beam	Gaussian beam	Mode purity after MOU	98%
Window	BN	Collector	Depressed
Cathode voltage	-53kV	Anode voltage	+20.8kV
Beam current	19.2A	Efficiency	52.5%

The six gvrotrons are very near to HL-2A tokomak, only about 7m from the device. The transmission lines are less than 10m, which are consisted of corrugated circular waveguide, two or three miter bends, a polarizer, two sliding waveguides and a DC break, showed in figure 2. After carefully re-alignment of the old two transmission lines, 500kW/500ms high power EC wave can be transmitted through the line with the polarizer in air. The new transmission lines are conditioning now. Because only two Φ350mm ports are available for six wave beams, one launcher for two beams should be changed for four beams. The new launcher is consisting of four focusing mirrors and one plane mirror, which is showed in Figure1. The plane mirror has two planes, which has 91.2^0 angles for reflecting the beams to the center of the plasmas. Because of the limited space, it is very difficult to design the launcher to change the injection angle of the beams. The power density of each beam is 158MW/m^2, and the radius is 31.7mm.For one mirror, the ohmic loss and diffraction loss are 0.27% and 0.64%. By the finite analysis software ANSYS, thermal analysis of the mirror has been done. The temperature increases less than0.47℃, so cooling system is not necessary.

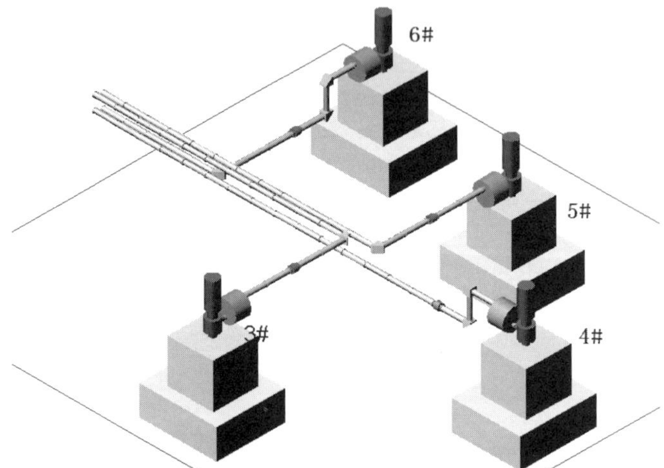

Fig1 Setup of four gyrotrons and four transmission lines

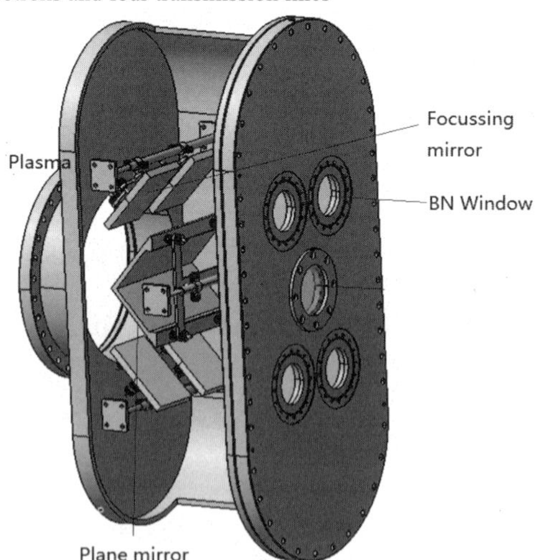

Figure 2.The launcher for four 68GHz wave beams.

3. High powerful ECRH experiments

Based on the ECRH system, series of O1 or X2 experiments have been explored in HL-2A tokamak. The total output power with four gyrotrons is up to 1.65MW and pulse duration is up to 600ms. During toroidal magnetic field 2.43T, plasma current 300kA, plasma density $1.7\times10^{19}cm^{-3}$ and EC power 1.57MW, the central electron temperature Te was increased from 1.2keV to 4.93keV, which is showed in Figure3.

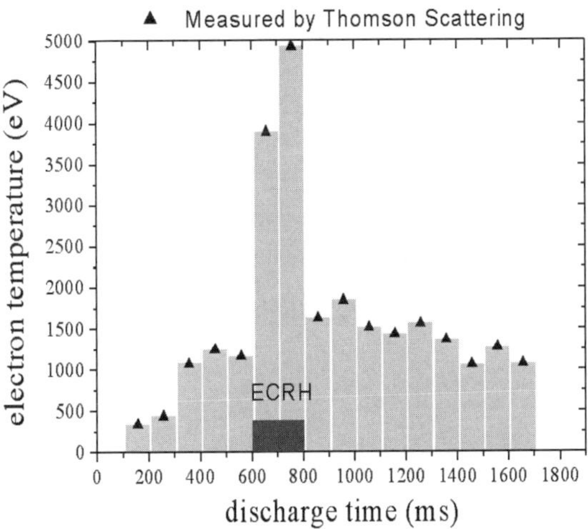

Figure3 Te increased from 1.2keV to 5keV during 1.57MW ECRH

MHD instabilities during ECRH have been studied. Non-standard sawtooth, such as saturated sawtooth, compound sawtooth, gaint sawtooth, a humpback or a hill, were found. Snake-like oscillation of the soft X-ray was suppressed by ECRH. ITB triggered by ECRH switch-off have been observed. When the heating location is from q=1 and q=2 surface, the central temperature measured by ECE increased and edge temperature decreased after ECRH/ECCD switch-off. As reported in [3], normal and modulated ECRH/ECCD in X2 mode experiments to suppress tearing mode have been explored. Experimental results show that m=2/n=1 tearing mode was suppressed, meanwhile the plasma density, temperature, stored energy and energy confinement time increase steadily throughout the modulated ECRH period. It seems that the modulated ECRH/ECCD is more powerful to suppress the m=2/n=1 mode. The m/n=2/1 NTMs were observed in a single null divertor discharge with low density and low beta plasma heated by 1MW ECRH, leading to a fall of 20% in βp.

Figure4 shows the H-mode discharge with the NBI power of 0.8MW and EC power of 1MW, plasma current of 180 kA, plasma density of $2.2\times10^{19}\,m^{-3}$ and toroidal magnetic field of about 1.3 T. The L-H mode transition occurs after 120 ms from ECRH injection and 20 ms from NBI injection. Energy, particle and impurity confinement improve simultaneously. Type III ELMs was driven with frequency of 500Hz..

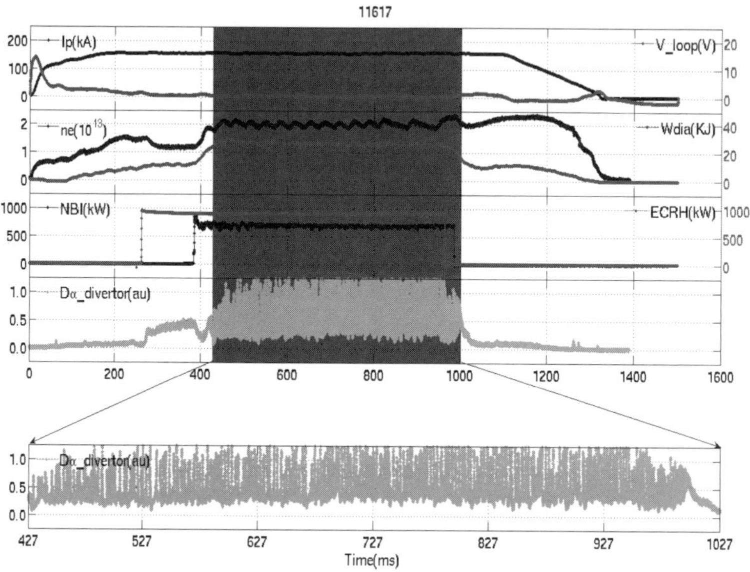

Figure 4 H-mode discharge on HL-2A with ECRH and NBI

4. Summary

3MW ECRH with six 68GHz/500kW gyrotrons has almost been developed on HL-2A. The six gyrotrons are going to operate together this year. With high powerful ECRH system, some MHD control, transport study, and confinement experiments have been explored on HL-2A. H mode discharges have been achieved with ECRH and NBI. For operation of HL-2A in higher parameters, 140GHz ECRH system is under discussion. H mode discharge triggered by ECRH is proposed.

Acknowledgments

Authors appreciate the colleagues of GYCOM Ltd, who provide the good gyrotrons. The authors would like to thank R Magne,G. Giruzzi,X Zou (CEA Cadarache), F Leuterer, D Wagner (IPP Garching), S Kubo,T Shimozuma

(NIFS) H Idei(Kyushu Univ.), K Nagasaki(Kyoto Univ.).This work could not have been done without their help, stimulating suggestions and useful discussions. The authors would like to thank the HL-2A staff, particularly the ECRH team, for conducting the experiments.

References

1. X.R.Duan et al.*Nucl.Fusion*, **49**, 104012(2009).
2. Zhou Jun et al., IAEA Technical meeting on ECRH Physics and Technology for Large Fusion Devices Gandhinagar, India., 2009
3. Yi Liu et al. 22nd IAEA Fusion Energy Conference, Geneva, Switzerland, EX-/P9-2, 2008

COMMISSIONING AND OPERATION RESULTS OF 110 GHZ ECH SYSTEM IN KSTAR

MI JOUNG, WON-SOON HAN, JONGSU KIM, YOUNG-SOON BAE, HYUNG-LYEOL YANG

National Fusion Research Institute, Gwahangno 113, Yuseong-gu Daejeon, 305-333, Korea

SEUNGIL PARK, HEEJIN DO, WON NAMKUNG, MOOHYUN CHO

Department of Physics, POSTECH, San 31, Hyoja-dong, Nam-gu Pohang, 790-784, Korea

YURI GORELOV, JOHN LOHR, JOHN DOANE

General Atomics, P.O. Box 85608 San Diego, CA92186, USA

A 110 GHz GYCOM gyrotron, which was loaned from DIII-D in General Atomics (GA) including a matching optics unit (MOU) and a magnet, was successfully commissioned and used for the second harmonic ECH-assisted startup in 2009 Korea Superconducting Tokamak Advanced Research (KSTAR) campaign. The gyrotron was aligned at the magnetic field axis of the magnet by adjusting aligners on top and bottom flanges of the magnet during the installation procedures. The measured images of the output beam profile at the window using a burn paper showed a good agreement with those measured in DIII-D. The maximum available RF power was just about 250 kW, which was measured at the terminal dummy load, due to the limitation of the existing power supply of which nominal beam voltage and current were 63 kV and 20 A, respectively. The 110 GHz, 250 kW EC wave, which corresponds to the second harmonic resonance wave to the toroidal field of 2 T, was injected to KSTAR for the plasma start-up and current ramp-up. The plasma start-up was successfully and reliably achieved through the second harmonic ECH pre-ionization. The injection mode was the X-mode with oblique launch angle to the toroidal magnetic field. This paper reports the initial commissioning and operation results of the 110 GHz ECH system. Moreover, the experimental results of the ECH-assisted start-up during the 2009 KSTAR campaign are presented and discussed.

1. Introduction

A second harmonic electron cyclotron heating (ECH) system has been requested as an essential tool to start-up in large superconducting tokamak devises with low loop voltage such as ITER [1-3]. A reliable plasma start-up was also achieved on Korea Superconducting Tokamak Advanced Research (KSTAR) with the low loop voltage ranged from 3 V to 4 V at the inner post, which is corresponding to an electric field $E_\varphi \sim 0.37$ V/m to $E_\varphi \sim 0.49$ V/m [4]. EC wave injection before the onset of the inductive voltage had provided the enough pre-ionization and additional heating during the current ramp-up phase.

In 2008 KSTAR campaign, the 2nd harmonic ECH-assisted startup was attempted using the CPI gyrotron operated at 84 GHz. The available EC power at the output window of the gyrotron was 500 kW, and the launched beam power to the KSTAR was 350 to 400 kW. For the second KSTAR campaign, a 110 GHz 2nd harmonic ECH-assisted start-up is newly implemented using the 110 GHz GYCOM gyrotron system which was loaned from General Atomics (GA) and adapted to the existing KSTAR ECH system because the 84 GHz gyrotron was found to have failures and sent to the factory for the repair due to the vacuum leak at the collector.

This paper presents commissioning and operation results of 110 GHz gyrotron system in the existing KSTAR ECH system and especially, the transmission losses for 110 GHz EC wave in the transmission line system optimized to 84 GHz EC wave. And the experimental results of the second harmonic 110 GHz ECH-assisted start-up are reported.

2. Commissioning of 110 GHz ECH System

2.1. *110 GHz GYCOM Gyrotron System*

A 110 GHz gyrotron, matching optics unit (MOU), superconducting magnet, heater transformer, and several power supplies except high voltage power supply and transmission line components were transferred from GA before the 2009 KSTAR campaign as a collaboration program between KSTAR and GA. The 110 GHz GYCOM-made gyrotron was specified with an output power of 1 MW for 2 s pulse length and 0.5 MW for 5 s pulse length with the efficiency of about 38 %. It uses a boron nitride disc for the output window. The electron gun is a diode magnetron injection gun with metal porous cathode. A schematic diagram of the 110 GHz gyrotron is shown in fig. 1. According to GA experimental results, the gyrotron without depressed collector showed RF power of 800 kW for the nominal voltage and current of 70 kV and 30 A, respectively.

Installation of 110 GHz gyrotron and the cryostat has been successfully completed with help of two GYCOM experts. Most important work was to reassemble the cryostat with two nitrogen screens and to align the center of the mounting flanges both on the top and bottom of the cryostat to the maximum magnetic field line. As the nitrogen screen made in copper sheet was installed between the helium tank and the wall of the cryostat at room temperature, the screens prevent helium tank from directly contact with the wall at room temperature. After the magnet was cooled down, the superconducting magnet was carefully inspected by applying a current. The nominal operation current of the main magnet was 61 A.

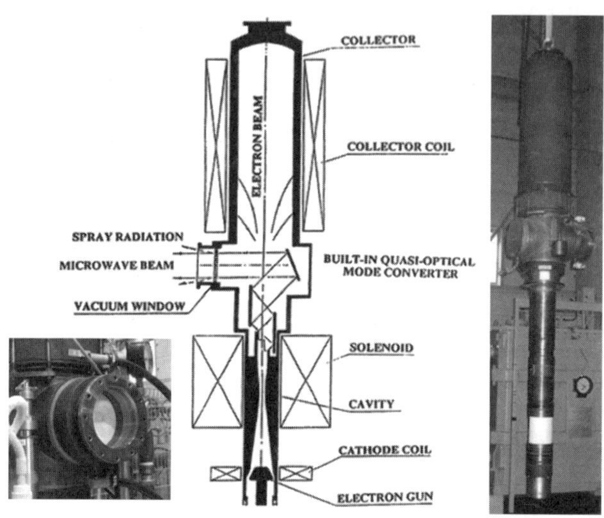

Figure 1. Schematic diagram and pictures of 110 GHz GYCOM gyrotron.

The MOU consists of two phase correction mirrors. The output RF beam with TEM_{00} mode from the gyrotron is transformed to HE_{11} mode by the MOU mirrors and matched to the 31.75 mm ID circular corrugated waveguide inserted into the MOU. Two mirrors are adjusted by using the knobs. The MOU system includes adjustable micrometers that are attached on the back of the mirrors to align the beam. The MOU chamber is held by the springs fixed on the supporting structure to protect the chamber against external shocks and has couples of view ports both for arc detection and for RF signal measurement.

2.2. *Commissioning Results*

The initial operation of the gyrotron was performed using an octanol load installed at the gyrotron output window in the atmospheric space [5]. The octanol load was put in the cubic chamber which is made of RF absorbing eccosorb material. The RF output from the gyrotron window was observed by a diode detector installed at the backside of the dummy load. In addition to the detector, a thermally sensitive paper was put at the gyrotron output window flange to diagnose the beam profile. Figure 2 shows the measured beam profile that is a good agreement to the result obtained in DIII-D as shown in the color image. The beam profile was also measured at the input end and the output end of the MOU output waveguide to check the mirror alignment. It was observed that the beam was slightly shifted up at the input and down at the output. Nevertheless, the mirrors were not moved because the beam shift was not considered as a significant problem.

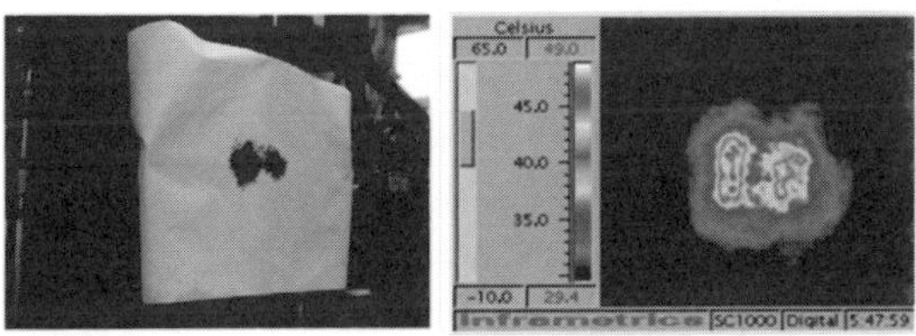

Figure 2. The beam profiles at the output window flange of the gyrotron.

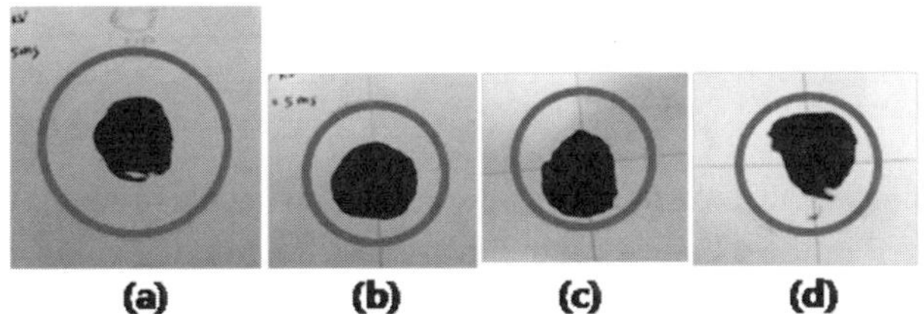

Figure 3. The beam profiles (a) at the input and (b) at the output of the MOU waveguide and (c) at 360 mm and (d) at 740 mm from the end of the MOU output waveguide.

With the beam voltage of 63 kV and beam current of 20 A which are maximum ratings in the existing power supply, the measured RF power at the SS load was 200 kW in averaged over 200 ms pulse length and the corresponding peak power was 250 kW. The power absorbed in the MOU was about 75 kW. Considering the theoretically calculated transmission loss of 25 %, the averaged RF power from the gyrotron was expected to be about 340 kW and the peak power about 410 kW.

2.3. *Transmission Line System*

The EC beam is transmitted from the gyrotron to the antenna by use of a circular corrugated waveguide with inner diameter of 31.75 mm. The ECH transmission line components, which were fabricated by GA, were optimized to transmit 84 GHz EC wave [6]. In order to know the feasibility of the transmission line for 110 GHz EC wave, we investigated the transmission efficiency of each component for 110 GHz EC wave.

Table 1. The transmission losses for 84 GHz and 110 GHz EC wave in the existing transmission line components [7]. These values are calculated by Dr. J. Doane, GA.

<table>
<tr><th>Items</th><th colspan="2">Loss</th><th>Value for 84 GHz</th><th>Value for 110 GHz</th></tr>
<tr><td>Waveguide</td><td colspan="2">Ohmic (70 GHz to 110 GHz)</td><td>~ 5 % (100m)</td><td>~ 5 % (100m)</td></tr>
<tr><td rowspan="4">Miter bend</td><td rowspan="2">Ohmic ~ $f^{1/2}$</td><td>H plane</td><td>~ 0.07 %</td><td>~ 0.07 %</td></tr>
<tr><td>E plane</td><td>~ 0.15 %</td><td>~ 0.15 %</td></tr>
<tr><td rowspan="2">Mode conversion of plane mirror</td><td>With mode mixture</td><td>~ 0.28 %</td><td>~ 20 %</td></tr>
<tr><td>Without mode mixture ~ $\lambda^{3/2}$</td><td>3 % per bend</td><td>~ 1.4 %</td></tr>
<tr><td>Diamond window</td><td colspan="2">Reflection</td><td>< 1 %</td><td>~ 45 %</td></tr>
<tr><td>Gap type*</td><td colspan="2">Radiation and mode conversion in a gap (0.4 ~ 2.5 mm) ~ $\lambda^{3/2}$</td><td>0.01 ~ 0.05 %</td><td>0.01 ~ 0.05 %</td></tr>
<tr><td>W/G up-taper</td><td colspan="2">Mode conversion</td><td>< 0.1 %</td><td>~ 4.3 %</td></tr>
</table>

* DC break, pump-out tee, RF gate valve, and waveguide switch

Table 1 shows the transmission losses of both 84 GHz and 110 GHz EC wave transmitted in the existing components. The diamond vacuum window and the mode mixture with the large transmission loss were removed. The total transmission loss from the output waveguide of the MOU to launcher is estimated to be about 25 % according to the theoretical values. The launcher did not affect the total loss. Although there are two dummy loads in the transmission line, the compact waveguide load was not utilized and only the large stainless steel (SS) load near the KSTAR tokamak, which has a long response time for 110 GHz EC wave, was used. The RF power absorbed in the terminal stainless steel tank load, the gyrotron cavity, the MOU chamber, and the MOU mirrors are measured using calorimetric method.

The polarization of the EC beam is controlled with two grooved mirrors adopted in the polarizer miter bends allowing pure ordinary (O) or pure extraordinary (X) mode power launching for arbitrary angles. However, the

Fourier transformation coefficients are changed for the 110 GHz wave that was and confirmed by the maker, GA. As a result, we could found it covers the required polarization range at 110 GHz.

3. Second harmonic ECH-assisted Start-up in KSTAR

For 2009 plasma campaign, the toroidal magnetic field was set to 2 Tesla at the major radius of 1.8 m that corresponds to the second harmonic resonant magnetic field for 110 GHz EC wave. ECH power launched into the vacuum vessel was about 250 kW. Figure 4 shows the results obtained by perpendicular injection of a linearly polarized X2 mode EC wave for 2.5 s pulse length. The power e-folding radius of 110 GHz EC wave at the second harmonic EC resonance position (R_{X2} = 1. 8 m) was about 5.7 cm by Gaussian beam optics in free space. It gives a power density 2.4 kW/cm^2 for the 250 kW, and it is higher than the power density of the 84 GHz EC beam.

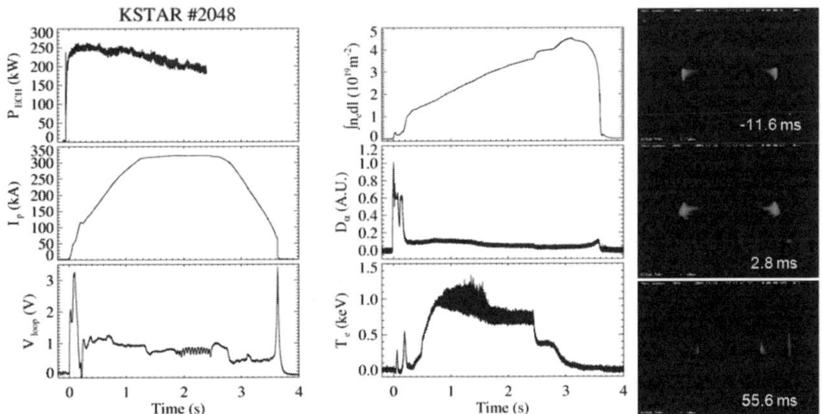

Figure 4. The experimental plasma parameters and the fast CCD camera images for shot number 2048

The experimental results during 2009 KSTAR campaigns showed the feasibility of the second harmonic 110 GHz ECH-assisted startup. The interesting observation was that the pre-ionization would not occur with a pure toroidal magnetic field, that is, when poloidal magnetic field null (FN) is not applied. Moreover, the small amount (~1 kA) of the toroidal plasma current was observed in the pre-ionization phase. It is considered that the vertical field component of the FN structure plays a role in confining the electrons during the pre-ionization phase. During the 2009 plasma campaign, the optimized condition of ECH pre-ionization was investigated with parameter scans of deuterium pre-fill gas pressure, resonance position, polarization, and vertical magnetic field without Ohmic discharge. The 2009 results showed that the pre-ionization was observed in most scanned conditions except X3 mode injection and the absence of initial FN configuration. The breakdown is increasingly delayed at the low pre-fill pressure, the inboard resonance position, O-mode injection, and the higher B_z at the FN center, as evidenced by the D_α emission and the central ECE signal. Despite optimized ECH pre-ionization, the plasma startup was not reproducible possibly due to the marginal ECH power, the unsatisfactory wall conditioning, and the lack of the gas controllability during the phase change from the pre-ionization to the Ohmic phase. A second gas puff introduced just before the onset of the Ohmic pulse enhances the plasma startup.

4. Conclusions

The 110 GHz GYCOM gyrotron was loaned from DIII-D in collaboration with GA and it was successfully commissioned at the KSTAR in time before the plasma operation in the 2009 campaign of the KSTAR. The 110 GHz ECH system was very essential for the plasma startup using the second harmonic EC resonance at the toroidal magnetic field of 2 T. It was also very helpful in the current ramp-up and the extension of the plasma duration.

Acknowledgments

Authors would like to thank the GYCOM experts, Dr. L. Popov and Mr. V. Orlov, for the installation and the initial operation of the gyrotron and magnet at the KSTAR.

References

1. K. Kajiwara, Y. Ikeda, M. Seki, *et al.*, Nucl. Fusion, **45**, 694 (2005).
2. J. Bucalossi, P. Hertout, *et al.*, Nucl. Fusion, **48**, 054005 (2008).
3. G. L. Jackson, M. E. Austin, *et al.,* Fusion Sci. and Tech., **57**, 1 (2009).
4. Y. S. Bae, *et al.*, Nucl. Fusion, **49**, 022001 (2009).
5. J. Lohr, D. Ponce, et al., General Atomics report GA-A22420 (1996).
6. Y. S. Bae, J. H. Jeong, *et al.*, Fusion Sci. and Tech., **52**, 321 (2007).
7. Private communication with Dr. J. Doane in GA.

ECH AND ECE APPLICATION FOR SPECTRAL ANALYSIS OF THE GLOBAL PLASMA OSCILLATIONS AT TOKAMAK. EXPERIMENTS ON T-10

V.I. POZNYAK, YU.V. GOTT, A.M. KAKURIN, V.V. PITERSKII, G.N. PLOSKIREV
Russian Research Center "Kurchatov Institute", 123182, Kurchtova Sq. 1, Moscow, Russia

O. VALENCIA
Russian People Friendship University, 117198, Miklukho-Maklaya St. 6, Moscow, Russia

T.V. GRIDINA
Bauman Moscow State Technical University, 105005, 2-nd Baumanskaya St. 5, Moscow Russia

The spectral characteristics of the large scale (*MHD*) plasma oscillations were investigated. ECH was used for a modification of plasma parameters. Main tool of the analysis was the high space resolution ECE method. The selection of different modes of oscillations was fulfilled. The simple relations between the eigen-frequencies of oscillations and the radii of the corresponding rational magnetic surfaces were discovered. The absolute values of all measured eigen-frequencies coincide well with the calculation by the introduced formula.

1. Introduction

As rule plasma theory considers the global oscillations, which we investigate, within of MHD models (*tearing, kink* and others). In an experiment those classes of oscillations are singled out by certain dynamic feature. However in our experiments we revealed for the same plasma object simultaneously different factors which must be inherent to the different classes of oscillations. This gives ground to find a general nature of process. It was discovered earlier that electron distribution function in those regimes is the essentially non-equilibrium and longitudinal and perpendicular velocities perform strong variations [1]. The restricted frameworks do not allow recite here our kinetic conception of the observed phenomena. However we suggest the simple enough semi-empirical expression for the eigen frequencies of the global plasma oscillations. This work represents the spectral analysis of the oscillations and the comparison of the measured eigen frequencies with the calculated values.

2. Methodology

On-axis ECH (1 – 3 gyrotrons 140 GHz, power up to 1.2 MW) is applied for an amplification of the global oscillations [2]. Off-axis ECH (1 – 2 gyrotrons 129

GHz, power up to 0.8 MW) is used for a gain of signals. All launches are across to the magnetic field. That method enables to investigate both discharges with the single driving mode (m/n=1/1) and with a family of modes (here m/n=1/1, 3/2, 2/1, 3/1) by a little change of a current penetration velocity to the plasma center. In the presented examples: B_t=25 kGs, I_p=250 kA, $\tilde{n}_e$~$1.7\cdot10^{13}$cm^{-3}, large radius R=150 cm, carbon rail limiter r_l ~ 30 cm, q_l~3.1.

Two kinds of oscillations: modulating and helical are distinguished methodically. They arise already in the OH stage and grow in the ECH regime. The first becomes apparent in oscillations of the ECE amplitudes. Their frequency range is 3 – 7 kHz. The frequencies of the second kind are determined as the quantities inverse to the time delay between ECE signals from external (LFS) and internal (HFS) sides of the torus (fig. 1a, here the time delay is 90 μs). Those delays are connected with the non-uniform motion of plasma along the spiral trajectory. The frequencies of both kinds can be registered also by means of the magnetic probes dislocated at the limiter shadow. The presented spectral analysis had required to use the ECE method with the high resolution along the radial coordinate for every channel (Δr/a~0.03). Such method becomes inapplicable when antenna pattern is more wide (fig. 1b) apparently owing to an averaging in space of an emission of electrons with different energy.

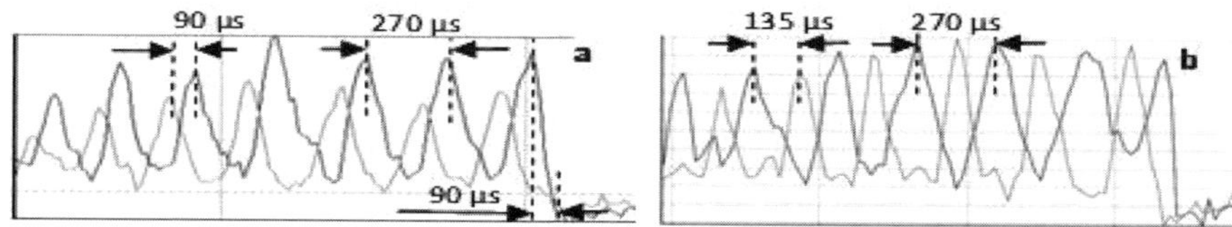

Figure 1. Shape of ECE signals (IF band is 600 MHz) in the regime with single *kink like* mode m/n=1/1 from LFS (light-gray color) and HFS (dark-gray): a – high space resolution, **the disturbance moves nonuniform**; b – low resolution, ECE signals are similar to the chord SXR signals, **the disturbance motion looks like uniform one**.

It should know the positions of the corresponding rational magnetic surfaces for an identification of the structure (mode) of oscillations. Their coordinates, averaging in time, are determined by an appearance of the flattened zones in the electron temperature profile, by the sharp change of its gradient, by the abrupt alternation of the phases of oscillations and some others.

3. Experiment

Data of two types of discharges are represented in this work. In first kind under the powerful on-axis ECH (fig. 2a), the m/n=1/1 *kink like* oscillations with the high amplitudes up to 50% at the center relatively to the slow changing background are observed. Under the low ECH power and shift of the absorption zone out of half q=1 radius, maximum of the global oscillations shifts from the center also and the oscillations acquire the sings of *tearing* mode. First type of discharges has a good repetition. Here we shall complement data of the works [1, 2].

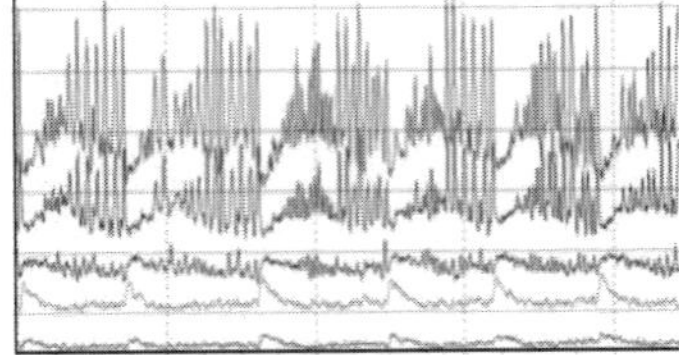

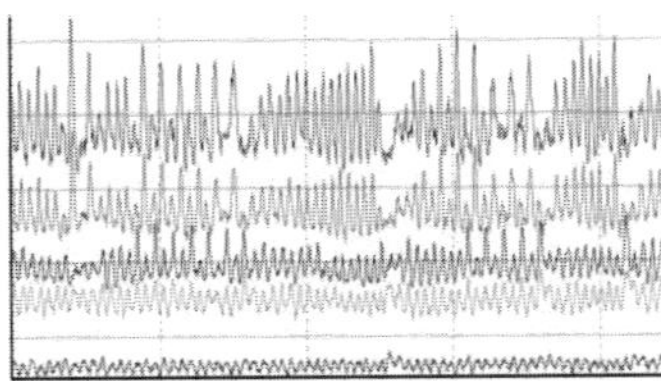

Figure 2. ECE signals in the regimes: a - #36057, *kink like* oscillations of plasma in m/n=1/1 mode with a periodical internal disruptions, frequency of amplitude oscillations in the beginning of every sawtooth cycles is ~ 6 kHz, after - 3.5 - 3.7 kHz; b - #36056, *fishbone like* oscillations (the first stage of ECH) in the regime with several driving modes.

In other type of discharges, the modes with more complicate structure become excited (for example fig. 2b). An appearance of the peripheral modes leads to an active interaction of plasma with the wall and scenarios can be essentially different. We represent information by the discharge in which several classes of oscillations exist simultaneously and process itself possesses by very complicate dynamics. It should mark that those data corresponds fully to the results which were obtained in a set of other discharges with simple scenarios.

By the physical nature, the global oscillations are divided to the eigen (resonant) and the forced which are a response to the primary eigen. The most important peculiarity of the investigated eigen oscillations is the constancy of their frequencies $f^k_{m/n}$ (k – number of harmonic, m/n – mode of oscillations) in regimes with the steady plasma current including a time of ECH. Their frequencies do not depend on electron temperature (1 – 3.5 keV), plasma density ($1.5 - 3 \cdot 10^{13} cm^{-3}$), ECH power (0.2 - 2 MW). The frequencies determined by the delays of ECE signals are the eigen always. At many cases the frequencies obtained by the amplitude modulation have also similar properties: in the OH stage with m/n=2/1 and 3/1 modes, directly after a start of ECH, during the stationary stage of ECH – in the first phase of every *tooth of saw* and sometimes during the total period of a saw. Oscillations with the variable frequencies will be named later as the modulating f^k_{mod}.

Fig.3a illustrates the structure and dynamics of the plasma oscillations in the regime with a multitude of modes in the first stage of ECH. Oscillations fill the total plasma volume what enables to determine the positions of all simple rational zones. Sawtooth oscillations are not pronounced. After the switching-off of one on-axis gyrotron electron temperature decreases on ~20% and frequencies of all modes except the eigen for m/n=1/1 rise synchronously doubling after ~70 ms(fig. 3b). In this moment all peripheral modes decay (fig. 3c). Only the frequencies inherent to the m/n=1/1 mode preserve. The discharge becomes identical to that of the first type. It should mark that the ECH absorption zone does not contact with any rational surfaces both under a strong onset and under a suppression of oscillations.

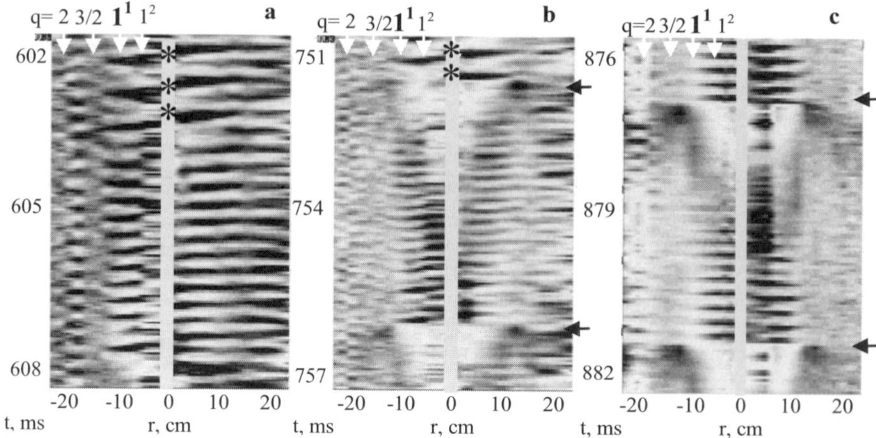

Figure 3. Cartograms of the global oscillations (equatorial plane, 24 channels, extraordinary polarization of 2nd ECE harmonic) in the discharge with several driving modes: a – the first *fishbone like* stage of ECH, three on-axis and one off-axis gyrotrons; b – the second stage after the switching-off of one central gyrotron and just before the decay of the peripheral modes, the motion of energy to the center exhibits (pinch); c – the third stage – single *kink like* m/n=1/1 mode. The white arrows are the positions of the rational magnetic surfaces, the black arrows – the moments of internal disruptions, the asterisks – high amplitudes on the first modulating subharmonic $f^{1/2}_{mod}$. The gray belts cover the frequency band of three on-axis gyrotrons.

Fig. 4 shows the phase relations between ECE signals from LFS and HFS. All signals are always synchronous on LFS (bottom). Signals from LFS and HFS have the same phase shift ~90 μs inherent to the m/n=1/1 mode. This shows that oscillations of this mode spread all over the column. The exchange of the sign of phases shows to the zonal character of the process at plasma volume. At the third ECH stage the oscillations out q=3/2 fully decay and the sign of the phase shift becomes positive as in the first type of discharge. 90 μs delay is not observed during the appearance of oscillations on the subharmonics of the modulating frequency $f^{k/2}_{mod}$ (fig. 4, top). In such case this method enables to detect the time delays corresponding to the eigen frequencies of the m/n=3/2 and 2/1 modes measured earlier during the OH stage. Although the velocity of the disturbance with the frequency $f^{1/2}_{mod}$ rises with a decrease of the corresponding surface radius certain relaxation process (here on HFS) synchronizes this motion maintaining its periodicity into the q=3/2 area which is common for the m/n=2/1 and 3/2 modes.

Fig. 5 shows the dynamics of the electron temperature profile inherent to the first kind of discharge. This dynamics is identical to the third stage of ECH after a reduction of the on-axis heating. Oscillations have the feature of all discussed classes of oscillations: the steps on the profile snapped to the rational zones (*tearing*), the swinging of all central part of the distribution (*kink*), the slowing down of disturbance motion in every period of oscillations (*locking*), the appearance of the structures with a size less than radius of q=1 zone (*filamentation*). All denoted peculiarities appeared also at the previous *fishbone like* stage. It should mark out especially the excitation of the second space harmonic of the m/n=1/1 mode inside the radius $r_{q12} \sim 4.5$ см. The motion of this disturbance ensures the second harmonic of frequency $f^2_{1/1} \sim 12$ кГц. Just this mode is the driving one in the regime with the on-axis ECH. Also essential that the steps (*"islands"*) appear in every period of oscillations and do not disappear during the disruption and after it.

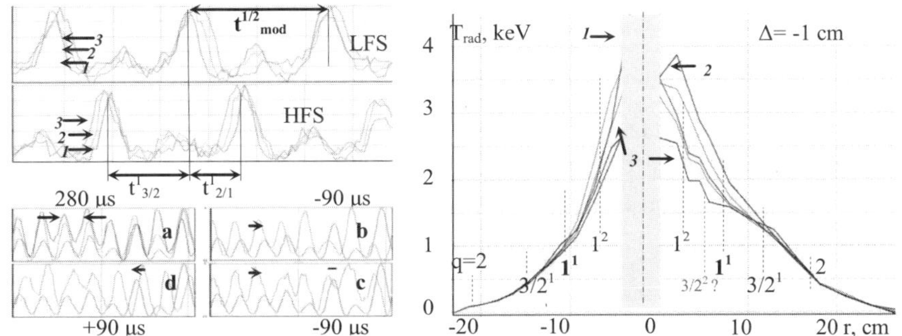

Figure 4. Phase relations of ECE signals in the first *fishbone like* stage of ECH. Bottom: **a** – on LFS; **b** – here and further - between LFS и HFS, dark-gray curve – the reference signal from LFS, here - inside q=3/2; **c** – out q=2; **d** – between q=3/2 и q=2. Top – under appearance of subharmonics of modulating frequency $f^{1/2}_{mod}$, *1* – r = ±5, *2* - ±7, *3* - ±9 cm.

Figure 5. $T_{rad}(r)$ profile dynamics in the third stage with single driving m/n=1/1 mode which have simultaneously signs of *tearing*, *kink*, *locking* and *filamentation* processes. *1* – maximum of amplitude - on HFS, *2* – maximum on LFS, *3* – just after disruption, the remainder curves – intermediate states.

The analysis of the amplitude distributions with the detached frequencies permits to localize their excitation areas and determine the modes. At the regime with single driving mode (fig. 6a) both modulating harmonics have a similar bell-like shape of the distribution. However the relation of those distributions shows to the particular role of two surfaces $q^1=3/2$ and $q^2=1$ in a creation of the forced oscillations f^2_{mod}. In the regime with several driving modes (fig. 6b) the distribution has the three-humped shape that corresponds to a impact of two sources: the power flow by ECH to the central area and the poloidal field flow moving from a periphery and misappropriating itself energy on oscillations near q=2 zone. The white curve on the graph is a typical shape of distribution in a pure Ohmic regime. The appearance of the odd subharmonics $f^{1/2}_{mod}$ and $f^{3/2}_{mod}$ is the result of a simultaneous activity of the m/n=2/1 and 3/2 modes inside certain common area. The power density inside $q^2=1$ in this regime is lower than that in the first one and the forced oscillations do not observed here.

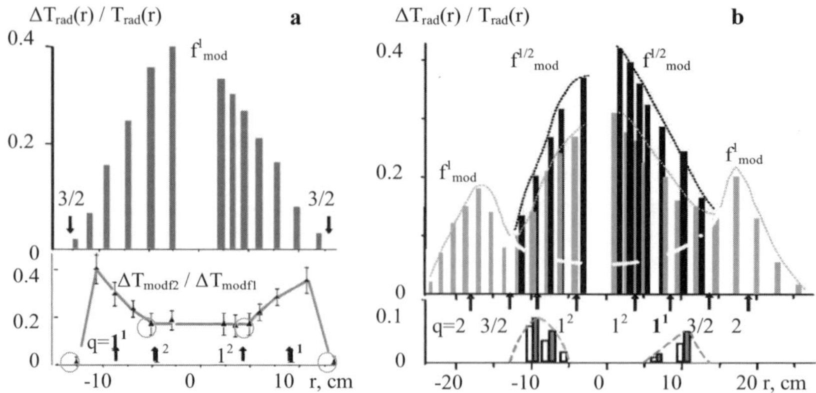

Figure 6. Radial amplitude distributions of the global oscillations (by ECE): a – *kink like* regime with single driving mode, top – first modulating frequency (~3.5 kHz), bottom – the relation between second (~7 kHz) and first modulating modes; b – *fishbone like* regime with several driving modes, top – first modulating mode (3.5 kHz, light-gray color) and first subharmonic of modulating frequency $f^{1/2}_{mod}$ (~1.75 kHz, black), bottom - third subharmonic of modulating frequency $f^{3/2}_{mod}$ (~5.2 kHz, white) and second modulating mode (~7 kHz, dark-gray).

Fig. 7 shows the spectrum dynamics of the oscillation for two types of discharges. In the first case (a) the values of the frequencies determined by the delays of ECE signals coincide with the frequencies measured by the magnetic probes and really are the eigen frequencies of the plasma current oscillations. The start of ECH demonstrates that the modulating frequencies occur from the eigen owing to the slowing down of the fundamental disturbance motion connected with the q^1=1 zone (*locking*). We assume that during a heating of the central area a part of the plasma current once for the period of the f^1_{mod} concentrates inside r_{q12} ~ 4.5 cm creating oscillations on the double frequency $f^2_{1/1}$ ~ 12 кГц. The residual part of the current disturbance inside q^1=1 zone moves more slowly and the visible frequency decreases. The amplitudes of the second space harmonic can exceed the amplitudes of the $f^1_{1/1}$ ~ 6.3 кГц in the Ohmic stage in 4 – 5 times. Such process repeats in every period of the f^1_{mod} setting on the stationary stage the odd rhythm (~3/4) in the succession of signals from LFS and HFS. The area of f^k_{mod} oscillations is limited by the q=3/2 zone and the probes do not register them.

At the second type of regimes (fig. 7b), the instabilities with evident feature of *lock mode* take place during the OH stage. The trains of the amplitude ECE oscillations in the m/n=3.2, 2/1 and 3/1 modes happen from time to time. Those instabilities decease just under ECH but oscillations in different modes fill a total plasma volume (fig. 3a). As opposed to the first kind of discharge, the frequency f^1_{mod} ~ 3.5 kHz appears just with the ECE start so far as oscillations with the first and second eigen frequencies $f^{1,2}_{3/1}$ of the m/n=3/1 mode already presents at this discharge. In what follows those oscillations discontinue to be the eigen and become the forced by the modulating component $f^1_{mod} \approx f^2_{3/1}$ of the m/n=1/1 mode. The reduction of ECH power diminishes an energy transport to a plasma periphery and hampers a self-excitation of all peripheral modes. The capture of oscillations by the m/n=1/1 mode occurs at the limited region inside q=3/2 where a current density is maximal. The frequencies increase tending to the three nearest attractors (gray rings) which are intrinsic to the m/n=1/1 mode. A weak rise of the frequency $f^2_{1/1}$ in the case (b) can be explained by a slow penetration of a current wave to the q^2=1 zone.

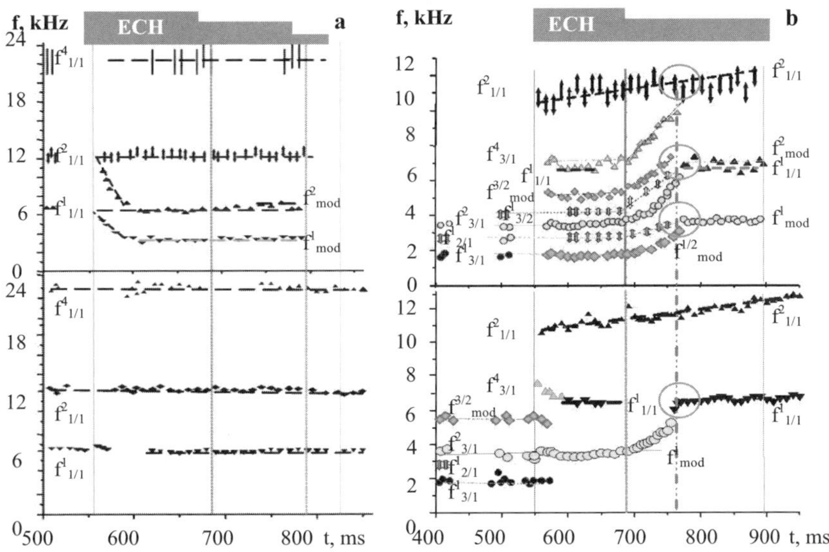

Figure 7. Spectra of the global oscillations: a – regime with single driving mode $f^k_{1/1}$; b – regime with several driving modes $f^k_{m/n}$. Top graphs – by the ECE measurements, bottom – by the magnetic probe measurements. The data marked by the vertical arrows – by the delays between ECE signals; the figure signs – by the amplitudes of signals.

4. Summary and discussion

The eigen oscillations can become excited at all area inside the corresponding rational magnetic surface. Their amplitude distribution depends on the energy deposition (fig.6). The process acquires the zonal properties under the excitation of several driving modes (fig. 3). The triggering of the oscillations occurs, as we assume, by the relaxation of the electron distribution on the potential electron plasma waves at LFS or HFS (fig. 1a or 4) that leads to the zonal plasma current quench. The current oscillations of the major modes can spread through the total column (fig. 4, 7). The spiral motion of the current disturbances together with the drift plasma motion follows to certain resonances. The frequencies of the eigen (resonant) global plasma oscillations are determined by the expression:

$$f^{k}_{m/n} = \min < [(v_e, v_i)_{\blacktriangledown B\ drift}] >_{r^{k}_{m/n}} / 2r^{k}_{m/n}, \ k = 2^{s-1}, s = 1, 2, 3\ldots, r^{k}_{m/n} = r^{1}_{m/n} / k$$

The table represents necessary data for a comparison with the formula. Ion temperature does not vary during all current plateau stage including the ECH time. This defines the constancy in time of the eigen frequencies (fig. 7).

m/n	1/1	3/2	2/1	3/1
$r^{1}_{m/n}$ (cm) (±5%)	8.5	13	19	29
$<T_i>$ (eV), inside q=3/2 zone (±10%)	500			
$f^{1}_{m/n}$ (kHz), experiment (±5%)	6.3	4.3	2.8	1.85
$f^{1}_{m/n}$ (kHz), calculation	6.5	4.2	2.9	1.9

A good agreement with the experiment confirms our assumption. More over it was found that the relations of the corresponding eigen frequencies and radii are the relations of the prime numbers inside an 5% accuracy - $f^{1}_{1/1} : f^{1}_{3/2} : f^{1}_{2/1} : f^{1}_{3/1}$ = **3/2**; $r^{1}_{3/1} : r^{1}_{2/1} : r^{1}_{3/2} : r^{1}_{1/1}$ = **3/2**.

The areas of the excitation put in as *Russian dolls*. The frequencies of the oscillations, determined by the corresponding zonal velocities, are set by the common parameter – the ion temperature whose distribution is almost constant inside the q=3/2 zone [3] as that of the electron density. In such case the stationary matching of the global oscillations is possible only with certain arrangement of the master magnetic surfaces in the common space.

The constancy in time of the zone positions does not mean of course the current distribution invariability but shows apparently to a non-monotonic current profile. Local peculiarities of it are connected with the small shear zones and strengthen them. So far as the rational zones positions averaging by a period of oscillations do not depend on amplitudes we believe that this can be a ground of so called *profile consistency phenomena.*

We thank the T-10 staff for the fruitful long time cooperation in experiments.

This work supported by the Nuclear Science and Technology Department of RUSATOM RF.

References

1. V.I. Poznyak, V.V. Piterskii, G.N. Ploskirev, E.G. Ploskirev, O. Valensia, Proc. of the 15th Joint Workshop on ECE and ECH, Yosemite National Park, California, USA, 2008, p. 136
2. V.I. Poznyak, Proc. of the 14th Joint Workshop on ECE and ECH, Santorini, Greece, 2006, p. 123
3. Yu.V.Gott, Yu.D.Pavlov, 34th EPS Conference on Plasma Phys., Warsaw, 2007, ECA Vol. **31f.** P-04.107

FEASIBILITY OF AN ECRH SYSTEM FOR JET: PROJECT OVERVIEW

G. GIRUZZI[1], M. LENNHOLM[2,10], A. PARKIN[3], F. BOUQUEY[1], H. BRAUNE[4], A. BRUSCHI[5],E. DE LA LUNA[2,11], G. DENISOV[6], T. EDLINGTON[3], D. FARINA[5], J. FARTHING[3], L. FIGINI[5],S. GARAVAGLIA[5], J. GARCIA[1], T. GERBAUD[12], G. GRANUCCI[5], M. HENDERSON[7], M. JENNISON[3],P. KHILAR[3], N. KIRNEVA[8], D. KISLOV[8], A. KUYANOV[8], X. LITAUDON[1], A.G. LITVAK[6], A. MORO[5], S. NOWAK[5], V. PARAIL[3], G. SAIBENE[9], C. SOZZI[5], E. TRUKHINA[8], V. VDOVIN[8]

JET-EFDA, Culham Science Centre, Abingdon OX14 3DB, UK

[1] CEA, IRFM, 13108 Saint-Paul-lez-Durance, France
[2] EFDA Close Support Unit, Culham Science Centre, Abingdon OX14 3DB, UK
[3] CCFE, Culham Science Centre, Abingdon OX14 3DB, UK
[4] Max-Planck-IPP, Euratom Association, D-17491 Greifswald, Germany
[5] Istituto di Fisica del Plasma CNR, Euratom Association, 20125 Milano, Italy
[6] Institute of Applied Physics, Nizhny Novgorod 603155, Russia
[7] ITER Organization, 13108 Saint-Paul-lez-Durance, France
[8] RRC 'Kurchatov Institute', Moscow, Russia
[9] Fusion for Energy, 08019 Barcelona, Spain
[10] European Commission, B-1049 Brussels, Belgium
[11] Laboratorio Nacional de Fusion, Asoc. EURATOM-CIEMAT, 28040, Madrid, Spain
[12] LPTP, Ecole Polytechnique, 91128 Palaiseau, France

A study has been conducted to evaluate the feasibility of installing an ECRH system on the JET tokamak. This paper presents an overview of the studies performed in this framework by an EU-Russia project team. The motivations for this major upgrade of the JET heating systems and the required functions are discussed. The main results of the study are summarised. The usefulness of an EC system for JET is definitely confirmed by the physics studies. No strong limitations for any of the functions envisaged have been found. This has led to a preliminary conceptual design of the system.

1. Introduction

The future JET programme [1], after the installation of the ITER-like wall [2], will be mainly focused on the consolidation of the physics basis of the three main ITER scenarios, i.e., the ELMy H-mode, the hybrid scenario and the advanced steady-state scenario [3]. In ITER, these scenarios will make substantial use of Electron Cyclotron (EC) waves, for heating as well as for control of both the MHD activity and the current density profile [4,5]. Therefore, a programme for preparation, validation and optimization of the ITER scenarios in present tokamaks would strongly benefit from, and even require, an ECRH/ECCD system. This gives a strong motivation for examining the feasibility of the construction and implementation of such a system in JET by 2014-2015, for an intensive exploitation before the start of ITER. The possibility of implementing an ECRH system on JET has already been considered in the past [6]. The 2000-2002 project is of course a good basis for the new feasibility study, however, changes in the general context and evolutions of motivations, physics and technology in this area had to be taken into account. Involvement of the Russian Federation in the construction and the operation of the system is considered a key element for the success of the project, and Russian scientists have actively participated in the feasibility study. Synergies with the development of the ITER ECRH system [7], and especially with the elements corresponding to European procurements, have been systematically pursued, in particular through the involvement of scientists from ITER IO and F4E (the European ITER Domestic Agency).

This paper reports on the results of these feasibility studies for a JET ECRH system, carried out under bilateral agreement between the European Union and the Russian Federation. Both physics studies, aiming at a definition of the basic parameters of the system (wave frequency, launching geometry, power required), and exploration of the main technical options (wave generation and transmission, antenna design, power supply, auxiliaries, layout and port allocation, control systems, diagnostics) have been performed in parallel. A preliminary assessment of cost, time schedule, manpower and risks has been made.

2. Physics requirements studies

The main functions of an ECRH system on JET can be summarized as follows: i) Central electron heating – to equilibrate electron and ion temperature and to control impurities. ii) Current profile control – to improve the performance of advanced and hybrid scenarios. iii) Sawtooth control – to shorten sawteeth through local current drive near the q=1 surface and hence avoid triggering Neoclassical Tearing Modes (NTMs). iv) NTM suppression through local current drive at the relevant rational q surface. These objectives were translated into the requirement to heat the plasma or drive currents at various plasma minor radii. In order to fulfil all the objectives, the system should be able to deposit the ECRH power over almost the full range of plasma minor radii. For MHD control, the width of the deposition profile is also an important parameter. As the ECRH power is absorbed at or near the Electron Cyclotron Resonance, the physics objectives can only be fulfilled for a selected range of toroidal magnetic fields.

Several state-of-the-art codes have been used to perform these studies, in particular the beam-tracing code GRAY [8] and the integrated modelling suite CRONOS [9]. Two recent JET discharges, both with parameters compatible with the ITER-like wall, have been identified and chosen as reference discharges for the simulations: discharge 73344 (for H-mode and hybrid scenarios) and 77895 (for advanced scenarios).

Wave propagation and absorption. The first aim of the study was to determine the best choice of operating frequency [10]. Two gyrotron options were considered: either 1) using dual frequency gyrotrons (113GHz, 150GHz) as used on ASDEX Upgrade with minor modification to increase the frequency; or 2) using the 170GHz gyrotrons developed for ITER. The study produced a comprehensive evaluation of the minor radii accessible as a function of toroidal field for the two options. The results showed that the 113GHz, 150GHz option allowed good performance for toroidal fields below 2.7 T (2nd harmonic extraordinary mode at 150GHz) and for fields above 3.3 T (fundamental ordinary mode at 113GHz) while operation around 3 T would be difficult with this frequency choice. Choosing 170GHz (2nd harmonic, extraordinary mode) would on the other hand allow excellent performance in the interval 2.7-3.1 T with off axis current drive available in a significantly wider range (see Figure 1). Given that the largest fraction of JET operation is foreseen to use toroidal fields around 3 T it was decided to focus the further studies on the use of 170GHz gyrotrons. Furthermore, this frequency choice allows JET to take full advantage of the significant efforts which have been and are being made in designing components for the ITER ECRH system. With this choice of gyrotron frequency the injection angles required to fulfil the objectives were found to be: toroidal: -25° - +25° , poloidal range: 30° around an appropriate mean value.

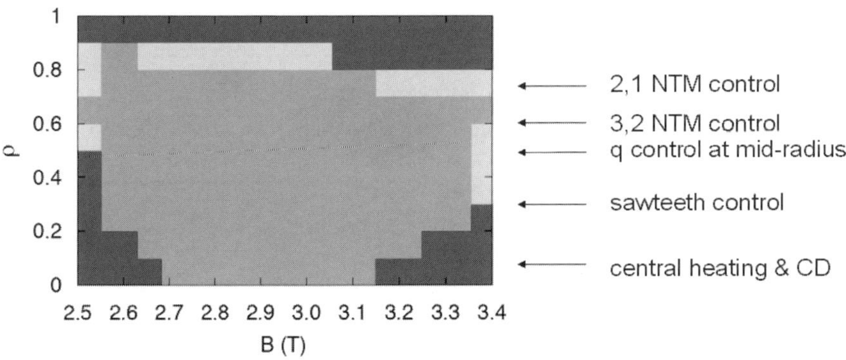

FIG. 1: Operation diagram for a JET ECRH system at 170 GHz, computed for a reference H-mode scenario. Green means that the envisaged funtions (listed on the right) are possible, red not possible, yellow marginally possible.

NTM stabilisation. Use of localised ECCD has been considered for stabilisation of (3,2) and (2,1) modes in H-mode scenarios [11]. The beam-tracing results have been used in connection with the generalised Rutherford's equation [12] to study the evolution of the magnetic island width in the presence of ECCD. It is found that EC power < 5-10 MW is sufficient for full island suppression in most cases of interest, without power modulation, provided the driven current is localised in a radial range < 10 cm. Note that beam-tracing calculations predict typical CD radial width of 5 cm for a single beam, however the superposition of multiple beams will make this requirement challenging.

Sawtooth control. Beam-tracing calculations combined with ASTRA transport simulations [13] and the Porcelli model [14] have been used to estimate the power required in order to have a significant impact on the sawtooth period for the H-mode scenario. It is found that ECCD localised close to the q = 1 surface can be used, counter-CD being more effective than co-CD. A power of the order of 2.5 MW is sufficient to double the sawtooth frequency.

Integrated modelling of JET scenarios with EC waves. The CRONOS code has been used for integrated modelling of full JET scenarios. Both interpretative and predictive simulations (using the Bohm/gyro-Bohm transport model [15]) of the reference discharges have been performed, in order to test the predictive capability of the code with respect to the current and temperature profiles. Benchmarks with the JETTO code has been done for the H-mode reference discharge. Scenarios at higher heating power and with ECRH/ECCD have then been simulated. It is found that a wave power of the order of 10 MW, absorbed in the central plasma region, is adequate in order to equilibrate electron and ion temperatures in the H-mode scenario. For advanced scenarios, 10 MW of off-axis ECCD is marginally adequate to significantly modify the q profile (see Figure 2). These calculations also indicate that fully non-inductive discharges at high power and density can be achieved. Nevertheless, a full inversion of the q

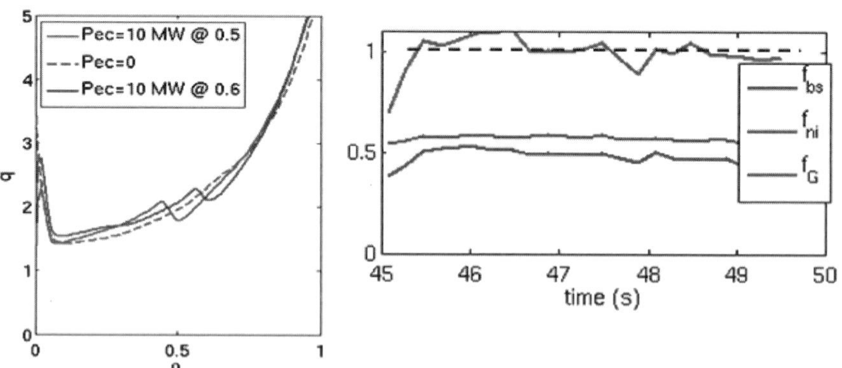

FIG. 2: Predictive simulations of a JET advanced scenario by the CRONOS code. Left: safety factor without and with ECCD at two different locations (ρ ~ 0.5 and 0.6). Right: bootstrap, non-inductive and Greenwald fractions vs time, with ECCD at ρ ~ 0.6.

profile inside the radius of ECCD absorption does not seem possible, owing to the dominant contribution of the current driven by the NBI power.

3. System components

Gyrotrons. Availability of adequate power sources (1 MW / 20 s) produced by the Russian partners has been investigated. IAP/GYCOM can provide gyrotrons satisfying the requirements for both frequency options [16]. A 170 GHz gyrotron of this type already exists: it is the tube developed for ITER, which has already attained performances exceeding 1 MW, 200 s pulse duration, 50 % efficiency [17]. A double-frequency 113-150 GHz gyrotron would require a modest development with respect to the existing 105-140 GHz tubes developed for Asdex-U. The present industrial production capability of GYCOM is of 5 tubes per year. The delivery of 12 gyrotrons for JET would be favourably phased with respect to that of 8 gyrotrons for ITER. Therefore, the production of 12 gyrotrons in 4 years appears a reasonably achievable target, with minimum risk.

Transmission lines. The basic question of choice between quasi-optical and waveguide transmission has been addressed. Although in the considered frequency range the two options could be equivalent for a system and a machine environment to be designed from scratch, the quasi-optical approach seems difficult to implement in the very busy environment of JET and complying with the safety requirements connected with tritium operation. The ITER-relevant evacuated waveguide solution is now considered preferable from the technical, safety and cost points of view and has been used as a reference solution for the JET ECRH system. A study of the waveguide routing has been performed [18]. The best solution is a transmission line of length ~ 75 m, with only 6 bends and 4.1 % expected attenuation.

Windows. Diamond windows are the only viable solutions, owing to the power / pulse length combination. Since 170 GHz is chosen, the same windows used for the ITER ECRH system could be employed, with substantial synergy with ITER and possible support by F4E (design and procurement). The use of two windows per line (for tritium segregation) is now assumed as the reference solution [18]. In this case, two separate windows (one at the entrance of the Torus Hall and the other at the vacuum vessel) are preferable to a double window, from the safety point of view. The time to produce such a large number of windows (24) is presently difficult to determine and could be a source of delay. The behaviour and lifetime of these windows is also a significant element of risk, but also a contribution of risk reduction for the ITER ECRH system.

Launcher. Possible options for a launcher design have been analysed, with the following requirements: plug-in launcher fitting in an equatorial port, 12 waveguides, toroidally and poloidally steerable actively cooled mirrors, use of the ITER steering mechanism [19] for at least one of the two movements. Available space in the port and the port shape suggest considering a two-module system, i.e., one mirror in the upper part of the port and one in the lower part, each launching 6 beams, with the 12 waveguides grouped in the middle of the port. On this basis, three main options have been considered, all of them using the ITER steering mechanism for the poloidal movement (which should be controlled in real time) and possibly a simpler solution for the toroidal movement, on a shot-to-shot basis. The feasibility of an EC antenna with these requirements seems very likely, but detailed design efforts continue to decide on the best option and assess its limitations.

Power supplies. The possibility of using an existing NBI power supply, manufactured by Siemens 27 years ago, has been analysed. Unfortunately, it has been found that only half of this system can be used with minor modificatons while the other half requires major refurbishement. The modification of both halves nevertheless seems both technically and economically attractive. An approach based on the use of the refurbished NBI powers supplies (type 1) to supply 8 gyrotrons and 4 new solid state power supplies (type 2) for the remaining 4 gyrotrons has been developed [20]. The Type 1 Power Supply consists of a rectifier transformer and a tetrode tube connected in series to the transformer DC - output: the old transformers are re-used here. Type 2 consists of new solid-state technology power supply, either PSM (Pulse Step Modulation) or primary clocked switch mode. A new building would be necessary for this type of power supply, although a solution using containers is also possible and cheaper. Corresponding requirements for the body voltage power supplies have also been determined.

Auxiliaries. A significant amount of auxiliary equipment and services are required to make the ECRH plant operational: cooling water, superconducting magnets for the gyrotrons, auxiliary power supplies (for the superconducting magnets, filaments, collector sweeping magnets, ion pumps), vacuum pumps and gauges, compressed air, motors for moving polarisers, mirrors and switches, arc detection systems, microwave measurement systems, etc.). The requirements for all these auxiliary systems have been analysed [21], using gyrotron specifications provided by IAP/GYCOM. No major feasibility issue has been encountered.

Layout and port allocation. Finding an adequate port and location for the EC system in the present JET environment is not easy. Four port options have been identified: Octants 1, 2, 5, or 6. Octant 2 could be a simple solution in the case of removal of the ITER-like ICRH antenna. Octant 6 would require relocation of several diagnostics, a problem that would be minimised in the case of Octants 1 or 5. Four different locations for the EC system (gyrotrons with auxiliaries and power supplies) have been identified, most of them requiring the construction of new buildings. A floor area requirement of at least 400 m^2 has been estimated, in 2 storeys. All of the possible locations require removal of existing equipment. Octant 2 has been chosen as the reference port allocation for the ECRH antenna ; consequently, a gyrotron building located north of J1 is the best solution [21].

Control systems and diagnostics. This project would yield a good opportunity to design a control system that adopts and tests a number of ITER-CODAC standards. A number of diagnostics, in particular microwave based ones, will require adequate protection from stray radiation produced by the gyrotrons. Stray radiation caused by ECRH power is assumed to be a few mW in normal operation, and may rise by a factor of 100 to 1000 under fault conditions. For the latter case, if components can handle 10 dBm (10 mW) a 40 dB rejection at 170 GHz is needed. A very narrow frequency width (~1GHz) is assumed for the 170 GHz source, to take into account any frequency jitter - particularly during switching. An upgrade of the ECE radiometre real-time network will also be necessary, for MHD control applications.

4. Conclusions

An important output of the project is that the usefulness of an EC system for JET is definitely confirmed by the physics studies. No strong limitations for any of the functions envisaged have been found. However, the power of 10 MW should be considered as a minimum, and some limitations in the magnetic field values at which the system can be used appear unavoidable, in particular for a single frequency system. However, the chosen frequency 170 GHz would maximise the synergies with the ITER system and possibly the contributions by F4E. This choice takes into account physics requirements and properties, together with the availability of existing GYCOM gyrotrons, the much simpler window concept (with respect to a double-frequency system) and the possibility of components and solutions being developed jointly with the ITER and F4E teams.

In conclusion, an ECRH system for JET seems feasible, but it will be a complex project, with significant associated costs. From a technical point of view, the most critical issues concern the launcher design and the behaviour of new components such as the gyrotrons, the diamond windows and the mirror steering mechanism. However, these components are the same as for the ITER ECRH system therefore their full scale test within the JET ECRH system would constitute a valuable risk reduction contribution to the ITER system. The total cost of the system is estimated ~ 51 - 53 M€. The manpower necessary for design, construction, tests, system commissioning on plasma and project management is estimated to 114 ppy; a dedicated team of at least 10 professionals and technicians seems necessary. A preliminary planning is compatible with the delivery of 10 MW ECRH power in the plasma in approximately 5 years, provided adequate funding and highly qualified manpower are available.

This work was supported by EURATOM and carried out within the framework of the European Fusion Development Agreement. The views and opinions expressed herein do not necessarily reflect those of the European Commission.

References

1. F. ROMANELLI, Fus. Sci. Techn. **53** (2008) 1217.
2. J. PAMELA et al., J. Nucl. Mat. **363-365** (2007) 1.
3. M. SHIMADA et al., Nuclear Fusion **47** (2007) S1.
4. C.E. KESSEL et al., Nucl. Fusion **47** (2007) 1274.
5. C. GORMEZANO et al., Nucl. Fusion **47** (2007) S285.
6. A.G.A. VERHOEVEN et al., Nucl. Fusion **43** (2003) 1477.
7. M.A. HENDERSON et al., Nucl. Fusion **48** (2008) 054013.
8. D. FARINA, Fusion Sci. Techn. **52** (2007) 154.
9. J.F. ARTAUD et al., Nucl. Fusion **50** (2010) 043001.
10. D. FARINA, L. FIGINI, this conference.
11. S. NOWAK, this conference.
12. E. LAZZARO, S. NOWAK, Plasma Phys. Contr. Fusion **51** (2009) 035005
13. G.V. PEREVERZEV, P.N. YUSHMANOV, *ASTRA-Automated System for TRansport Analysis*, IPP 5/98, February 2002.
14. F. PORCELLI et al., Plasma Phys. Contr. Fusion **38** (1996) 2163.
15. M. ERBA et al., Plasma Phys. Contr. Fusion **39** (1997) 261.
16. G.G. DENISOV, this conference.
17. G.G. DENISOV, A.G. LITVAK et al., Nucl. Fusion **48** (2008) 054007.
18. S. GARAVAGLIA et al., this conference.
19. R. CHAVAN et al., Fusion Engineering and Design **82** (2007) 867.
20. H. BRAUNE et al., this conference.
21. M. LENNHOLM et al., this conference.

Feasibility of an ECRH system for JET: wave propagation, absorption and current drive

Daniela Farina, Lorenzo Figini and JET-EFDA contributors*

JET-EFDA, Culham Science Centre, Abingdon OX14 3DB, UK

Istituto di Fisica del Plasma "P. Caldirola", Consiglio Nazionale delle Ricerche, EURATOM-ENEA-CNR Association, via R. Cozzi 53, 20125 Milano, Italy

Investigation of electron cyclotron wave propagation, power absorption, and current drive has been performed for a set of JET scenarios, aiming to assess the optimal wave frequency, launching position, and injection angles to achieve the various physics goals of an EC system in JET. EC power absorption and current drive have been computed for three values of the EC frequency, namely, 113 GHz, 150 GHz, and 170 GHz. On the basis of extensive beam tracing calculations performed in a wide range of magnetic fields (2.5 T – 3.4 T), the frequency of 170 GHz has been chosen, since it covers the wider magnetic field range of operation in JET and corresponds to that foreseen in ITER.

1. Introduction

A feasibility study for an ECRH system at JET has been carried out in 2009 to explore the potential physics capabilities and the technical options [1,2]. An EC system working in JET before ITER operations would make the foreseen JET programme [3] more ITER relevant both from the viewpoint of the physics and of the related technology issues. The main physics tasks to be fulfilled by an EC system in JET have been identified as: i) electron heating in the plasma core, i.e., for $\rho < 0.3$ (ρ being the square root of the normalized toroidal flux); ii) well localized current drive for NTM control ($\rho \simeq 0.5-0.75$), for current density profile tailoring ($\rho \simeq 0.5-0.6$), and sawtooth control ($\rho \simeq 0.3-0.4$). The choice of the optimal wave frequency and the evaluation of the current drive performance in the main JET scenarios have been based on extensive and systematic EC beam-tracing calculations in a wide range of magnetic fields ranging from 2.5 T to 3.4 T. In addition,

*See the Appendix of F. Romanelli et al., Proceedings of the 22nd IAEA Fusion Energy Conference 2008, Geneva, Switzerland

the inputs for NTM's stabilization studies [4], and for the antenna design [5] have been provided by the analysis.

In the feasibility study, three frequency values have been considered in detail, 113, 150, and 170 GHz, each of these covering a different range of magnetic field operation. The first and last values of the frequency are those chosen in the previous JET project [6,7] and in ITER, respectively. The choice of the optimal EC frequency depends both on the range of magnetic field at which JET is expected to operate, and on the physics objectives of the project itself, which determine the radial region in which EC plasma interaction must take place. Being the EC interaction a resonance process that allows localized absorption and current drive depending on the magnetic field value, an EC system at a single frequency can not be designed to operate at all magnetic fields of interest in JET. A compromise between available sources and required physical objectives has to be found.

To lowest order, the features of the EC system can be identified investigating the occurrence of the EC resonance within the plasma volume varying the JET magnetic field. At 113 GHz, first harmonic interaction occurs in the high field side for $B_0 > 2.8\,\mathrm{T}$, while second harmonic resonance occurs in the low field side for $B_0 < 2.6\,\mathrm{T}$. Interaction occurs at second harmonic both at 150 and 170 GHz in the whole magnetic field range considered, and, at a given field, in inner regions at 170 GHz. Third harmonic absorption may affect the EC efficiency for $B_0 < 2.8\,\mathrm{T}$ at 170 GHz, if the electron temperature is large enough. For $B_0 > 3\,\mathrm{T}$, interaction occurs in the low field side, thus reducing the current drive efficiency, due to trapped particle effects. In the following we shall focus on the analysis at 170 GHz, while the results at 113 and 150 GHz can be found in [1].

Two representative JET plasma discharges at $B_0 = 2.65\,\mathrm{T}$ have been considered which correspond to a standard H-mode regime at high delta and high density (#73344), and to a non-inductive advanced regime (#77895). These two scenarios are among those developed at JET as being ITER relevant. Note that the advanced scenario has a higher temperature and lower density with respect to the standard H-mode scenario, with the ratio T_e/n_e $3 \div 4$ times larger. Investigation of the ECCD performance has been performed in a wide range of magnetic field values rescaling the reference scenarios varying the magnetic field, the plasma current, and the plasma pressure keeping the edge safety factor q_{95} and the plasma β constant, and the plasma density and temperature keeping $n/n_{GRW} = const$, being n_{GRW} the Greenwald density, so that $I, n, T \sim B$.

Two different launching positions have been considered, with the same radial and two different vertical positions, $R = 4.3\,\mathrm{m}$, $z = \pm 0.345\,\mathrm{m}$, roughly corresponding to the upper and the lower extreme of a JET equatorial port. A divergent beam with waist $w_0 = 3\,\mathrm{cm}$ at the mirror has been considered. The poloidal and toroidal injection angles α and β are varied in a wide range, $-40° \leqslant \alpha \leqslant 40°$, and $0° \leqslant \beta \leqslant 26°$, with β positive in order to drive a co-current.

Calculations are performed using the GRAY code [8], a quasi-optical beam tracing code for EC Gaussian beam propagation. Power absorption is computed solving the fully relativistic dispersion relation for EC waves [9], while the current drive is computed via the adjoint method. All the evaluations are done for a nominal EC injected power of 1 MW, and for $Z_{eff} = 2$. While the values of the current and of the deposition radius are almost independent of the beam parameters, the current density profile may be strongly affected by the assumptions made for the beam width and divergence, except at large toroidal angles when Doppler broadening determines the J_{cd} profile shape. Optimization of the wave beam might lead to a better performance of the EC system.

2. ECCD performance at 170 GHz

EC waves at 170 GHz are injected polarized as extraordinary mode, and are absorbed mainly at the second harmonic. Figure 1 shows the driven current obtained scanning the whole range of injection angles under consideration versus the radial position of the peak current density for three values of the magnetic field and the two scenarios. This kind of plots provides the following information: i) the envelope of the scattered points gives the maximum current that can be driven at a given radius, ii) the "density" of the points gives a qualitative estimate of the sensitivity with respect to variations of the injection angles. A much higher current is found in the case of the advanced scenario (by a factor up to 3, depending on the radius), since this scenario is more favourable to ECCD being a high temperature and relatively low density scenario. The driven current per unit power is inversely proportional to density and increases (almost linearly) with temperature for the JET parameters.

Third harmonic absorption may strongly reduce the driven current in scenarios at low magnetic field, this effect being more relevant at high temperature. The third harmonic resonance is located within the plasma region for $B_0 \lesssim 2.6\,\mathrm{T}$, so that, depending on the magnetic field value and on the plasma electron temperature, power available at the second harmonic resonance may be reduced due to downshifted third harmonic absorption in

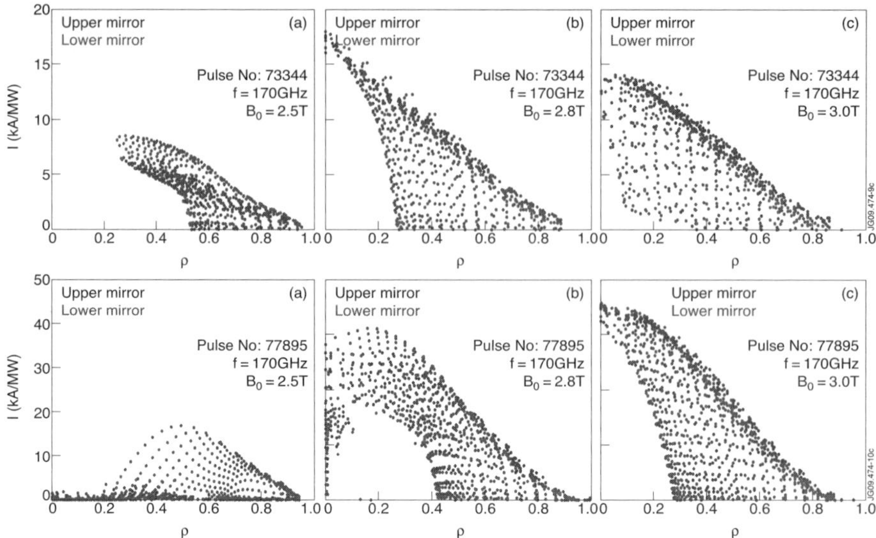

Fig. 1. EC driven current versus normalized radius for f=170 GHz in the H-mode scenario #73344 (top row) and in the advanced scenario #77895 (bottom row), obtained by means of a scan in poloidal and toroidal injection angles, for power absorption larger than 95%. Cases (a), (b), and (c) refer to $B_0 = 2.5, 2.8, 3.0\,\mathrm{T}$, respectively.

the low field side even for $2.6\,\mathrm{T} \lesssim B_0 \lesssim 2.8\,\mathrm{T}$.

The driven current and the maximum value of the EC current density as a function of the injection angles, together with the contours of the radial location of the driven current density peak are plotted in Fig. 2. It is shown that it is possible to maximize either the driven current or the current density on a specific magnetic surface by means of a proper choice of the injection angles. For the considered launching conditions, the effective poloidal steering range required to cover all the plasma positions in the whole magnetic field range considered is about 40° in the case of #77895, and somewhat lower, $\sim 30°$, in the case of #73344.

In order to synthesize the large amount of ECCD results and to estimate the performance at different magnetic fields, the maximum value of the driven current attained in a given radial interval $\Delta\rho = 0.1$ is calculated and represented graphically as a density plot in the (ρ, B_0) space in Fig. 3 for shots #73344 and #77895. The ECCD efficiency is about 2.5 times larger in the case of the advanced scenario #77895 with respect to the H-mode #73344, mainly due to the favourable high temperature and low density. The maximum ECCD is found at $B_0 = 2.8\,\mathrm{T}$ for #73344 and at

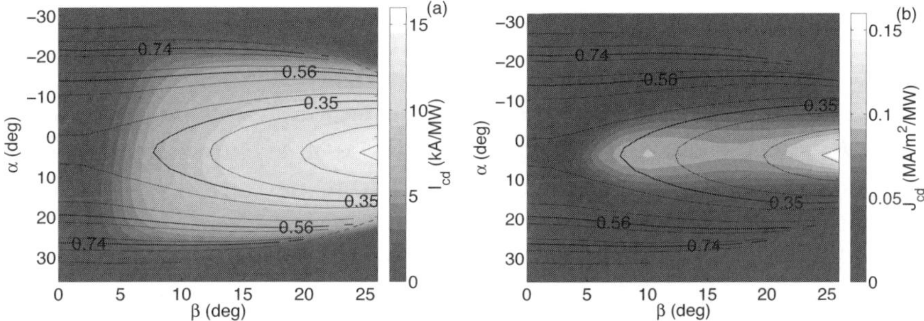

Fig. 2. EC driven current (a) and peak current density (b) as a function of the launching angles from the upper mirror, for shot #73344 at $B_0 = 2.65\,\mathrm{T}$. The contour lines give the radius at which J_{cd} is peaked.

3.0 T for #77895, because of the larger outward shift of the plasma centre for #77895. The inner region of the plasma ($\rho \leqslant 0.2$) at high and low B_0 is not accessible in the H-mode scenario, since the EC resonance is located far from the center in the low and high field side respectively. Although Fig. 3 provides a simplified picture of the possible use of an EC system at 170 GHz, we can conclude that a single frequency gyrotron at 170 GHz would cover the range $2.7\,\mathrm{T} - 3.2\,\mathrm{T}$ in both scenarios, and up to $3.3\,\mathrm{T}$ in the case of the advanced scenario for which the ECCD efficiency is peaked at higher B_0. Further extension of the magnetic field range covered by the system can be obtained by using a double frequency gyrotron (e.g., 136 GHz [1]).

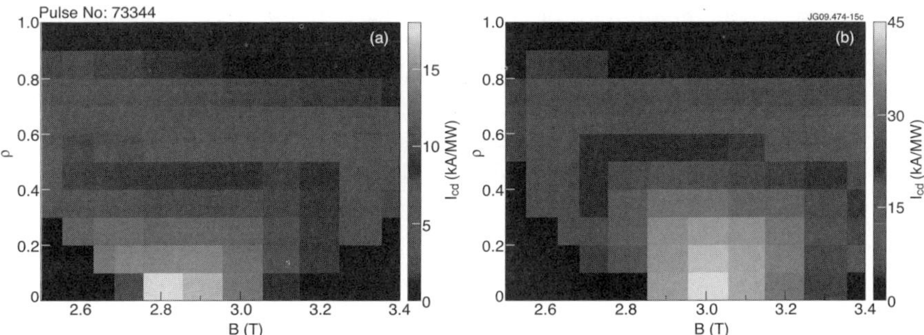

Fig. 3. EC maximum driven current at 170 GHz within the radial interval $\Delta\rho = 0.1$, for the H-mode scenario(a) and the advanced scenario (b). The plotted results are the average of the upper and lower mirror results, and are filtered with the following criteria: absorbed power larger than 95%, $\beta < 25°$, and current profile width $\Delta\rho < 0.1$.

3. Conclusions

As a result of the overall analysis, the frequency 170 GHz has been taken as the reference value for the present ECRH project in JET. The choice has been made on the basis of the estimated performance, showing that the range $B_0 = 2.7 - 3.2\,\mathrm{T}$ is well covered, with possible extensions to slightly higher fields. Lower magnetic fields suffer from parasitic third harmonic absorption that increases as the electron temperature increases. However, since also the current drive efficiency increases with temperature (at least in the range considered here), a large ECCD efficiency is still found in the high T_e advanced scenario.

Acknowledgments

This work was realised under EFDA JET Task Agreements n. JW9-NEP-ENEA-84, and JW9-OEP-ENEA-86, in the framework of the Project "Feasibility study of an ECRH system for JET". The views and opinions expressed herein do not necessarily reflect those of the European Commission.

References

1. G. Giruzzi *et al.*, *Feasibility study of an ECRH system for JET, Final Project Report*, tech. rep., EFDA JET (2009).
2. G. Giruzzi *et al.*, Feasibility of an ECRH system for JET: Project overview, in *Proceedings of this conference*, 2010.
3. F. Romanelli, *Fusion Sci. Technol.* **53**, p. 1217 (2008).
4. S. Nowak, Feasibility of an ECRH system for JET: Neoclassical tearing modes stabilization, in *Proceedings of this conference*, 2010.
5. C. Sozzi *et al.*, Feasibility of an ECRH system for JET: Options for an ECRH/ECCD launcher design, in *Proceedings of this conference*, 2010.
6. H. Zohm *et al.*, *JET-EP ECRH Physics*, tech. rep., EFDA JET (2002).
7. A. G. A. Verhoeven *et al.*, *Nucl. Fusion* **43**, p. 054013 (2003).
8. D. Farina, *Fusion Sci. Technol.* **52**, p. 154 (2007).
9. D. Farina, *Fusion Sci. Technol.* **53**, p. 130 (2008).

FEASIBILITY OF AN ECRH SYSTEM FOR JET: OPTIONS FOR AN ECRH/ECCD LAUNCHER DESIGN

C. SOZZI[1], A. BRUSCHI[1], D. FARINA[1], L. FIGINI[1], M. LENNHOLM[2,3], A. MORO[1]
JET-EFDA, Culham Science Centre, Abingdon OX14 3DB, UK

[1]IFP-CNR, EURATOM-ENEA-CNR Association, Milan, Italy
[2]EFDA Close Support Unit, Culham Science Centre, Abingdon OX14 3DB, UK
[3]European Commission, B-1049 Brussels, Belgium

A study has been conducted to evaluate the feasibility of installing an ECRH plant on the JET tokamak. The possible options for the wave launching system for JET have been investigated, assuming a frequency of 170 GHz, the use of an evacuated transmission line and the availability of an entire JET mid-plane port for the launching system. Applications include NTM and sawteeth control, core heating and current drive, current profile tailoring .

1. Guidelines for the preliminary design

While in principle ECRH/ECCD systems in toroidal fusion devices are nowadays nearly "standard" equipment, serious issues for the launcher design come when high performance, extended flexibility and lack of space are among the constraints of the project. This is the case of JET where the most powerful ECRH plant worldwide in an existing machine, with wide poloidal and toroidal steering, real time capabilities, and tritium-grade vacuum integration is being considered for installation.

The technical requirements for the launcher have been specified after an extensive wave propagation study related with the range of the ITER-oriented physics applications desirable for JET [1]. The main requirements to be fulfilled are:

- Power capability of 10 MW delivered into plasma
- 20sec pulse duration repeated every 15 minutes
- Steering range of $\Delta\beta=\pm30°$ (toroidal, shot to shot positioning), $\pm25°<\Delta\alpha<\mp5°$ (poloidal, real time controlled)
- Optical performances (power and driven current density) good enough for NTM control (J_{CD} current channel 5-10 cm wide ($\Delta\rho<0.1$) [2,3] in an extended absorption range (1-1.5 m long, depending on the plasma

magnetic field and island location)
- Inclusion of the ITER poloidal steering mechanism [4,5] with minimum modifications in terms of design and functionalities
- Port integration compatible with plug-in installation (no or minimum use of remote handling and/or fixed in-vessel installations)

For what concerns the antenna, three design options (Figure 1) have survived the initial stage of the feasibility study. All the options use two nearly identical modules, vertically displaced, each one occupying half of the available volume, including the ITER steering mechanism (see Table 1) and using 2x6 corrugated waveguides of 63.5 mm inner diameter as input. Each beamline is equipped with a focusing mirror positioned in front of the waveguide redirecting the beam towards the steering unit. The other mirrors in the system are flat. The toroidal mechanism in not yet identified in detail and is only considered in term of required functionalities (i.e. angular range and needed clearance).

Option L1: the whole steering system is placed in the narrow part of the JET port close to the plasma boundary. A modified ITER-like mechanism mounts the poloidal mirror tangentially. The toroidal steering is realized rotating the whole poloidal mechanism or the mirror only around a nearly vertical axis.

Option L2: an unmodified ITER-like mechanism in the rear part of the JET port provides the poloidal steering. The toroidal steering is obtained moving the whole mechanism in such a way that the beams can enter the plasma either directly or through 1 or 2 bounces on two large, fixed, flat (non parallel) side mirrors. This layout provides a range of toroidal injection angles with forbidden angular windows.

Option L4: an unmodified ITER-like mechanism in the rear part of the JET port provides the poloidal steering. The toroidal injection is obtained steering one of the side mirrors with 1 or 2 bounces towards the plasma. This layout provides a nearly continuous range of toroidal angles.

Table 1. Main characteristics of the ITER steering mechanism

	Poloidal steering mechanism
Size	~250(L)x245(D)mm, cylindrical shaped
Actuator	Pneumatic (He gas), servo-valve 3.5 - 21.5 bar
Range	+/- 7° (rotation around symmetry axis)
Speed	Full range < 3 secs
Accuracy	0.05 °

The last option (L4), which decouples the two steering mechanisms, offers significant advantages in terms of achievable steering range and appears suitable for easier port integration. The remainder of the paper is therefore mainly focused on option L4.

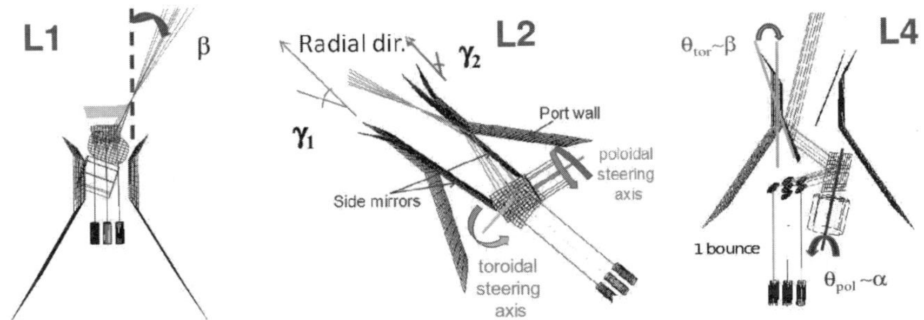

Figure 1: Options for the steering unit layout (top view, not in scale). β is the toroidal injection angle, γ_1 and γ_2 are the fixed angular positions of the side mirrors (with respect to the radial direction), θ_{tor} (θ_{pol})is the adjustable angle of the toroidal (poloidal) steering mechanism,.

2. Steering range

The steering range is expressed in terms of the poloidal angle α (between the horizontal plane at constant z and the poloidal component of the beam) and the toroidal angle β (the angle between the beam and the poloidal plane).

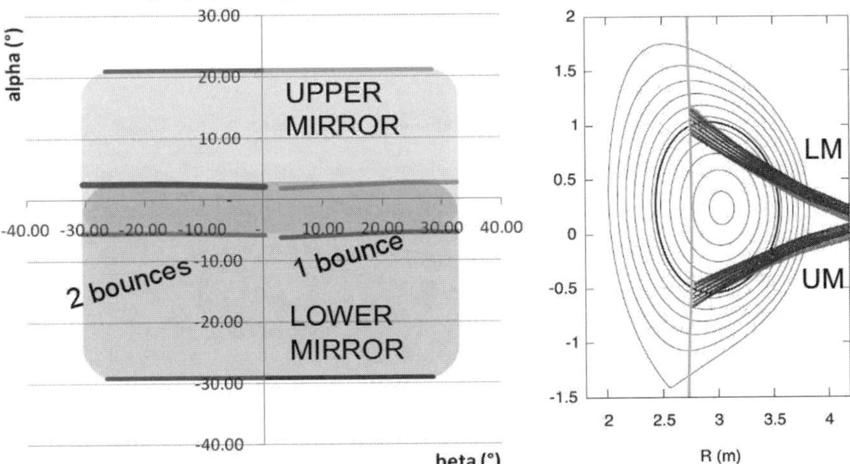

Figure 2: On left: steering range obtained with the L4 layout design. Note that positive α's are downward. On right: beams paths for the upper and lower steering units (B=2.8T, resonance near the q=3/2 magnetic surface). Beams paths from UM and LM crosses outside the last closed magnetic surface.

The achievable steering range for option L4 in the α,β plane is shown in

Figure 2 (shaded area). It spans about ±30° toroidally and ±30° poloidally. In order to cover most of the plasma volume, the upper and lower ranges are generally separate, except for a superposition region around the equatorial plane represented by the darker area. A small forbidden gap of about ±0.5° due to the 1 to 2 bounces range transition is located around 1° toroidal injection angle.

3. Focusing optics and beam size optimization

The focusing mirror parameters have been optimized with the criterion of minimizing the beam size in a reference absorption position in the plasma, corresponding to R=2.5m with radial injection. The input parameters for the optimization are the input beam waist (w_{0in}=20.4mm at the waveguide's tip) and the input distance from the waveguide to the focusing mirror d_{0in}. The output parameters (for each value of d_{0in}) are the focal length f of the mirror and the ellipsoidal surface equation obtained matching the phase front curvature on the mirror surface.

Figure 3 shows the output beam waist w_{0out} and the beam dimension w_{pl} at R=2.5m versus d_{0in}. As expected, larger input distance d_{0in} gives better focusing in the plasma, with little effect on w_{0out}. Figure 4 compares the beam size versus the propagation distance from the input waveguide for the different layouts. For L4 option d_{0in}=800mm, f=1037mm and mirror diameter D=100mm (≅3w at d_{0in}) have been considered, as longer input distances seem to be impractical. Dots in the figure are physics optics computation [6] including beam truncation effects.

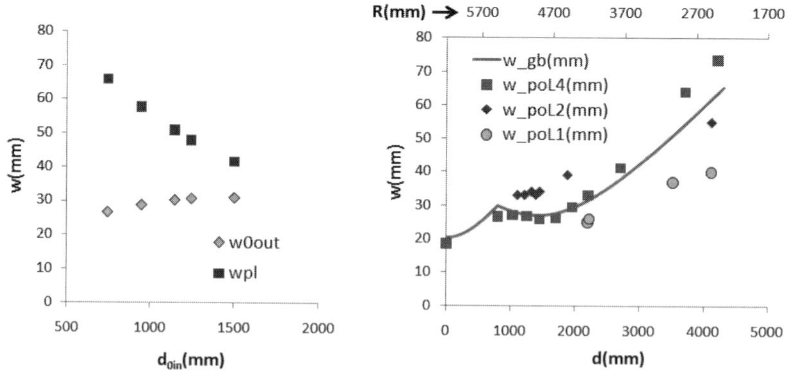

Figure 3 On left: Beam waist (w_{0out}) and size (w_{pl}) at the reference absorption location as a function of the distance from the input waveguide to the focusing mirror.
Figure 4 Onright: Beam size along the propagation distance d (and plasma radius R) from the waveguide exit computed with a physics optics tool (dots) for the various layout's options. Line represents the corresponding gaussian beam propagation for L4 option.

The resulting beam size is in line with the requirements, but safer margins can be obtained with a mode converter artificially increasing the effective d_{0in}. Such device is currently being designed.

4. Power distribution on mirrors

Physics optics computation have been performed to compute the power density and beams distribution on the mirrors. Figure 5 shows the beams footprint on the poloidal steering mirror for the L4 options, with the optical

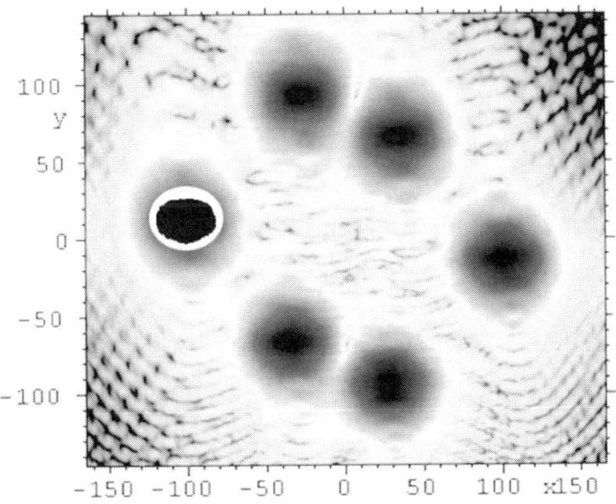

Figure 5: Power distribution on the poloidal steering mirror (see text). Fringe pattern in the corners are artifacts. Units on axis are mm.

parameters specified above. The increased contrast area on one of the beams marks the $1/e^2$=-8.7dB contour including 86.46% of the equivalent gaussian beam power. The peak power density is 97 kW/cm^2 assuming 1MW beam power. The average power density in the beam radius (white contour) is 55 kW/cm^2. These values are rather high but seems acceptable for a transient system [7].

5. Optimisation of the beams superposition in the plasma

Applications such the NTM control and suppression are critically dependent from the driven current density in the magnetic island location and therefore the superposition of the beams along an extended region is a key factor in a multi-beam system operating over a range of magnetic fields. Assuming the spacing among the input waveguides of 10cm and the L4 layout the maximum achievable current density has been computed for the resonance on the q=3/2 magnetic surface at B=2.8T using the beam tracing code GRAY

[8]. An optimal geometrical convergence of the beams corresponding to $\delta\alpha=2°$ has been derived. Then the driven current profile has been evaluated for magnetic fields $2.6<B<3.2$ T. In this range the current channel broadening (FWHM) is below 20% of the optimized case, being smaller at the lower extreme of the magnetic field range. This broadening corresponds to a reduction of about 12% of the peak current density.

The convergence of $\delta\alpha$ between beams launched from different z position of the same module is achievable by introducing a similar tilt in the orientation of the focusing mirrors.

6. Conclusions

A few layout options for the ECRH/ECCD launcher of JET have been studied. From the feasibility point of view, at least one of them (L4) is compliant with the requirements and does not present major technical drawbacks.

References

1. G.Giruzzi et al., this conference
2. S.Nowak, this conference
3. D.Farina, L.Figini, this confererence
4. R.Chavan et al., Fusion Engineering and Design **74**, 437–441, (2005)
5. A.Collazos et al., Fusion Engineering and Design **84**, 618–622, (2009)
6. K.Pontoppidan, GRASP® Tech. Description, TICRA, Denmark (2005).
7. F.Sanchez et al., Fusion Engineering and Design **84**, 1702–1707, (2009)
8. D. Farina, Fusion Sci. Technol. **52** 154 (2007)

Acknowledgments

This work, carried out under the European Fusion Development Agreement, supported by the European Communities and "Istituto di Fisica del Plasma P.Caldirola – CNR ", has been carried out within the Contract of Association between EURATOM and IFP. The views and opinions expressed herein do not necessarily reflect those of the European Commission or IFP.
Authors are indebted to the colleagues of the EC/ECRH community at ITER/IO, F4E, CRPP Lausanne, IPP Greifswald and IPF Stuttgart for their valuable suggestions.

FEASIBILITY OF AN ECRH SYSTEM FOR JET: NEOCLASSICAL TEARING MODES STABILIZATION

S. NOWAK [1] AND JET EFDA CONTRIBUTORS[†]
JET-EFDA, Culham Science Centre, OX14 3DB, Abingdon, UK

[1]Istituto di Fisica del Plasma CNR, Euratom Association, 20125 Milano, Italy

In the framework of the study of the feasibility of an ECRH system for JET, the aim of this work is to verify that the foreseen 10 MW–170 GHz Electron Cyclotron (EC) input wave power would be sufficient for Neoclassical Tearing Modes control, crucial issue for operation at high β and to determine how much the required power might be further reduced in the case of the microwave source modulation. The stabilization of low order (m,n) modes is discussed using a Generalized Rutherford equation and JET ELMy H-mode scenarios are considered in a wide range of operational magnetic field.

1. Introduction

A study has been conducted to evaluate the feasibility of installing an ECRH system on the JET tokamak [1]. In this framework, the stabilization of Neoclassical Tearing Modes (NTMs) by Electron Cyclotron Heating (ECH) and Current Drive (ECCD), a crucial issue for operation at high β, is investigated in different plasma configurations.

The basis for the stabilization of low order (m,n) NTMs is discussed in terms of a Generalized Rutherfor Equation (GRE) dynamic model [2] coupled to the mode rotation frequency evolution [3], the control criterion being to reduce the mode amplitude to its stable marginal width by balancing the ECH and ECCD terms with all the other ones in the GRE.

The stability of the main (2,1) and (3,2) modes, often rotating at very high frequency in JET (10-20 kHz), is investigated in JET ELMy H-modes scenarios in a wide range of operating toroidal magnetic field between 2.5 and 3.4 T. The aim is to verify that the foreseen 10 MW–170 GHz EC input wave power, injected at the 2nd harmonic from the low field side, would be sufficient for NTM control in all the considered plasma configurations and to determine how

---

[†] See the Appendix of F. Romanelli et al., Proceedings of the 22nd IAEA Fusion Energy Conference 2008, Geneva, Switzerland

much the required power might be further reduced in the case of the microwave source modulation.

2. Mode evolution equations and stabilization criterion

The island width (W) evolution of NTMs at the rational surface q=m/n located at the minor radius r_s is well described by a model based on GRE including stabilizing (-) and destabilizing (+) terms in the form:

$$dW/dt \propto -\Delta'_0 + \Delta'_{bs} - \Delta'_{GGJ} \pm \Delta'_{pol} - \Delta'_h - \Delta'_{cd} - \Re e(\Delta'_w) \qquad (1)$$

with the usual stability parameter Δ'_0 and the terms due to the Bootstrap current $\Delta'_{bs} \propto \beta_p \varepsilon^{1/2} L_q/L_p W/(W^2+W^2_d)$, to the curvature $\Delta'_{GGJ} \propto \beta_p\ \varepsilon^2 L^2_q/L_p W/(W^2+W^2_d)$, to the ion polarization current $\Delta'_{pol} \propto \beta_p\ \varepsilon^{3/2} L^2_q/L^2_p \omega(\omega\text{-}\omega_T)/W^3$, to the ECH $\Delta'_h \propto P_{EC} L_q \eta_h[W,\delta_h]/n_e T_e$ and ECCD $\Delta'_{cd} \propto P_{EC} L_q I_{cd} \eta_{cd}[W/\delta_{cd}, W^2/\delta^2_{cd}]$ and to the eddy currents in the resistive wall $\Re e(\Delta'_w) \propto 2\ m\ \omega^2$. The typical quantities in these terms are the local poloidal beta β_p, the local inverse aspect ratio ε, the local safety factor q and pressure p gradient lengths $L_{q,p}$, the mode frequency ω, the natural plasma frequency ω_T, the EC injected power P_{EC}, the EC driven current I_{cd}, the full e^{-1} power current density widths δ_h, δ_{cd}, the helical function $\eta_{h,cd}(W/\delta_{h,cd})$ related to how much the heating and the driven current are efficient in the island, the small width W_d due to the finite perpendicular transport reducing the bootstrap drive for the incomplete flattening of the pressure. The mode rotation frequency evolution is given through the balance of the electromagnetic, viscous and inertial torques as:

$$d\omega/dt = [\ -n\ (\ T_{em} + T_{visc}\ )\ -\ T_{in}\ ]\ /\ I \qquad (2)$$

with the torque due to the eddy currents $T_{em} \propto \omega\ W^4$, $T_{visc} \propto (\omega\text{-}\omega_T)/W$, $T_{in} \propto (\omega\text{-}\omega_T)\ dW/dt$ and the moment of inertia of the plasma $I \propto W$.

The NTMs can evolve from an initial seed island to their saturated sizes when the local β_p is larger than a minimum marginal critical value $\beta_{p,cr}$ corresponding to the marginal island width below which the NTMs are unconditionally stable. The stabilization criterion is to reduce the mode at its marginal width W_{mar} by balancing Δ'_h and Δ'_{cd} with all the other terms in GRE. The Δ'_{pol} term, stabilizing if rotating in the direction of the ion drift, shifts the W_{mar} at larger values with respect to those calculated without this term and its stabilizing effect is much stronger if ω increases. In JET neutral beam injection (NBI) heated plasmas the modes are often rotating with frequency up to 10 kHz for

(2,1) and 20 kHz for (3,2), the natural frequency ω_T being comparable with ω, because in this case it includes both the ion diamagnetic frequency (1-2 kHz) and the plasma rotation one. Therefore, for scenarios with NBI we choose $\omega/\omega_T=1.2$ with Δ'_{pol} not negligible compared to the other terms in GRE. The needed power should decrease for $\delta_h,\delta_{cd}<W_{mar}$ and for large W_d (~ 4 cm in JET).

3. (2,1)-(3,2) mode stability by P_{EC} CW injection in H-mode scenario

The stability of (2,1)-(3,2) NTMs is investigated for the H-mode scenario of the JET reference discharge #73344 at 61.02 s (2.65T/2.5MA/3.9KeV/8.8$10^{19}$m^{-3}) for P_{EC} continuously injected from the low field side from 2 different launching points above (Upper Mirror-UM) and below (Lower Mirror-LM) the midplane, located at z=±0.345m. The main plasma parameters used in GRE are in Table 1:

Table 1. main parameters used in GRE for (2,1)-(3,2) modes

q	ψ_n	ε	r_s [m]	L_q [m]	L_p [m]	β_p	n_e [10^{19}m^{-3}]	T_e [KeV]
2/1	0.68	0.3	0.91	0.4	0.28	0.38	6.9	1.6
3/2	0.49	0.25	0.77	0.55	0.35	0.60	7.6	2.4

The EC power needed for stabilization of these modes is calculated by using the ECH/ECCD profiles provided by the GRAY beam tracing code [4] with different toroidal and poloidal injection angles. The launching geometry is in agreement with the launcher conceptual design [5]. The results for various values of I_{cd} and δ_{cd} are plotted in Fig,1 (q=2) and Fig.2 (q=3/2), showing the role of all the terms in GRE. I_{cd} and δ_{cd} are ranged between 1.74-3.33 kA/MW and 0.04-0.12 m at q=2 and between 2.6-7.2 kA/MW and 0.035-0.1 m at q=3/2.

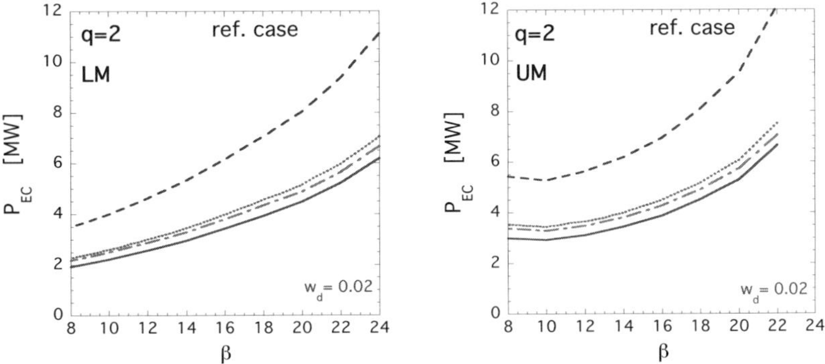

Figure 1. P_{EC} needed to stabilize the (2,1) mode, using the beam tracing results for I_{cd} and δ_{cd}, for both lower (left) and upper (right) mirrors. The roles of different terms in GRE are shown: $\Delta'_0+\Delta'_{bs}$ (dashed line), $\Delta'_0+\Delta'_{bs}+\Delta'_{GGJ}$ (dotted line), $\Delta'_0+\Delta'_{bs}+\Delta'_{GGJ}+\Delta'_w$ (dash-dotted line) and $\Delta'_0+\Delta'_{bs}+\Delta'_{GGJ}+\Delta'_w+\Delta'_{pol}$ (solidl line).

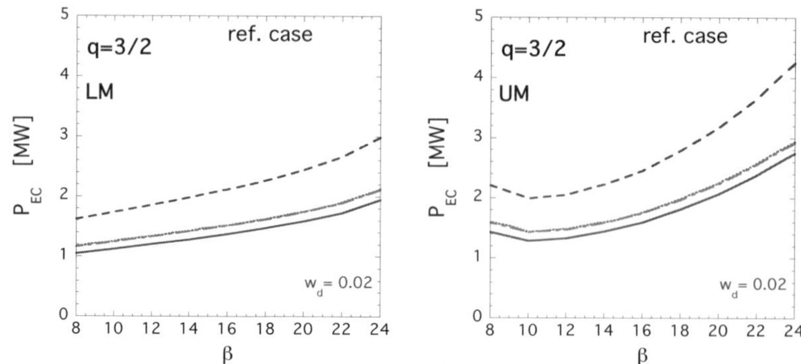

Figure 2. Same as in Fig.1 for the (3,2) mode. The effect of the resistive wall (dash-dotted line) is negligible at the location q=3/2.

EC power less than 6 MW seems sufficient to stabilize the (2,1) mode at its marginal width for toroidal angles < 20° corresponding to the lower driven current density widths ~ 0.04 m. It is to be noted that in this scenario $\Delta'_h << \Delta'_{cd}$. The mode (3,2) can be controlled using $P_{EC}<3$ MW, because the local EC peak current density is higher than in the previous case. The δ_{cd} is also smaller ~ 0.03-0.35 m. A general trend of EC power needed for stabilization vs driven current I_{cd} is given in Fig.3 for both the modes considering 2 fixed widths δ_{cd}=0.05-0.1m: (2,1) is stabilized with P_{EC}<8MW for I_{cd}>2.5kA/MW and (3,2) with P_{EC}<5MW for I_{cd}>4kA/MW. The stabilizing role of the ion polarization term Δ'_{pol} is shown. Less power is required as the δ_{cd} decreases and I_{cd} increases.

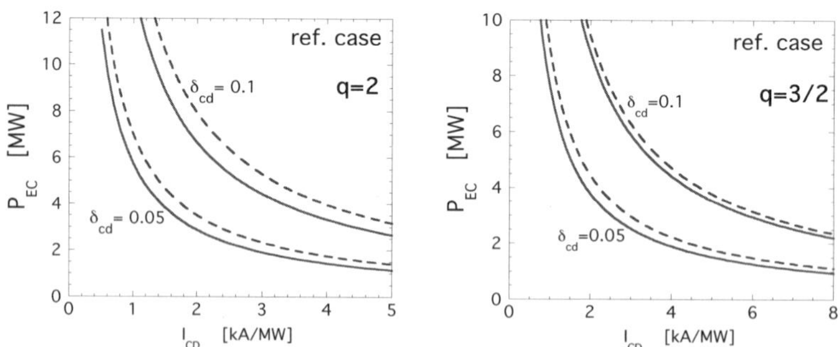

Figure 3. P_{EC} vs I_{cd} for 2 values of δ_{cd} =0.05-0.1m for (2,1) on the left and (3,2) on the right. The dashed lines refer to combination of terms in GRE without Δ'_{pol}, the solid traces with it.

In this reference scenario P_{EC}>10 MW is required for I_{cd} <1.5kA/MW (q=2) and I_{cd} <2kA/MW (q=3/2) for both the chosen widths δ_{cd}.

4. NTM stability by P_{EC} CW injection in ref. scenario at various B

The stability for (2,1) and (3,2) modes is calculated for various magnetic fields between 2.5T<B<3.2T, rescaling the plasma current, the electron density and temperature by a factor $\alpha = B/B_{ref}$ keeping constant q_{95} and β. In Fig.4 P_{EC} is plotted against the magnetic field B, calculated for the optimum value of the injection toroidal angle in general different for each value of B. The values of power needed for stabilization using I_{cd} and δ_{cd} from the beam tracing are compared with those obtained with the same driven current for δ_{cd} =0.05-0.1 m.

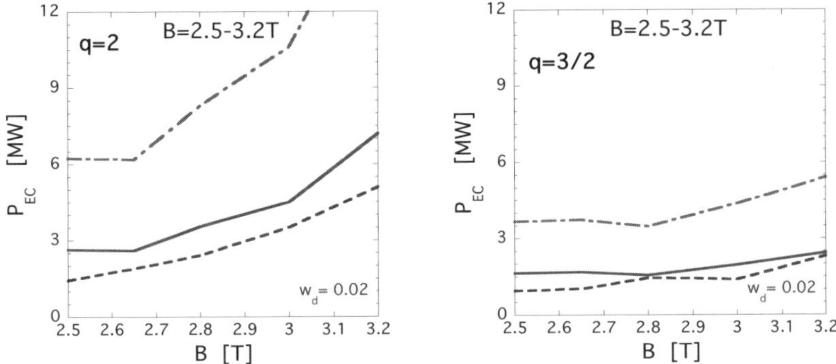

Figure 4. P_{EC} vs B using δ_{cd} =0.05-0.1m for (2,1) on the left and (3,2) on the right. The dashed lines are from beam tracing, the solid and the dash-dotted lines for δ_{cd} =0.05m and =0.1m respectively.

P_{EC}<6 MW are foreseen to stabilize the (3,2) mode at any magnetic field, while more than 8 MW are required to control (2,1) at B>2.8T for a driven current channel width δ_{cd}=0.1 m. At high magnetic fields the (2,1) control becomes marginal because of the poor current drive efficiency. All the traces plotted in Fig.4 correspond to the data from the lower mirror both for q=2 and q=3/2.

5. (2,1) stability by P_{EC} modulated injection in ref. H-mode scenario

The aim of the modulation is to minimize the ECCD power needed for NTM stabilization [6,7] by a gyrotron 50% duty cycle. Since for $W \ll \delta_{cd}$ the helical efficiency η_{cd} depends on W^2/δ_{cd}^2 without modulation and on W/δ_{cd} with modulation, Δ'_{cd} is more efficient in the second case and less P_{EC} is required for stabilization. For a current diffusion time much larger than the island rotation period the phasing of the ECCD peak with the centre of the mode is advantageous and therefore modulation could be useful, as in JET where the time modulation is small (30μs-100μs) respect to the current diffusion time (~100ms). Benefits come for W/δ_{cd}<0.5 . As in the previous cases we have $0.4 < W/\delta_{cd} < 1$, we can conclude that in JET the EC power modulation is not strongly competitive with CW injection, which is sufficient to stabilize the NTMs with less than 10 MW. In Fig.5 is shown an example of stabilization by unmodulated (ECCD phasing between -π and π) and modulated (ECCD phasing between -π/2 and π/2) EC power injection for the JET discharge #76082 at 50.5s, 20 MW NBI heating and (2,1) frequency of 12 kHz.

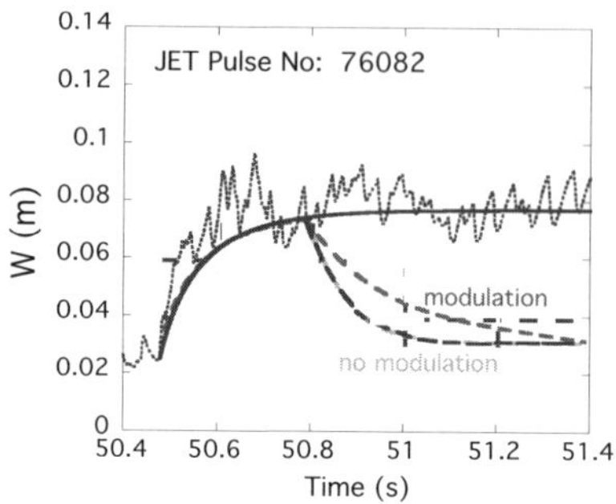

Figure 5. (2,1) amplitude evolution stabilized at its w_{mar}~0.032m by unmodulated and modulated EC power injecting 1.5MW and 0.7MW respectively for W/δ_{cd}~.029. Experimental line is dotted.

6. Conclusion

The stabilization of rotating NTMs in JET, by application of localized EC power, is discussed in term of a generalized Rutherford equation coupled to the mode rotation frequency evolution. Detailed analysis of the power needed for stabilization of (2,1) and (3,2) modes has been performed for JET ELMy H-mode scenario in a wide range of operating magnetic field. $P_{EC}\leq 6$ MW can control (3,2) mode in all the cases, while $P_{EC}\leq 8$ MW stabilizes (2,1) except for B>2.8T and $\delta_{cd}\geq 0.1$m . Modulation can be required only for $W/\delta_{cd}<0.5$ to stabilize the (2,1) mode using less power than that one needed in CW injection.

This work was supported by EURATOM and carried out within the framework of the European Fusion Development Agreement. The views and opinions expressed herein do not necessarily reflect those of the European Commission.

References

1. G. GIRUZZI et al., *this Conference.*
2. E.LAZZARO, S.NOWAK, *Plasma Phys. Contr. Fusion* **51**(2009) 035005.
3. G.RAMPONI, E.LAZZARO, S.NOWAK, *Phys. Plasmas* **6** (1999) 3561.
4. D.FARINA, L.FIGINI, *this Conference.*
5. C.SOZZI et al., *this Conference.*
6. G.GIRUZZI et al., *Nucl. Fusion* **39** (1999) 107.
7. O.SAUTER, *Phys.Plasmas* **11** (2004) 4808.

FEASIBILITY OF AN ECRH SYSTEM FOR JET: PLANT LAYOUT, AUXILIARIES AND SERVICES

M. LENNHOLM[1,2], F. BOUQUEY[3], H. BRAUNE[4], J. FARTHING[6], S.GARAVAGLIA[5], G. GIRUZZI[3], G. GRANUCCI[5], M. JENNISON[6], A. PARKIN[6]

JET-EFDA, Culham Science Centre, OX14 3DB, Abingdon, UK,

[1] EFDA Close Support Unit, Culham Science Centre, Abingdon OX14 3DB, UK
[2] European Commission, B-1049 Brussels, Belgium
[3] CEA, IRFM, 13108 Saint-Paul-lez-Durance, France
[4] Max-Planck-IPP, Euratom Association, D-17491 Greifswald, Germany
[5] Istituto di Fisica del Plasma CNR, Euratom Association, 20125 Milano, Italy
[6] CCFE, Culham Science Centre, Abingdon OX14 3DB, UK

A study conducted over the last year to asses the desirability and feasibility of installing an ECRH system on the JET tokamak has concluded that such a system is indeed both desirable and feasible. Details of physics studies, launcher and transmission line design, and power supplies are presented elsewhere in these proceedings. This paper concentrates on the logistical implications of installing this system at JET. The paper addresses issues such as port allocation and plant location. The study has concluded that a new building will be needed to house the ECRH plant. Building layout proposals are presented together with considerations regarding the required auxiliary equipment.

1. Introduction.

Having an ECRH system would strongly enhance the ability of JET to fulfil its mission in preparing ITER operational scenarios. Therefore a study evaluating the feasibility of installing such a system on JET within a timeframe allowing results to be beneficial for ITER has been carried out during 2009 in a collaboration between JET, various European Fusion associations and the Russian Federation with substantial input from both Fusion For Energy and ITER [1]. A previous JET ECRH design [2] formed the basis for the study, though the evolution of both requirements and technical and scientific experience has lead to a proposal which differs significantly from the previous design. Investigating different system frequencies and evaluating their ability to fulfil various physics requirement such as NTM stabilisation[3], Sawtooth control, Current Profile Control and central electron heating, 170GHz was found to be the best compromise [4,5]. A significant additional advantage of choosing 170GHz, which is the ITER frequency [6], is that components, not least the gyrotrons themselves, can be taken directly from the ITER design allowing verification not only of ITER ECRH physics but also of ITER ECRH technology. A plant of twelve 1 MW gyrotrons [7,8], capable of injecting 10MW of ECRH power for 20s into the plasma was found adequate, though more power would be desirable to achieve substantial current profile control [1,3,4,5]. In the following the investigation leading to the choice of launcher and plant location to be used in the reference design are described followed by considerations regarding building layout and a brief evaluation of the required auxiliary equipment. Power supply and transmission line considerations are presented in separate contributions [9,10].

2. Port Allocation.

One of the biggest challenges facing the feasibility study was the search for a port where an ECRH launcher could be located and it rapidly became clear that no such position could be made available without sacrificing or, at best, relocating existing equipment. With the proposed power of 10 MW transmitted through 12 separate transmission lines, a launcher with the required steering range requires the use of an entire JET main horizontal port [11]. With this boundary condition all 8 main horizontal ports were considered. Figure 1 show the present use of these ports. Ports 3, 4, 7 and 8 were rapidly ruled out as they are completely occupied by neutral beam injectors (Ports 4 and 8), The lower hybrid current drive launcher (port 3) and the interferometer used to measure plasma density (port 7). The diagnostic systems occupying ports 1 and 5 (Neutron monitor and LIDAR respectively) are hard if not impossible, to relocate. This in conjunction with the fact that these ports are used for access into the machine meant that neither port 1 nor 5 were considered ideal candidates. Port 6 presently contains a large number of smaller diagnostic systems. Relocating any single diagnostic from this port would probably not

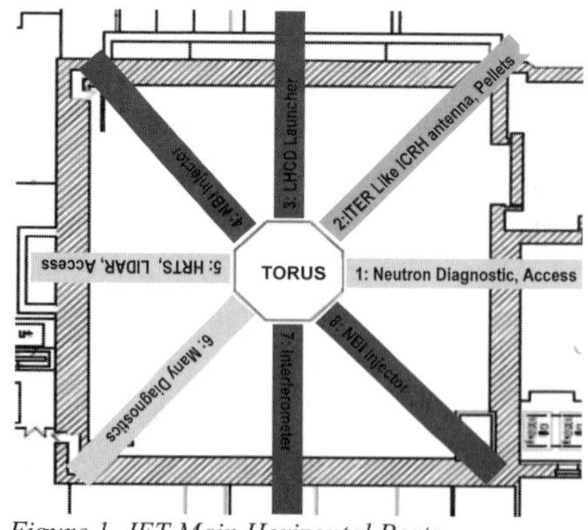

Figure 1. JET Main Horizontal Ports – Present Use

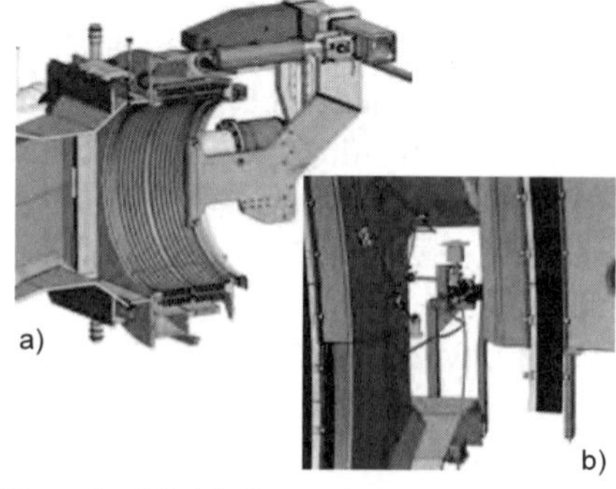

Figure 2. a) ILA bellow and support structure, b) View of port 2 from inside the JET vessel showing the pellet injection pipes.

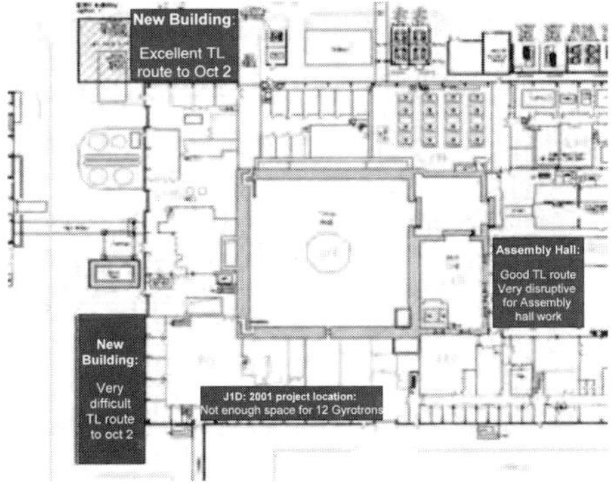

Figue 3, Plant location alternatives

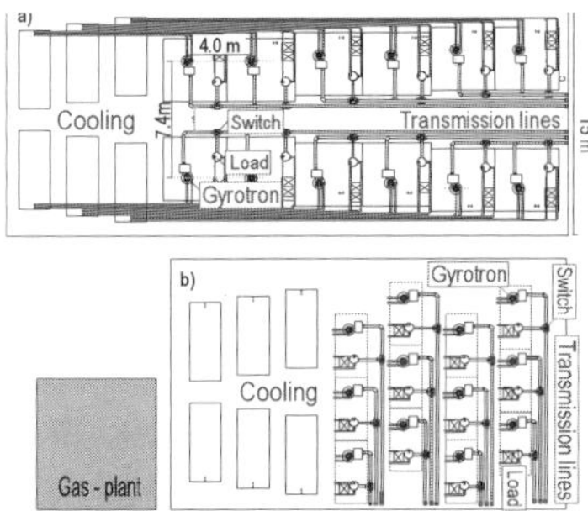

Figure 4. Plant building alternatives
a) initial 6x2 gyrotron layout.
b) compact 4x3 'honeycomb' layout

pose a significant problem but relocating all of them was found to be very difficult with significant risk of reducing the availability of essential diagnostics. The final port (port 2) is presently occupied by the ITER like ICRH antenna (ILA). This antenna is not operational at present and it is unlikely that it will be refurbished in the near future. Therefore replacing this antenna by an ECRH launcher would present a simple solution. If port 2 is used, the ECRH launcher could use infrastructure presently used by the ILA such as the main support and bellow seen in figure 2a. Port 2 also hosts some pellet injection pipes as seen in figure 2b, which would have to be accommodated by the ECRH launcher. These pipes are not very large and it is not considered a problem to leave space for them. Given these considerations it was decided that the feasibility study should concentrate on the use of port 2 for the ECRH launcher. This being said the use of one of the ports (1,5,6) could be reconsidered if the decision was taken to repair the ITER like ICRH antenna.

3. Plant location – and preliminary building design.

Having decided on the location of the ECRH launcher, the search for a convenient location for the ECRH plant started. The required space is mainly defined by the fact that the distance between adjacent gyrotrons has to be at least 4 metres in order to avoid interference between the magnetic fields of the gyrotrons. Figure 3 shows various potential plant locations. Installing 12 gyrotrons with their auxiliary equipment inside existing JET buildings was found to be, if not impossible, highly disruptive and very costly. This led to the decision that a dedicated ECRH building would be required. The final choice of a building location at the North West corner of the torus building was governed by the ability to find a simple routing for the ECRH transmission lines [9]. A preliminary building design using all the available space in this location and allowing the required distance between gyrotrons is presented in figure 4a. In

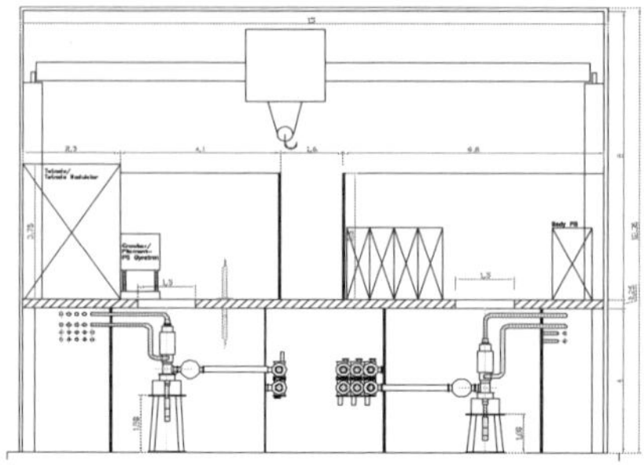

Figure 5. Elevation view of plant building – initial 6x2 layout

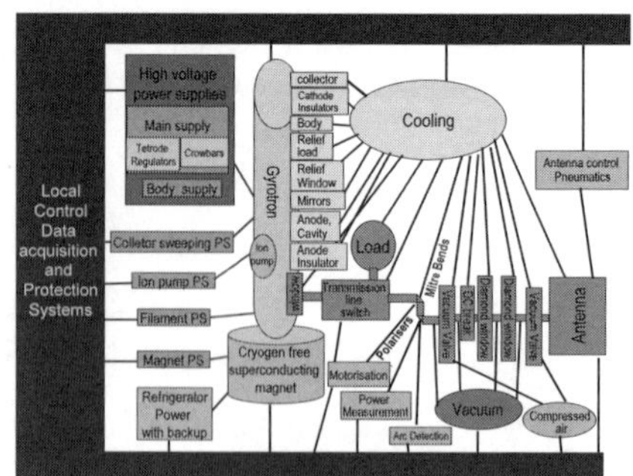

Figue 6. Plant schematic overview

this design the gyrotrons would be distributed in two rows of 6 gyrotrons. The gyrotrons and their associated cooling circuits and cooling plant would be on the ground floor while power supply equipment would be installed on the first floor. Removable walkways on the first floor would allow easy access to install and remove gyrotrons with a modest size overhead crane capable of lifting 300kg loads (fig. 5). A more compact layout, which avoids having to relocate an existing gas plant, is proposed as seen in figure 4b. The 4x3 honeycomb layout of the gyrotrons in this proposal would maximise the distance between gyrotrons for a given building size and it would assure the most symmetric distribution of stray magnetic fields.

4. Auxilliary Equipment.

Figure 6 shows an overview of the ECRH plant for one gyrotron. Of the systems shown in this figure, the launcher [11], transmission line [9], gyrotron [7] and high voltage power supplies [10] are described elsewhere in these proceedings. In the following the remaining equipment is considered:

Cooling: A number of system components needs to be actively cooled. The main energy to be removed is ~1-1.5 MW from each gyrotron collector and 1 MW from each test load. It is unlikely that it will be necessary to operate all gyrotrons simultaneously for 20 seconds into test loads and hence the total energy to be removed during one 20s pulse from collectors and loads should remain below 400 MJ. The total energy to be removed from all remaining components is modest in comparison. All cooling water has to be demineralised and a heat exchanger is required to take the heat from the demineralised water into the site water. Including a sizeable tank in the system will smooth the cooling of 20s pulses over a longer period though in doing this it has to be considered that the maximum water inlet temperature for the gyrotron cooling is 35°C. The gyrotron is cooled by several cooling circuits with different pressures and flow rates, some of which have to be electrically isolated from each other. In addition the diamond windows have to be cooled by an independent system with a special anti corrosive additive in the cooling water. The required transmission line cooling is under investigation and given the modest pulse length, it is likely that figure 5 is pessimistic in assuming that all transmission line components will require active cooling. The final component likely to need active cooling is the launcher. The launcher design [11] uses the ITER upper launcher steering mechanism with a slightly modified mirror for poloidal steering [6]. This assembly is designed to be water cooled. The cooling for the remaining launcher mirrors remains to be designed. In JET, for reasons of tritium compatibility, it would be desirable to avoid water cooling inside the torus vacuum. Whether interpulse gas cooling can fulfil the requirements is under investigation. A possible compromise could be to retain the water cooling for the poloidal steering assembly while designing the remaining structures for gas cooling.

Superconducting Magnets: Gyrotrons require a magnetic field of approximately 7 Tesla inside their cavity provided by external superconducting magnets. Two types of magnets have been considered: i) Traditional superconducting magnets cooled to liquid Helium temperatures using external cryogenic supplies. ii) So-called 'cryogen free' magnets, which are self-contained units that do not require any external cryogenic plant. The 'cryogen-free' magnets contain internal refrigerators and require only mains power to maintain the magnets at superconducting temperatures. As the existing JET cryoplant was found to be insufficient to supply liquid helium for a 12 gyrotron system, option ii) is considered to be the most economical and convenient solution. As the 'cryogen free' system is also the system chosen by ITER the study concluded that option ii) would also be the correct choice at JET.

Control and Data Acquisition: In line with the attempt at following the ITER design as closely as possible it is proposed that the Control and Data Acquisition system should be based on the ITER-CODAC/I&C design. Accordingly 'slow' instrumentation and control should use the Siemens S7 PLC range while using a model of the CODAC Plant System Host (PSH) computer for communications with the PLC. Above this host level, standard JET CODAS 'black-box' software should be used. For the safety systems, CISS (central Interlock and Safety System) and PSACS (Personnel Safety and Access Control System), standard JET procedures will be followed.

Other auxiliary equipment: In addition to the equipment described in the previous paragraphs other required systems are briefly discussed in the following. As the proposed transmission line solution [9] is based on evacuated cylindrical waveguides, a number of vacuum pumping groups including valves and gauges will be required both in the torus hall and in the plant building. Mechanical actuation is required to move the launcher mobile mirrors. For the ITER steering mechanism used for poloidal steering these actuators are pneumatic, while the toroidal steering may be achieved either using similar pneumatic actuators or using a simple motor driven pushrod system. As the wave polarisation has to be varied when injection angles change, motorised control of polarising mirrors is also required. To allow safe and reliable gyrotron operation a system measuring incident and reflected power is needed. Finally an optical arc detector system is essential to prevent damage to diamond windows both in the gyrotron and in the transmission line.

5. Conclusions.

The feasibility of installing a 12MW ECRH plant at JET has been studied. A solution for the choice of main horizontal port to host the ECHR launcher has been proposed and a promising location for a new building for the ECRH plant allowing a very simple transmission line trajectory to the tokamak has been identified. An initial evaluation of the required auxiliary equipment has been undertaken. As a conclusion the study has found no major impediments to installing the proposed system into the JET environment. Given the use of a new building to house the plant the installation of such a system could proceed with minimal interference with JET operation.

Acknowledgments:

The authors would like to acknowledge the substantial work carried out by the rest of the 'ECRH for JET' team [1] and thank the EC community as a whole for always being ready to promptly provide information and advice which has been invaluable for the progress of this study.

References:

1. G. GIRUZZI, this conference.
2. A.G.A. VERHOEVEN et al., Nucl. Fusion **43** (2003) 1477.
3. S. NOWAK, this conference.
4. D. FARINA, Fusion Sci. Techn. **52** (2007) 154.
5. D. FARINA, L. FIGINI, this conference.
6. M.A. HENDERSON et al., Nucl. Fusion **48** (2008) 054013
7. G.G. DENISOV, this conference.
8. G.G. DENISOV, A.G. LITVAK et al., Nucl. Fusion **48** (2008) 054007.
9. S. GARAVAGLIA et al., this conference.
10. H. BRAUNE et al., this conference.
11. C. SOZZI et al., this conference.

FEASIBILITY OF AN ECRH SYSTEM FOR JET: TRANSMISSION LINES & WINDOWS

S. GARAVAGLIA[1], G. GRANUCCI[1], M. LENNHOLM[2,3], A. PARKIN[4], G. GIRUZZI[5]

JET-EFDA, Culham Science Centre, Abingdon OX14 3DB, UK

[1]Istituto di Fisica del Plasma CNR, Euratom Association, 20125 Milano, Italy
[2] EFDA Close Support Unit, Culham Science Centre, Abingdon OX14 3DB, UK
[3]European Commission, B-1049 Brussels, Belgium
[4] CCFE, Culham Science Centre, Abingdon OX14 3DB, UK
[5] CEA, IRFM, 13108 Saint-Paul-lez-Durance, France

Quasi-optical (QO) and evacuated waveguide (EWG) transmission lines (TL) have been analyzed as possible solutions to transfer the power from the gyrotrons to the launcher. The EWG solution has been chosen as the reference TL solution due to the achieved space saving, its flexibility and its ITER relevance. The proposed EWG routing is compatible with the use of standard components, a small number of mitre bends and <5% of expected attenuation. Concerning the vacuum window, the only material currently available giving the power/pulse length combination is based on the use of disks in synthetic diamond (CVD). This choice gives the opportunity to adopt the window design proposed for the ITER project. A technical solution exploiting two separate single disk windows rather than a double window has been analyzed in order to fulfill safety requirements and to take into account problems arising in transmitting high power.

1. Introduction

In 2009 a feasibility study to evaluate an ECRH system on the JET tokamak has been conducted[1]. A specific comparison between different transmission line concepts has been considered even without the detailed layout of the system. As input data for this study we have considered a tritium compatible system, composed of 12 gyrotrons injecting 10 MW of power for 20s.

2. Concept options

An analysis has been conducted to choose the appropriate solution to transmit the power from the gyrotrons to the launcher. Two concepts can be considered: a quasi optical (QO) TL, based on the use of several mirrors that transmit the TEM_{00} mode in air and an evacuated waveguide (EWG) that transmit the low loss HE_{11} mode. While in several existing ECRH systems the

TL is based on the EWG solution (TCV, TS, DIII-D, JT-60, LHD), only few of them use a QO solution (TJ-II, W7-X).

2.1. *Quasi Optical solution*

As a reference solution for Q.O. TL, no real experience exists at multi-megawatt level and for long pulse length, except for the tests of the W7-X system. The W7-X ECRH system can be profitably used to evaluate the requirements of a QO line and its applicability to the JET environment.

The space reserved in the W7-X plant for the transmission of 10 MW is a straight duct of 30 m with a cross-section of 2.5 x 3 m, directed towards the machine. Each TL is composed of 16 mirrors, 6 of which are multi-beam (5 beams on the same large surface). The overall transmission efficiency is ~86%[2]. The total length of TL is ~60 m (depending on the gyrotron), while the distance between W7-X and the ECRH system building is ~30 m. This solution, although well designed and well integrated in the W7-X system, is not viable in the case of JET, for a number of reasons:

- A great space is necessary in the JET Torus Hall and connected buildings and such a volume is not available.
- Periodical mirror cleaning is needed and can be in conflict with the access limitation in the Torus Hall
- A metallic shielding must be installed to absorb stray radiation and to maintain low humidity and low air temperature. At the same time it should permit the accessibility for the mirror cleaning.
- The Tritium segregation requires a double barrier to avoid any risk of contamination. A large diameter window would probably be required.
- This is not the ITER solution and cannot benefit from the ITER project development.

2.2. *Evacuated waveguide solution*

EWG consists of corrugated aluminium waveguide with an internal diameter of 63.5 mm, connecting the gyrotrons to the launcher with mechanical continuity. The 12 lines can be allocated in a bundle of 1 x 1.2 m with 0.3 m of inter-axis. The transmission efficiency is very high (~95% in TCV[3]). The advantages of using EWG are:

- The occupied volume is minimized.
- RF leakage is avoided, except for the DC break, where shielding is mandatory.
- Polarisers are located in the mitre bends, under vacuum, minimizing arc occurrence.

- Very low losses.
- No maintenance is needed after installation.
- Some conditioning pulses should be done at the beginning of the experimental campaign and just after opening the line or arc occurrence.
- The windows work with vacuum on both sides, which is the safest way.
- The EWG is the ITER solution.

The drawbacks of this solution are the rigidity with respect to torus displacements (thermal expansion and disruptions), the spurious mode conversion in repetitive alignment stands and the employment of auxiliary equipments.

2.3. *Proposed solution*

The Q.O. solution is not compatible with the JET environment because there is not enough space for the mirror routing and this is not relevant for ITER. EWG is the solution proposed. This solution allows the required space to be minimised, to have more flexibility in the routing, to be fully ITER relevant and therefore to exploit the development made for the ITER project being, at the same time, a test for ITER.

3. Routing

A separate study has been accomplished to choose the best positions of the gyrotron Hall in the JET plant and of the port in which the ECRH launcher can be installed[4]. Our study has therefore been based on the assumptions that the Gyrotron Hall is in a new building and the launcher is positioned in octant 2
Figure 1 shows the chosen solution: the gyrotrons will be located outside J1 north-west corner. The 12 WGs reach the vessel with only 6 mitre bends, crossing the Torus Hall wall via 12 small drilling holes.

4. TL components

A list of necessary components has been defined on the basis of the experience on similar systems.

- Corrugated waveguide (63.5 mm i.d., $HE_{1,1}$ low loss mode): aluminium + INCONEL section connected to the launcher.
- 90° Mitre bends, with plane mirrors; They are responsible for most of the losses, mainly due to diffraction and mode conversion.
- Beam Switch to direct the power into the load. It can be integrated in the Matching Optics Unit to simplify the layout and the distribution of components.

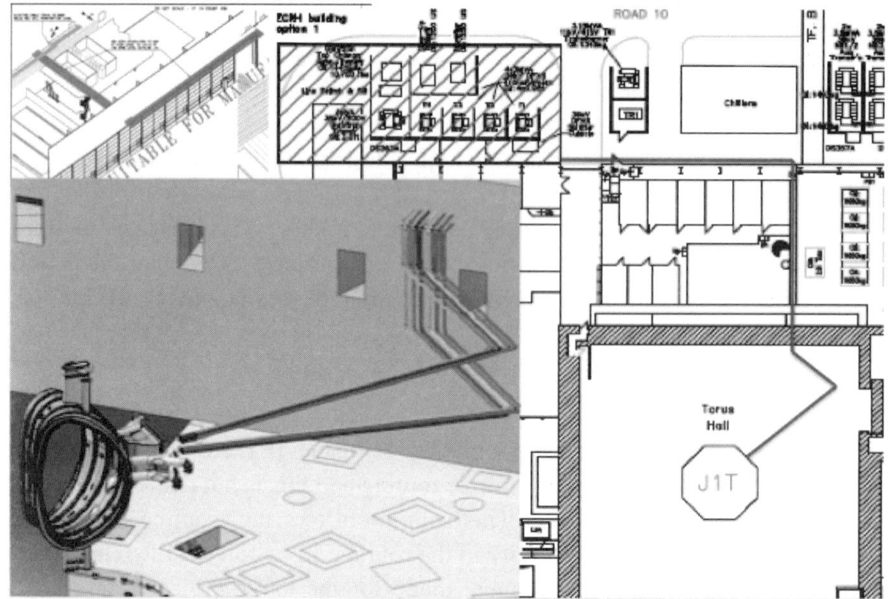

Figure 1. (Right) The suggested gyrotron building (the shadowed rectangle) and the approximate Transmission Line routing (red line). (Left): the route outside and through the J1H building (up) and the routing in the Torus Hall (down).

- Polariser: 2 Mitre Bends equipped with rotating corrugated surfaces (depth $\lambda/4$ and $\lambda/8$) remotely controlled in real time.
- Power monitor: a single Mitre Bend with detectors in both directions. Located as close as possible to the launcher.
- Sliding joints (bellows): to compensate length variation due to mechanical stress.
- DC break at the Torus Hall entrance.
- Gate valve: one before vessel, one after gyrotron+load
- Pumping unit: two groups to evacuate waveguides to 10^{-5} mBar through Pumping Tees.
- Load: 1MW/20s (cw equivalent) for every gyrotron.

The principal components exist and are routinely produced and tested at MW level.

Table 1 shows the quantity of each components for a single ECRH line, with an estimation of losses and the required cooling.

Table 1. Losses and required cooling. The total number of 90° bends is 6.

Components	Quantity	Losses (%)	Cooling (l/min)
WG63	70m	0.23	-
Mitre Bend (required in the Torus Hall)	3	1.5	5
Power Monitor (=1 MB with detectors)	1	0.5	7
Polarizer (=2 MBs with movable mirrors)	1	1.4	20
Bellows	2	0.006	-
DC break (at the entrance of Torus Hall)	1	0.002	-
Gate valve (in the vicinity of windows)	2	0.002	-
Pumping T	3	0.003	-
Beam Switch	1	0.5	5
Load (able to sustain 1MW@20s)	1		360
TOTAL		**4.13**	**407**

5. Vacuum window

Considering pulse length (20s) and transmitted power (1 MW) the only choice is a synthetic diamond window. A good opportunity is to use the ITER design as a reference, with substantial synergy with ITER and possible support by F4E (design and procurement). In the previous JET-EP project[5] a double window based on the use of two diamond disks with a very small inter-space was proposed. This solution has been abandoned because it doesn't resolve the problem about how to detect the failure of one disk and moreover it is not ITER relevant. The actual ITER project presents one single disk CVD diamond window, with 1.11mm of thickness, 75mm of diameter[6] and without arc detector looking at the window surface. The segregation of tritium requires the presence of a double barrier. In this project the use of two separate ITER windows has been considered, one positioned close to the launcher and the other one at the entrance of Torus Hall (Figure 2). This arrangement appears safer because it reduces the risk of damaging both windows in one incident.

A technical point to be solved is how to detect a failure of one of the windows. The recommended standard JET solution used in similar diagnostic lines is reducing as much as possible the volume between the barriers and installing a real time leak test, filling the inter-space with a known gas such as Neon at 0.5 Bar and monitoring the presence of Ne in the vessel and in the rest of the EWG. However 0.5 Bar is not compatible with high power operation of the EWG (required 10^{-5} Bar), since the risk of arcing would significantly increase.

A different strategy proposed is looking for Tritium and/or He in real time between the windows during the operation and making an automatic Ne leak test by filling the inter-space at 0.5 Bar after the end of operation each day. This point requires further investigation to find an acceptable solution.

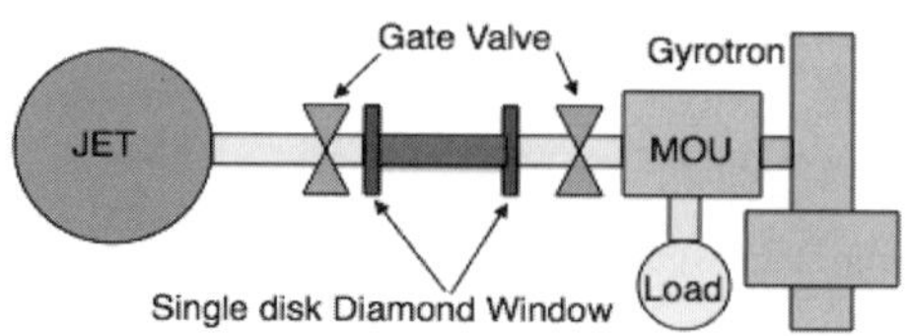

Figure 2. Proposed single line layout

6. Conclusions

The EWG is the selected TL solution since the QO option is not ITER relevant and seems impossible for JET. The suggested routing is fully compatible with an EWG, allowing full use and hence testing of ITER components. Two routing options have been proposed and analyzed: In the best solution the number of MB is reduced to 6, allowing low overall losses (<5%). Concerning the vacuum window, a good opportunity is to use two separate ITER CVD windows per line. As the standard JET procedure for detecting damaged windows is not compatible with high power TL operation an alternative strategy is proposed.

Acknowledgments

This work was realized under EFDA JET Task Agreements n. JW9-NEP-ENEA-84 and JW9-OEP-ENEA-86 in the framework of the Project "Feasibility study of an ECRH system for JET". The views and opinions expressed herein do not necessarily reflect those of the European Commission.

References

1. G. Giruzzi et al., This conference.
2. G. Michel et al., 18th Topical Conf. on Radio Frequency Power in Plasma, AIP Conf. Proc. 1187, pp. 559-562 (2009).
3. T. Goodman et al., Proc. 19th SOFT, Ed. C. Varandas, F. Serra, (North-Holland, Amsterdam, 1997), Vol. 1 pp. 565 (1996).
4. M. Lennholm et al., This conference.
5. A.G.A. Verhoeven et al., Nucl. Fusion 43 (2003) 1477
6. T. A. Scherer et al., This conference.

FEASIBILITY OF AN ECRH SYSTEM FOR JET: HIGH VOLTAGE POWER SUPPLIES REQUIREMENTS AND PROPOSED STRUCTURE

H. BRAUNE[1], G. GIRUZZI[2], J. HAY[3], P. KHILAR[3], M. LENNHOLM[4,5], L. MOREIRA[3], A. PARKIN[3], A. VADGAMA[3]

JET-EFDA, Culham Science Centre, Abingdon OX14 3DB, UK

[1] Max-Planck-Institut für Plasmaphysik, Euratom Association, D-17491 Greifswald, Germany
[2] CEA, IRFM, 13108 Saint-Paul-lez-Durance, France
[3] CCFE, Culham Science Centre, Abingdon OX14 3DB, UK
[4] EFDA Close Support Unit, Culham Science Centre, Abingdon OX14 3DB, UK
[5] European Commission, B-1049 Brussels, Belgium

The future JET programme, after the installation of the ITER-like wall, will be mainly focused on the consolidation of the physics basis of the three main ITER scenarios. These scenarios will make substantial use of Electron Cyclotron (EC) waves, for heating as well as for control of both the MHD activity and the current density profile. Therefore, a programme for preparation, validation and optimization of the ITER scenarios in present tokamaks would strongly benefit from an ECRH/ECCD system. A study has been conducted to evaluate the feasibility of installing an ECRH system on the JET tokamak. An important intention of the study was to investigate the feasibility to utilise some unused conventional NBI – power supplies for the ECRH project.

1. Microwave power – Microwave sources

The investigations concerning the required microwave power, absorbed in the central plasma region, led to the result that a microwave power in the region of 10 MW with a frequency of 170 GHz would be required. Commercially available microwave sources with the required frequency generate a radiation power of 1 MW at the output window. Therefore, taking into account the losses in the transmission lines, 12 microwave sources are intended to be used, so that 10 MW of microwave power can reliably be deposited in the central region of the JET plasma.

The microwave sources will be diode type gyrotrons operating in depressed collector mode in order to achieve efficiency up to 55 %. Diode type gyrotrons require two DC high voltage sources, the main voltage and the body voltage power supply.

Electrical requirements	
acceleration voltage	80 -85 kV
cathode voltage	-55 kV
depression voltage	< 35 kV
beam current	40 – 46 A
frequency	170 GHz
RF output power	1 MW
pulse length at JET	20 s

Figure 1. Gyrotron 1MW 170 GHz, GYCOM Russia.

2. Power supplies - requirements

The main voltage supply has to deliver 65kV, 50A with negative polarity on high potential. The body voltage source has to deliver 30kV, 50mA.

Both high voltage supplies generate the electron beam acceleration voltage of the gyrotron as shown at Fig. 2. The radiation output power of a gyrotron depends strongly on the acceleration voltage. Therefore highly regulated voltage sources are required for controlled gyrotron operation (accuracy ≤ 1%).

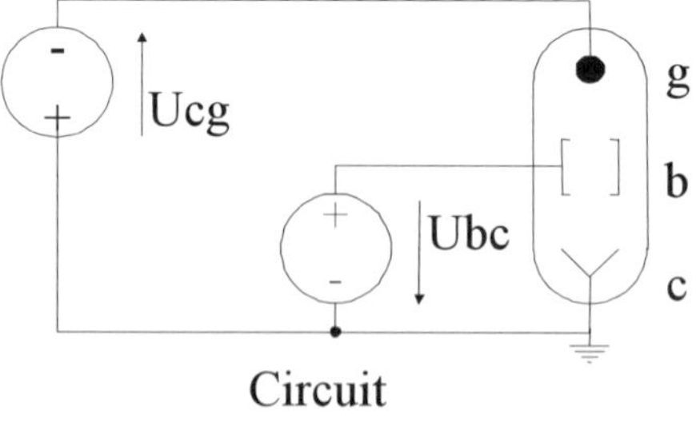

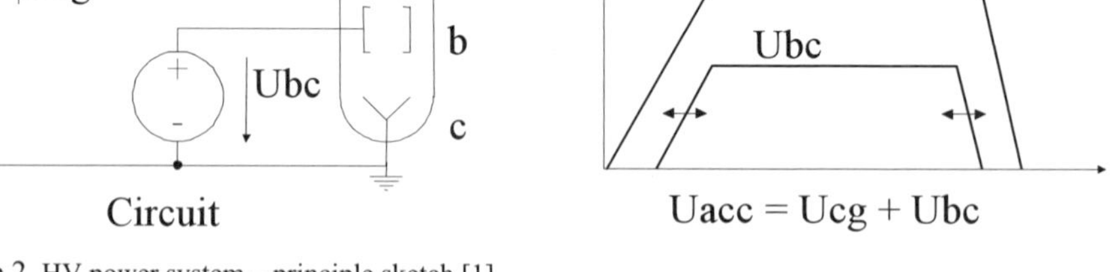

Figure 2. HV power system – principle sketch [1]

3. RF power regulation and modulation

Using body voltage variation for power regulation would lead to significant additional waste power at the collector surface with the risk of collector damage. Therefore main voltage regulation for power control should be used, to significantly reduce the collector power during modulation

The most flexible ECRH operation regime will be achievable if each gyrotron is fed by separate HV power supplies.

4. What is available? -> Four power supply units from the old Neutral Beam system!

Each PS units consists of two rectifier transformers with a so called protection unit in series and one matching transformer. Fig. 3 shows the out door part. The rectifier transformers deliver 120kV and 60A each. Thyristor star point controllers perform the DC output voltage adjustment. The rectifier transformers can operate separately or in series and are designed to deliver positive voltage. The two rectifier transformers per unit, referred to as master and slave, are different. The protection units include the tetrode and the required auxiliaries for exact DC – high voltage regulation. The tetrode installation has the wrong polarity and is situated at the wrong place for use in the proposed ECRH system.

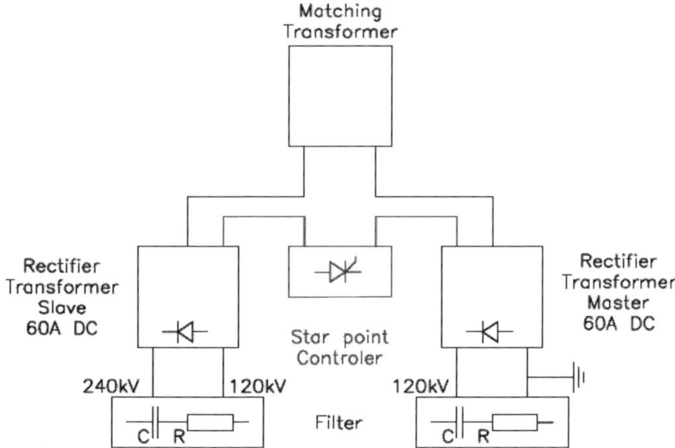

Figure 3. NBI power supply unit (out door part).

Master rectifier transformer

- output connector with the positive polarity is isolated up to 120 kV with respect to earth and the output connector with the negative polarity had been designed for a maximal voltage of 3.5 kV to earth.

Slave rectifier transformer

- both output connectors are isolated up to least 120 kV with respect to earth.

Only the slave transformer of each unit has currently the capability to deliver 80 kV DC in both polarities with very minor modifications, while more significant changes are required to allow the master transformer to deliver negative voltage.

5. Crowbar

The specified maximum released energy in the case of an arc inside the gyrotron demands an additional tube protection circuit. Each power supply therefore has to be equipped with a crowbar installed close to the gyrotron.

The design of the crowbar unit should be based on a thyratron tube in order to minimize the volume compared to a thyristor based crowbar unit.

A snubber must be installed between thyratron and gyrotron. The snubber ensures sufficient voltage at the thyratron terminals for safe triggering of the crowbar in case of voltage breakdown at the gyrotron due to arcing.

6. Proposed power supply structure

6.1. *Main power supply*

Two types of power supplies are proposed:

- **Eight** power supplies using the existing rectifier transformers with star point controllers and tetrodes in series, allowing fast voltage regulation and modulation up to 20 kHz (Type 1)
- **Four** power supplies based on solid state solutions without any added regulators and a limited modulation capability up to 1 – 2 kHz (Type 2)

6.1.1. *Main power supply Type 1*

The existing four NBI – power supplies can be split in 8 independent supplies capable of providing the required voltage and current for 8 gyrotrons. Using the NBI power supplies, only 4 new power supplies are required for the remaining gyrotrons. However, certain modifications and thorough refurbishment of these power supplies is needed in order to make them suitable for feeding gyrotrons.

- Modification (polarity change) of four master transformers
- Replacement of eight 12-pulse star point controllers and star point inductors
- Reconstruction of 8 protection units for operating with changed polarity
- Relocation of the reconstructed protection units to a new gyrotron building

The power supply has the capability to modulate the main gyrotron voltage up to 20 kHz. The body voltage supply becomes simple and compact.

6.1.2. *Main power supply Type 2*

For the purposes of the assessment we have assumed a power supply design based on pulse step modulator technology (PSM). The modules are equipped with IGBT switches rated 1700 V / 200 A. The complete system is air-cooled.

A short-circuit energy < 15 Joules is achievable with PS based on the PSM principle and the short-circuit switch-off time will be less than 8μs.

DC voltage generated by power supplies based on solid-state technology has a non-negligible ripple. The body power supply becomes more complicated in order to compensate the main voltage ripple. The power supply has the capability to modulate the main gyrotron voltage only up to 2 kHz.

6.2. *Body power supply*

The kind of body voltage power supply depends on the type of main power supply. The body voltage has to switch off fast in a case of a gyrotron interlock. Therefore a fast switch based on power MOSFET technology is connected in series between the body power supply and the gyrotron body at both types of main power supplies.

6.2.1. *Requirements when used in combination with main PS type 1:*

If the main high voltage is fast regulated by a tetrode tube based regulator without any ripple and noise, the requirements on the body voltage power supply are not very stringent. The body voltage supply would not need a fast high voltage amplifier.

6.2.2. *Requirements when used in combination with main PS type 2:*

The solid-state main high voltage power supply generates an exact voltage which may however be affected by a not insignificant ripple. Depending on the level of this ripple on the direct voltage, compensation could be required in order to achieve a stable RF power. The ripple compensation is feasible by using a body voltage supply equipped with a fast high voltage amplifier with slew rates up to 600 V/μs. Two kinds of amplifiers are currently in operation at comparable gyrotron installations:

- Tube based as used at the W7-X ECRH plant in Greifswald [2]. Unfortunately the production of the required tubes (TH 5188) has been stopped.
- Solid-state based as at the ITER gyrotron test facility in Lausanne [3]. This amplifier needs a lot of space compared to tube based solutions.

7. Conclusions

Several devices from the old NBI power supplies could be reused in the new ECRH system.

- In the proposed solution eight power sources will be able to perform a power modulation up to 20 kHz and four power sources will be basically used for continuous plasma heating without power modulation.
- From the physics point of view, this is a promising solution concerning the requirements of plasma heating and plasma stabilization.

Acknowledgement

This work, supported by the European Communities, was carried out within the framework of the European Fusion Development Agreement. The views and opinions expressed herein do not necessarily reflect those of the European Commission.

References

1. P. Brand, G. Mueller, *Circuit design and simulation of a HV-supply controlling the power of 140 GHZ 1MW gyrotrons for ECRH on W7-X*, 22nd Symposium on Fusion Technology, September 9 - 13, 2002, Helsinki, Finland
2. P. Brand. H. Braune, G. Mueller, *Design and test of a HV device for protection und power modulation of 140GHz/1MW CW gyrotrons used for ECRH on W7-X,* 23rd Symposium on Fusion Technology, September 20 - 24, 2004, Venice, Italy
3. D. Fasel et al., *Installation and commissioning of the EU test facility for ITER gyrotron*, 24th Symposium on Fusion Technology, September 11 – 15, 2006, Warsaw, Poland

RAY TRACING STUDY OF 170GHZ ELECTRON CYCLOTRON WAVES IN KSTAR PLASMAS

YOUNG-SOON BAE
National Fusion Research Institute, Gwahangno 113, Yuseong-gu Daejeon 305-333, Korea

M. JOUNG[1], H. L. YANG[1], W. NAMKUNG[2], M. H. CHO[2], H. PARK[2], R. PRATER[3], R. A. ELLIS[4], J. HOSEA[4]
[1]National Fusion Research Institute, Gwahangno 113, Yuseong-gu, Daejeon 305-333, Korea
[2]Pohang University of Science and Technology, San 31, Hyoja-dong, Nam-gu, Pohang 790-784, Korea
[3]General Atomics, P.O. Box 85608, San Diego, CA92186-5608, USA
[4]Princeton Plasma Physics Laboratory, P.O. Box 451, Princeton, NJ08543, USA

The electron cyclotron heating/current drive (ECH/ECCD) system has become an essential tool for the fusion plasma research in toroidal devices. In Korea Superconducting Tokamak Advanced Research (KSTAR) tokamak, development of high power and multi-frequency ECH/ECCD system is in progress. The frequencies employed in KSTAR are 84 GHz, 110 GHz, and 170 GHz. Multiple frequency sources can easily support the wide range of operating regimes from 1.5 T to 3.5 T in KSTAR tokamak. In particular, the 170 GHz source, that will be adapted to the ITER, corresponds to the second harmonic frequency of the KSTAR operating range from 2.6 T to 3.5 T. This frequency will be mainly used for the control of the local plasma current profile to manipulate the internal MHD instabilities such as the neoclassical tearing mode (NTM) critical in high-beta plasma operation. This paper presents simulated ray tracings of the 170 GHz EC waves for a various plasma conditions in KSTAR. The TORAY-GA ray tracing code is used, along with Interactive Data Language (IDL) procedures that create the input files, to study the effect of ECH/ECCD on the plasma equilibrium profiles as a function of the initial density and temperature profiles and of toroidal field.

1. Introduction

The new EC H&CD system operating at a frequency of 170 GHz, that will be adapted to the ITER and corresponds to the second harmonic frequency of the KSTAR operating range from 2.6 T to 3.5 T, is under plan for the control of the local plasma current profile to manipulate the internal MHDs such as the neoclassical tearing mode (NTM) critical in high-beta plasma operation. The initial 170 GHz, 1 MW EC H&CD system is under design and fabrication for the initial commissioning and operation in the fourth KSTAR campaign in 2011 using JAEA's ITER pre-prototype of 170 GHz, 1 MW CW (800 s) gyrotron, and an 1 MW (10s) steerable front-end mirror of the launcher. This passive-cooling launcher is under fabrication through the collaboration with Princeton Plasma Physics Laboratory (PPPL) and Pohang University of Science and Technology (POSTECH). Figure 1 shows a three dimensional view of 3 MW KSTAR heating and current drive systems in the future. The first 1 MW, 170 GHz ECCD launcher will be installed at Bay-E port on the mid-plane.

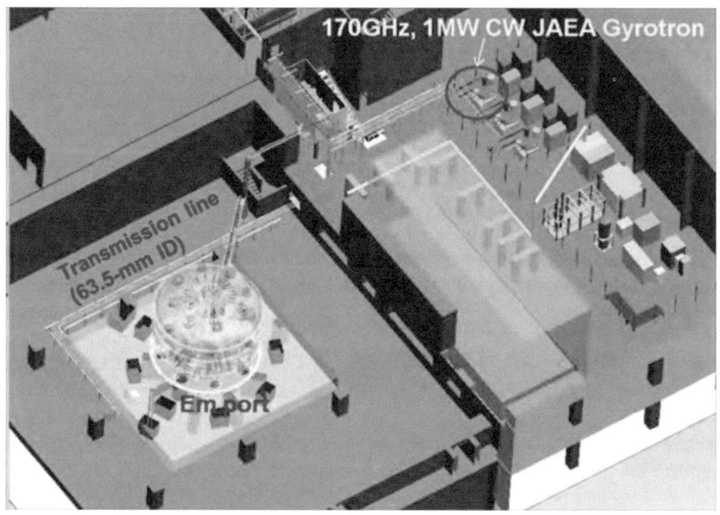

Figure 1. The layout of KSTAR 3 MW, 170 GHz ECCD system.

Presently, the EC frequencies adopted in the KSTAR are 84 GHz, 110 GHz, and 170 GHz. The present two frequencies of 84 GHz and 110 GHz are being used for the start-up and to assist the current ramp-up [1-3]. For the main heating and current drive, the 110 GHz and 140 GHz frequencies are also under consideration as well as 170 GHz. Figure 2 shows the operation window for three EC frequencies. The $R_{EC} = N28B_0R_0/f$ is the major radius of the EC resonance as a function of the toroidal magnetic field B_0 at the magnetic axis R_0 = 1.8 m and the harmonic number N. As summarized in table 1, the higher toroidal magnetic field operation requires 170 GHz. The 140 GHz frequency will be a good one if the high beta scenario with a reduced toroidal magnetic field in range of 2 to 3 T is required in the KSTAR.

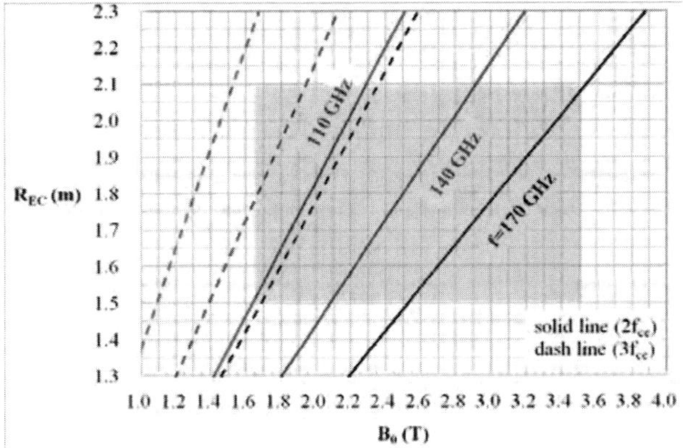

Figure 2. Operation window of EC frequencies in the KSTAR. The R_{EC} is the major radius of the EC resonance.

Table 1. Summary of the operation window of the resonance position and the toroidal magnetic field for three frequencies, 110 GHz, 140 GHz, and 170 GHz avoiding the third harmonic resonance.

f (GHz)	R_{EC} (m)	B_0 (T)	Remark
110	1.5 ~ 2.1	**1.7 ~ 2.1**	No 3rd harm. resonance
140	1.5 ~ 2.1	**2.1 ~ 2.9**	3rd harm. exists in B_T = 1.7~2.0
170	1.5 ~ 2.1	**2.5 ~ 3.5**	3rd harm. exists in B_T = 1.8~2.35

2. TORAY-GA code

The TORAY-GA code originates from the RAYS code developed by D. Batchelor et al [4]. The ray launching from an actual antenna is a set of rays with a Gaussian beam which propagates in the far field region from a single point with divergence, *D*. The wave absorption model used for 170 GHz wave for the KSTAR is "Matsuda-Hu" model near 2nd harmonic and weakly relativistic. The current version used in this paper is v1.8. It uses Interactive Data Language (IDL) procedures that create the input files toray.in and echin using an equilibrium data (G-EQDSK file).

The beam divergence in the TORAY-GA code is defined as $w = Dz$. The beam radius w is given by Eq. (1) with the focusing effect of the focusing mirror of the antenna.

$$w(z) = w_1\sqrt{1 + ((z - z_1)/d_1)^2}. \qquad (1)$$

Where, $z_1 = R_0(1+(R_0/d_0)^2)^{-1}$, $d_1 = \pi w_1^2/\lambda$, $d_0 = \pi(w_0')^2/\lambda$, $w_0' = w_0[1+(\lambda z_0/(\pi w_0))^2]^{1/2}$, $w_0 = 0.322D_0$, D_0 the waveguide inner diameter, z_0 the distance between the waveguide and the focusing mirror, and $R_0 = f/\sin 2\alpha$.

For the focusing mirror with $f = 500$ mm and $z_0 = 200$ mm, the beam radius $w = 30$ mm is calculated at the 2nd harmonic resonance position, $R_{X2} = 1800$ mm. Since the antenna pivot location, $R_{pivot} = 2800$ mm and $w(z) = Dz$, $D = w/z = 30$ mm/1000 mm = 0.03 rad. The full width at half maximum, FWHM = $D\sqrt{[\ln 2/2]}\times 180/\pi\times 2 = 2$ deg. If no focusing is adapted at the focusing mirror of the antenna, the beam radius equation is given by Eq. (2).

$$w(z) = w_0\sqrt{1+(\lambda z/(\pi w_0))^2}. \tag{2}$$

For the ray tracing and current drive calculation, the scaled plasma equilibrium data is generated using an IDL procedure with the KSTAR reference scenario of the plasma current of 2 MA, the toroidal magnetic field of 3.5 T, and other KSTAR plasma shape profiles. The IDL procedure scales the reference equilibrium data with new toroidal magnetic field and the plasma current with the same scaling factor of the toroidal magnetic field to keep the same safety factor, q. In this calculation, the toroidal magnetic field at the magnetic axis, R = 1.8 m, is 3 T. And, the profiles of the plasma density and temperature are assumed with the parabolic profiles.

3. Simulation results

For the KSTAR ECCD simulation, three EC beam launch pivot positions are considered. The pivot position which is being considered for the first 1-MW launcher is on the mid-plane of the KSTAR tokamak. The other pivot position, 30-cm above or 30-cm below from the mid-plane is also used for ECCD simulations. Figure 3 shows the ray tracing of the 170 GHz EC beams from the two launchers at different pivot positions for the KSTAR plasma equilibrium of 3 T. Figure 4 shows the EC-drive current density (J_{CD}) profiles for the EC beam launch angles for the mid-plane launcher (z = 0 cm). Figure 4 (a) shows the EC-driven current density profiles for the oblique toroidal launches with the poloidal angle, $\theta = 0$ deg injecting toward to the center of the plasma. In this case, the range of the EC-driven current density is very low in the normalized minor radius, that is, $\rho < 0.21$. For the toroidal launch angle greater than 20 deg, the EC-driven current density decreases. On the other hand, the poloidal launch scan shows the high range of the EC-driven current up to $\rho = 0.6$ for the toroidal launch angle, $\phi = -20$ deg as seen in Fig. 4 (b). The negative sign of toroidal angle means the clock wise direction from the top view of the tokamak. But, as the poloidal launch angle increases, the EC-driven current density decreases.

Since the launcher is pivoted on the mid-plane, J_{CD} profiles are up-down symmetric for the poloidal directions.

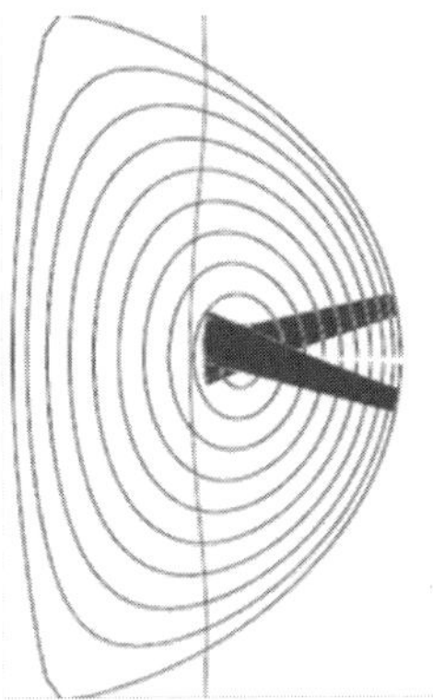

Figure 3. The ray tracing of 170 GHz EC beam from two launchers pivoted at +30 cm and -30 cm from the mid-plane in the KSTAR plasma equilibrium with toroidal magnetic field of 3 T.

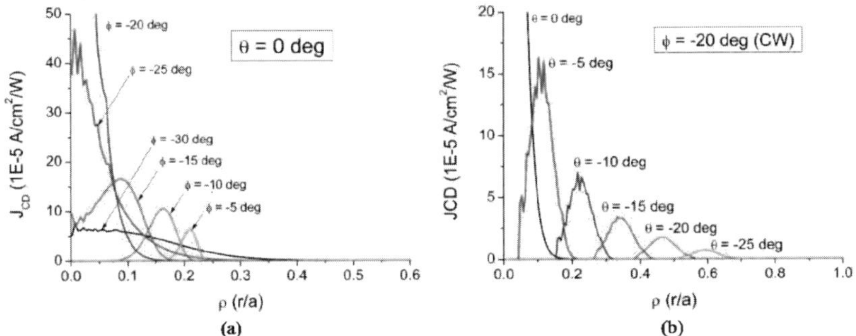

Figure 4. The EC-driven current density profiles for the launcher with pivot position on the mid-plane (z = 0 cm).

Figure 5 shows the J_{CD} profiles for the launcher with a pivot on the upper position from the mid-plane (z = +30 cm). In Fig. 5 (a), the downward launch with the poloidal angle θ = -15 deg is for the injection toward the plasma center. Similarly, the J_{CD} has low range as a function of the toroidal angle. The toroidal angle of ϕ = -20 deg gives the maximum current drive at the center of the plasma. But, the other poloidal launches with different angles give the wide-range of J_{CD} as seen in Fig. 5 (b).

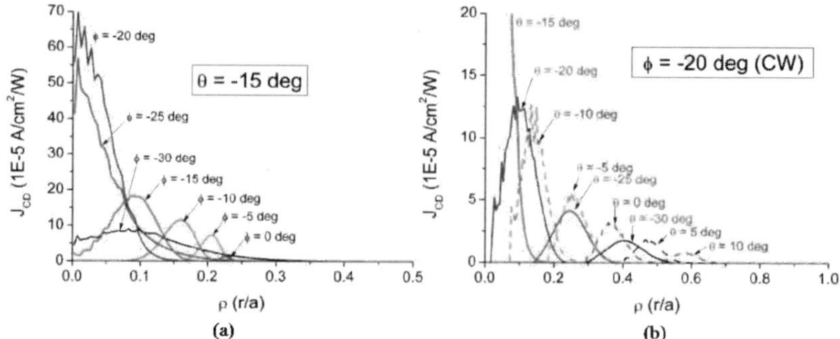

Figure 5. The EC-driven current density profiles for the launcher with a pivot on the upper position from the mid-plane (z = +30 cm).

Figure 6 shows the dependence of the J_{CD} on the full width at half maximum (FWHM) of the beam divergence for the launcher pivoted on the mid-plane. The high EC-driven current density is obtained with narrow width for the FWHM of 2.0 deg of EC-beam in the KSTAR 170 GHz launcher equipped with the focusing mirror with the focal length of 500 mm as described in previous section. The peak value of J_{CD} linearly increases as the FWHM decreases.

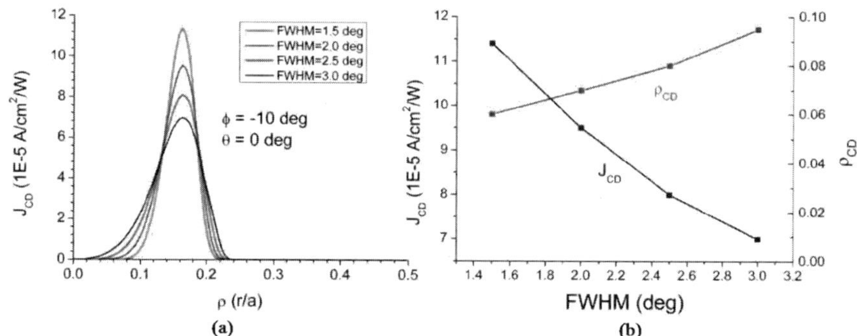

Figure 6. The dependence of EC-driven current density profiles on the beam divergences.

The dependence of J_{CD} and the current drive efficiency defined as $\eta = \langle n_e \rangle R_0 I_{CD}/P_{CD}$ is scanned for various temperatures T_{e0} and densities n_{e0} at the plasma center. The J_{CD} decreases as the peak density increases for the given the plasma temperature, but the current drive efficiency increases with the peak plasma density. Figure 7 shows the JCD and η for various plasma temperatures and the plasma densities for the EC beam launch toward the plasma center with the toroidal angle of 20 deg from the launcher pivoted on the mid-plane. Both the J_{CD} and η increases with the plasma temperature. For the $T_{e0} = 5$ keV and $n_{e0} = 0.9 \times 10^{20}$ m^{-3}, the current drive efficiency of 0.36×10^{19} Am^{-2}W^{-1} is obtained.

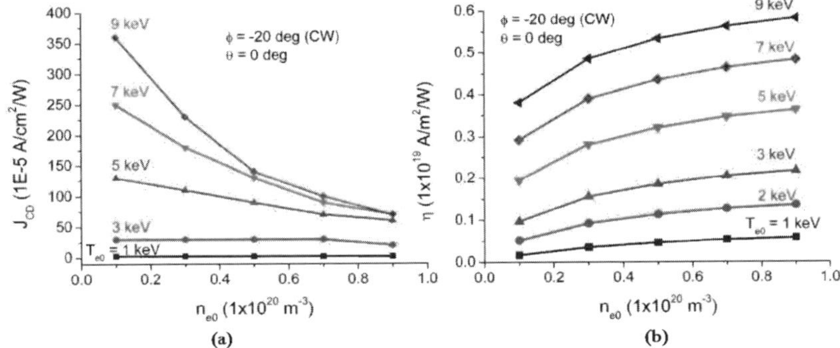

Figure 7. The EC-driven current density and its efficiencies as a function of the plasma density for the temperatures at the plasma center.

4. Conclusion

The ray tracing of 170 GHz second harmonic EC wave in KSTAR is studied for the toroidal magnetic field of 3 T. For $n_{e0} = 0.5 \times 10^{20}$ m^{-3} and $T_{e0} = 5$ keV, a total obtained maximum driven current is 160 kA for the core current drive with 3-MW input EC beam power launched from two launchers pivoted at the upper position (z = +30 cm) and on the mid-plane, respectively. Even when the mid-plane launcher, which will be the 1-MW first EC launcher, is pivoted at z = -30 cm, the nearly same total integrated driven current is obtained. The current drive efficiency η increases as the plasma density and the electron temperature increases, but the peak value of the current density decreases as the plasma density increases. For the electron temperature dependence; both the current drive efficiency and the peak value of the current density increases as the electron temperature increases. The first 1-MW non-cooled launcher is under fabrication in collaboration with Pohang University of Science and Technology (POSTECH) and PPPL. The first launcher will be installed with a fixed focusing mirror and a steerable mirror pivoted on the mid-plane for the initial CD at the plasma core for 2011 KSTAR plasma campaign.

References

1. Y. S. Bae et al., *Fusion Sci. Tech.* **52**, 321 (2007).
2. Y.S. Bae et al., *Nucl. Fusion* **49**,022001 (2009).
3. M. Joung et al., *The 16th Joint Workshop on Electron Emission and Electron Cyclotron Resonance Heating*, Sanya, April 12-15, 2010.
4. D. B. Batchelor and R. C. Goldfinger, *Nucl. Fusion* **20**, 403 (1980).

O2- and X3-Heating Experiments on ASDEX Upgrade

H. Höhnle*, W. Kasparek, U. Stroth
Institut für Plasmaforschung, Universität Stuttgart,
Stuttgart, 70569, Germany
**E-mail: hoehnle@ipf.uni-stuttgart.de*
www.ipf.uni-stuttgart.de

J. Stober, A. Herrmann, R. Neu, J. Schweinzer and the ASDEX Upgrade Team
Max-Planck-Institut für Plasmaphysik, EURATOM Association,
Garching, 85748, Germany

The improved H-mode in ASDEX Upgrade (AUG) is an operation regime with high confinement and densities near the Greenwald limit. Since the first wall in AUG is completely coated with tungsten, the ECRH is needed to control the tungsten accumulation in H-modes [1], therefore the routinely used X2 mode must be deposited in the plasma centre [2]. For ITER-relevant plasma discharges a safety factor of $q_{95} \approx 3$ must be achieved. At $B = 2.5$ T plasma currents of > 1.2 MA corresponds to a $q_{95} \approx 3$, which results in densities above the X2 mode cutoff. Another way to decrease q_{95} is to reduce the magnetic field, but then the X2 mode is not resonant in the plasma centre. For these two cases new ECR heating scenarios using X3 and O2 mode were developed and are presented here.

Keywords: electron cyclotron resonance heating, second harmonic ordinary mode, third harmonic extraordinary mode, holographic gratings

1. Introduction

On the ASDEX Upgrade tokamak (AUG) ECRH is typically used in the second harmonic mode at a frequency of $f = 140$ GHz and a magnetic field of 2.5 T. For extraordinary polarisation (X2 mode) of the wave a central deposition is complete and is routinely used to prevent tungsten accumulation [2]. Otherwise at ITER relevant parameters the concentration of the high Z material tungsten would lead to a radiation collapse of the plasma and a breakdown of the discharge. In a different scheme, off-axis current drive (ECCD) can be used to reduce neoclassical tearing modes (NTM).

This work addresses the options for ECRH in high density discharges. To avoid the X2 cutoff and the connected reflection of the microwaves at the cutoff-layer, new heating scenarios were developed for ITER relevant plasmas at safety factors of $q_{95} \approx 3$ ($q_{95} \propto B_t/I_p$ with B_t toroidal magnetic field and I_p plasma current). Low values of the safety factor can be achieved either at high plasma current I_p or reduced magnetic field. Both scenarios entail problems for the use of ECRH:

Increased plasma current will result in better confinement and a higher Greenwald density ($n_e^{GW} \propto I_p$), and therefore in a cutoff of the extraordinary mode. Here the ordinary mode (O2 mode) comes into play. This wave has twice the cutoff density of the X2 mode but suffers from incomplete absorption at the second harmonic resonance at accessible electron temperatures of few keV in AUG [3]. A method to handle the non-absorbed power is presented in section 2.

At reduced magnetic field the third harmonic extraordinary mode can be used for central heating, which also suffers from incomplete absorption. The solution of this problem is described in section 3.

At the end, the results are summarised and discussed, and an outlook for further investigations is given.

2. O2 Mode Heating

2.1. *Heating Scenario*

The beam tracing program TORBEAM [4] was used, to see whether an ECRH beam reaches the plasma core and to calculate the absorbed power. The calculations were done for different poloidal and toroidal injection angles. The central density was set to $1.4 \cdot 10^{20}$ m^{-3}, which is above the X2 mode cutoff; the central electron temperature was chosen as 3.5 keV. Both values are characteristic for improved H-mode discharges. In Fig. 1 on the left hand side the absorption of the O2 mode in dependence of the poloidal and toroidal injection angles of the new ECRH launcher is shown. The highest achievable absorption in this scenario is ≈ 75 % at poloidal angles of $18-25°$ and a toroidal angle of around $\pm 15°$. However, the shine through of 25 % is still problematic for microwave absorbing components at the inner wall of AUG. In order to take care of this shine through power a second pass of the beam through the plasma is needed. Therefore injection angles where chosen at which the beam hits directly a double tile at the inner column of ASDEX Upgrade, see Fig. 1 (right hand side).

On this tile, a mirror surface (thus conformal to the inner wall of ASDEX

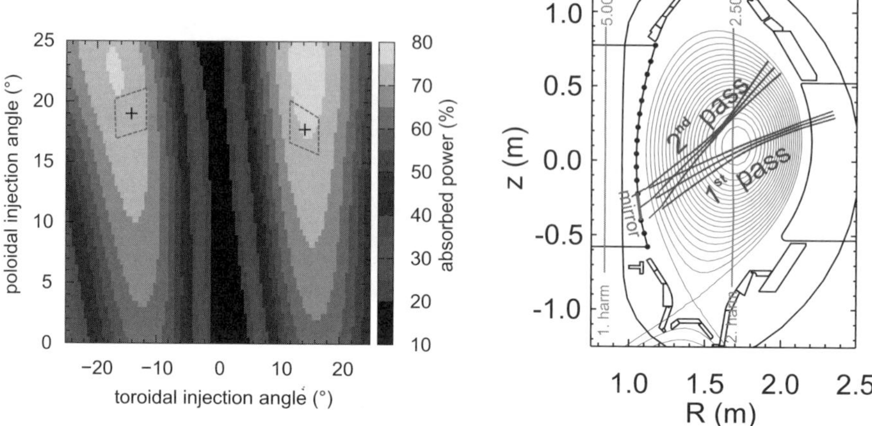

Fig. 1. O2 absorption from TORBEAM calculations for different injection angles (left). The crosses indicate the optimal injection angle to hit the holographic mirror (dotted lines). In the right figure the first and second pass of the beam through the plasma for the optimal injection angles are shown. Both pictures are calculated with typical density and temperature profiles from an improved H-mode discharge at AUG ($n_e(0) = 1.4 \cdot 10^{20}$ m^{-3}, $T_e(0) = 3.5$ keV).

Upgrade) was designed, which reflects the beam back to the core plasma in the correct direction ($\approx \mp 15°$) and correct polarisation (O2 mode). Thus the total absorption is enhanced to > 90 % and the remaining 10 % of the power hit the upper PSL (passive stabilization loop), where no critical microwave absorbing components are located. All these requirements are fulfilled by a holographic grating, which reconstructed the incoming phase of the beam to an outgoing beam phase. The design of such mirrors is described in References [5,6]. With this method, mirrors with a calculated efficiency of $\gtrsim 90$ % were manufactured out of graphite. Because of the tungsten program in AUG, the surface of these mirrors was also coated with tungsten. The efficiency of such mirrors was tested in resonator measurements and the results ($88 - 94\%$) agree very well with the calculated values.

2.2. *Experimental Results*

To keep the beam focused on the mirror, a new detection system was built. It consists of thin thermocouples (diameter: 0.25 mm, response time: 10 ms) embedded into the mirror, which allow to locate the beam in real time. Fig. 2 (left hand side) depicts the mirror with the thermocouples marked by

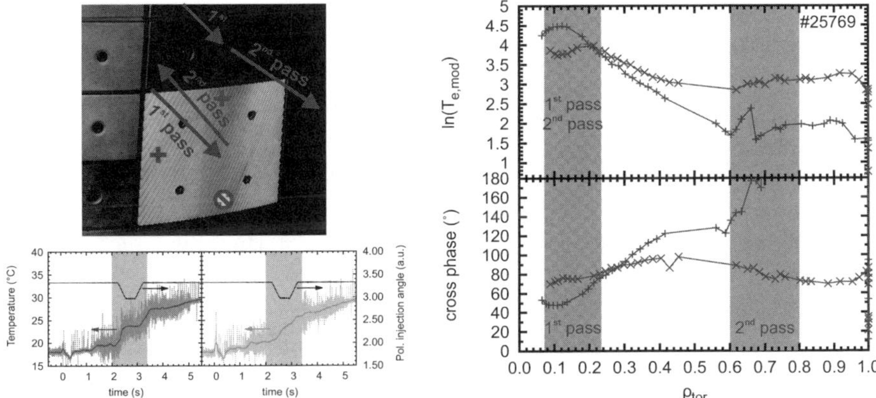

Fig. 2. Measurement setup for modulation experiments at ASDEX Upgrade. The ECRH beam in O2 mode was moved from above the holographic mirror to the centre and back above, in order to verify the improvement of the directed second pass of the beam. The thermocouples, shown here as coloured crosses, response only if the beam hits the mirror in their direct neighbourhood (red time trace belongs to the red cross). The modulation measurement shows an improvement of the directed second pass in comparison with the reflection at a standard tile.

coloured crosses. The couples were tested by moving the ECRH beam from above the mirror to the centre and back. The time traces of the temperature of two thermocouples are shown in the lower plots. The red and green curves belong to the thermocouple in the same colour above. The black curve indicates the mirror movement and the grey region displays the time of the O2 mode injection. It can be seen, that only the thermocouple in the direct neighbourhood of the beam have a clear response, so it is possible to locate the beam on the mirror.

In the same discharge the injected ECRH power was modulated with $f_{mod} = 30$ Hz. It was possible to show the influence of the second pass through the plasma centre in comparison with the reflection at a standard tile (2nd pass absorption in the plasma edge). The modulation amplitudes and the corresponding phases of the temperature modulation are shown in the right picture of Fig. 2. Because of high sawtooth amplitudes in the plasma centre a detailed analysis of the modulation amplitudes cannot be done, but a larger amplitude in the centre indicates enhanced absorption for the reflection from the holographic mirror. Also the phases for the directed 2nd pass are consistent with a heat pulse starting from the centre while in the case of the non-directed 2nd pass a second minimum of the phase indicates off-axis absorption.

3. X3 Mode Heating

3.1. *Heating Scenario*

The other way of achieving $q_{95} \approx 3$ is to reduce the magnetic field to < 2 T at a plasma current of 1.1 MA. The problem here is that the X2 resonance shifts to off-axis. If the magnetic field is decreased to 1.7 T the third harmonic can be used for on-axis heating. However, at achievable electron temperatures of 3 keV the absorption of the third harmonic is incomplete, too.

In this scenario the second harmonic resonance can help to avoid stray radiation. At a magnetic field of 1.8 T the third harmonic resonance is still central; at the same time the second harmonic lies in the plasma pedestal at the high-field side and can act as beam dump. This heating scenario was calculated with the TORBEAM code. The result of this calculation and the deposition profile are shown in Fig. 3. At the central X3-resonance an absorption of ≈ 60 % is found at temperatures of 2 keV and densities of $1.0 \cdot 10^{20}$ m^{-3}. The remaining 40 % of the power are efficiently absorbed at the X2 resonance at $\frac{r}{a} = 0.9$.

3.2. *Experimental Results*

With this heating scenario it was possible to generate the first ITER-relevant plasma with ECRH impurity control at ASDEX Upgrade. The

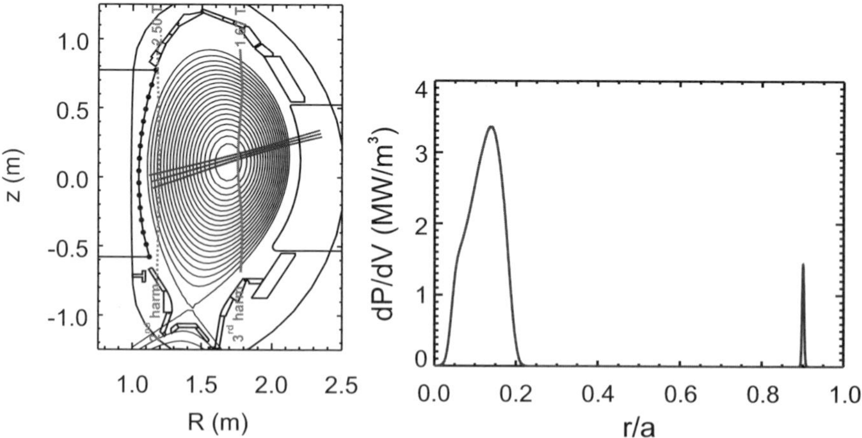

Fig. 3. In the X3 mode heating scenario the X2 resonance at the plasma edge acts as beam dump for the non complete (60 %) central X3 absorption. The use of the beam dump was demonstrated in [6].

time traces are shown in Fig. 4. The tungsten accumulation could be successfully suppressed with 1 MW of ECR in X3 mode, which is the minimum power needed for suppressing tungsten accumulation.

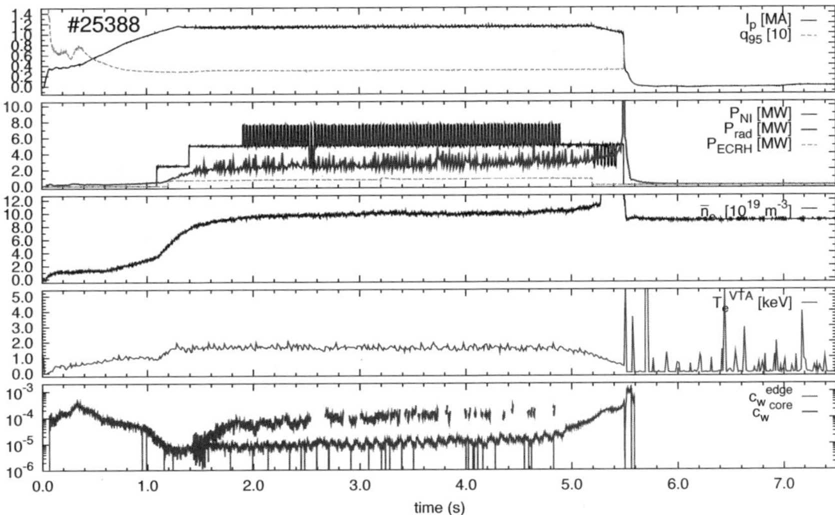

Fig. 4. Time traces of the first ITER relevant discharge at AUG with X3 mode ECRH. The tungsten accumulation is successfully suppressed with the central ECR heating. The safety factor q_{95} of 3 is due to the lower field and the plasma current of 1.1 MA.

4. Conclusion and Outlook

Two new ECR heating scenarios at ASDEX Upgrade were introduced. In both scenarios, ITER relevant plasma discharges in the full tungsten machine ASDEX Upgrade were possible without impurity accumulation.

The O2 mode heating works with a controlled second pass of the beam through the plasma. The benefit of the second pass could be demonstrated in modulation measurements. With 1 MW of ECRH in O2 mode into a 6 MW NBI heated discharge close at the Greenwald limit tungsten accumulation in the centre could be almost stopped (not shown in this paper). With more ECRH power available in the future, it should be possible to totally avoid the tungsten accumulation in high current discharges.

The first ECR heated stable ITER relevant plasma was made at lower magnetic field. The X3 mode heating with the X2 resonance on the high-field side as beam dump works very well. But also here more ECRH power will be beneficial.

References

1. A. C. C. Sips and O. Gruber, *Plasma Physics and Controlled Fusion* **50**, p. 124028 (2008).
2. J. Stober, ECRH on ASDEX Upgrade, *16th Joint Workshop on Electron Cyclotron Emission and Electron Cyclotron Resonance Heating* (Sanya, 2010).
3. V. Erckmann and U. Gasparino, *Plasma Physics and Controlled Fusion* **36**, 1869 (1994).
4. E. Poli, A. G. Peeters and G. V. Pereverzev, *Computer Physics Communications* **136**, 90 (2001).
5. O. Mangold, W. Kasparek and E. Holzhauer, Optimization of microwave reflection gratings for electron cyclotron resonance heating in O2 mode, *Conference Digest of the 2004 Joint 29th International Conference on Infrared and Millimeter Waves and 12th International Conference on Terahertz Electronics* 27. Sept.-1 Oct. 2004.
6. H. Höhnle, W. Kasparek, J. Stober, A. Herrmann, R. Neu, U. Stroth and the ASDEX Upgrade Team, Investigation of the O2- and X3-mode heating in ASDEX Upgrade, *36th EPS Conference on Plasma Phys. Sofia, June 29 - July 3, 2009 ECA Vol.33E, P-1.145* (Sofia, 2009).

ECRH and ECE in high T_e, low density plasmas of LHD

S. Kubo*, H. Takahashi, T. Shimozuma, H. Igami, H. Tsuchiya, Y. Nagayama,
S. Muto, I. Yamada, Y. Yoshimura, T. Mutoh

National Institute for Fusion Science, 322-6 Oroshi-cho, Toki, 509-5292, Japan
**E-mail: kubo@LHD.nifs.ac.jp*
www.nifs.ac.jp

The central electron temperature of more than 15 keV is achieved in LHD as a result of increasing the injection power and the lowering the electron density. Such collision-less regime is important from the aspect of the neoclassical transport and also potential structure formation. The creation of the high energy electrons and the transport of them can trigger the potential structure formation. Evidences of the presence of high energy electrons are indicated from hard X-ray PHA, and the discrepancy between the stored energy and kinetic energy estimated from Thomson scattering. ECE spectrum in such low density regime under the presence of high energy electrons are discussed by comparing the calculated one that solves the radiation transfer equation. Such investigation might be used for the estimation of high energy electrons as an inverse process of the ECRH.

Keywords: ECRH;ECE;collision less;high energy electrons;Gyrotron

1. Introduction

Recent upgrade of the ECRH system in LHD [1] enabled to study the plasma confinement properties at the far low collisional or collision-less regime in the helical system where specific confinement features are predicted from the neo-classical transport theory. These specific features includes confinement degradation in the so called "$1/\nu$" regime where the ripple loss become dominant but radial electric field can drastically alter the situation [2]. In order to discuss the confinement properties in such collision-less regime, the accurate estimation of the behavior of the high energy electrons as well as that of the bulk electrons temperature is required, since the high energy electrons can absorb injected ECRH power at the relativistically down shifted frequency and can deposit their energy to the bulk electrons out of the region where they absorb the energy from injected EC wave. Thus, the power deposition profile which is the key parameter in the transport analysis can be much affected by the presence of high energy electrons. One of the most reliable methods to experimentally deduce the power deposition profile is using the transient analysis of the electron cyclotron emission (ECE) [3]. ECE is known to be sensitive to the presence of the non-thermal electrons and furthermore tend to show a lower value as expected from the blackbody level of the bulk electron temperature [4,5]. In these sense, appropriate treatment and interpretation of the ECE spectrum are necessary, in particular in the high power EC heated plasma at collision-less regime.

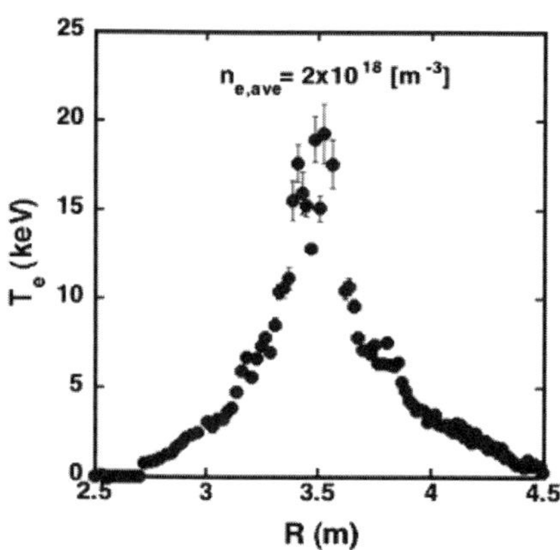

Fig. 1. Electron temperature profile measured by YAG-Thomson scattering when $n_{e,ave} = 2 \times 10^{18} m^{-3}$. This profile is obtained by accumulating Thomson scattering data of identical 10 shots. Central electron temperature exceeds 15 keV.

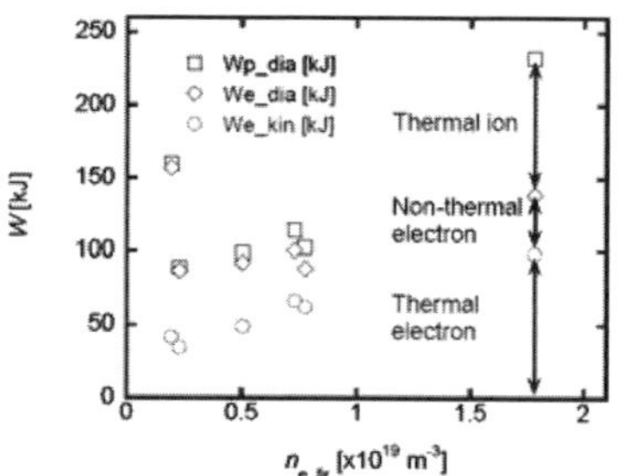

Fig. 2. Comparison of the stored energy and kinetic energy as a function of averaged electron density. Open squares indicate the stored energy measured by diamagnetism. Open diamonds indicate the total electron energy that is obtained by subtracting the total ion kinetic energy from diamagnetic energy. Open circles indicate the total electron kinetic energy obtained by volume-integrating the electron kinetic energy measured by the Thomson scattering.

2. Achievement of high electron temperature

ECRH system have been upgraded step by step to perform high power localized heating in the LHD. Recent topics are the introduction of 1MW class 77 GHz gyrotrons that are developed in collaboration with Univ. of Tsukuba. These gyrotrons are demonstrated to be operated stably up to 5 s with the power level of more than 1 MW [1].

Using this upgraded ECRH, the plasma parameter region of the low density and high temperature is explored. Figure 1 shows the electron temperature profile measured by YAG-Thomson scattering. The central electron temperature is confirmed to exceed 15 keV in the low density region ($n_e \leq 3 \times 10^{18}$ m^{-3}).

High temperature low density plasma produced by ECRH normally contains appreciable fraction of high energy electrons. The system and the analysis of the YAG-Thomson scattering in LHD are proved to be stable and give a good estimation of the bulk electron temperature even in such low density regime with high energy electrons. Simultaneous pulse hight analysis of hard X-ray (HX-PHA) indicates the presence of supra thermal electron component of the energy of the order of 100 keV. The dependence of the diamagnetic stored energy and the kinetic electron and ion energy as a function of the averaged electron density is shown in Fig.2. Ion kinetic energy fraction is small, especially in the low density region of $n_{e,ave} \leq 0.5 \times 10^{19}$ m^{-3}, since the ion temperature is less than 0.5 keV due to the long equi-partition time in such low collisional regime. Thus the high energy of the order of 100 keV electron fraction can reach near 5 %.

3. Behavior of ECE in high power ECRH low density plasma

In Fig.3 a)-c) are shown temporal behavior of the electron density, diamangetic stored energy and ECE radiation temperatures during a typical low density, high T_e discharge when ECRH(77 GHz) 2.63MW is injected into low density plasma. The averaged electron density is kept almost at 2 $\times$ 10^{18}m^{-3}. The ECE radiation temperatures shown in Fig.3 c) are cross calibrated by the bulk electron temperature measured by YAG-Thomson scattering at $t = 0.9$ s when high energy component is presumed to be well thermalized. The bulk electron temperature measured by Thomson scattering indicated 15 keV at the center and has sharp profiles as shown in Fig.3 d). The bulk electron temperature stays almost unchanged in time after t =0.3 s toward the end of ECRH injection. ECE radiation temperature profiles at the same time slices are plotted with open marks as a function of cold second harmonic resonance position. In Fig.3 e) are shown the similar radiation temperature spectrum, but as a function of the center frequency of each channel.

In order to understand the situation the ECE spectrum with the presence of high energy components, calculation method integrating the radiation transfer equation in LHD is developed [6]. In this calculation, the radiation transfer equation is integrated along the ECE line of sight

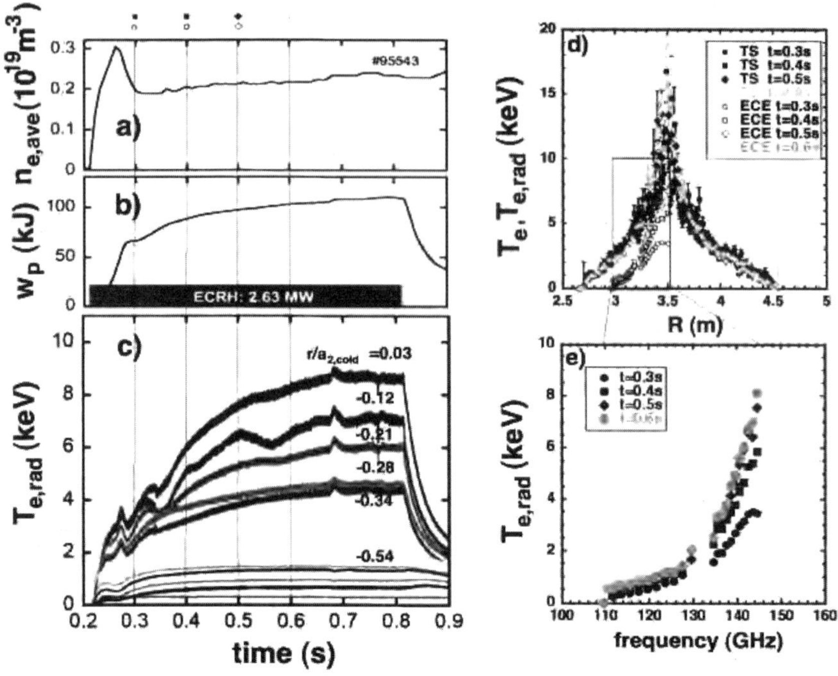

Fig. 3. Time evolutions of the a) averaged electron density, b) diamagnetic stored energy, and c) ECE radiation temperature during high power (2.63 MW) ECRH injection into low density plasma. ECE radiation temperature is deduced by the cross calibration with the electron temperature measured by YAG-Thomson scattering at t=0.9 s where non thermal electrons are thermalized. d) Electron temperature profile measured by YAG-Thomson scattering at the time slices t = 0.3, 0.4, 0.5 and 0.6 s in the shot shown in a)-c). Radiation temperature estimated from ECE are also plotted with open marks as a function of cold second harmonic resonance position of each channel. e) Radiation temperature spectrum as a function of frequency at these time slices. ECE radiation temperature stay less than the electron temperature from YAG-Thomson scattering. ECE radiation temperature profile keep increasing in time while the bulk electron temperature profile stays almost constant.

backward from the receiving antenna. Here the weakly relativistic quasi-perpendicular formula [7] at any harmonics are used. To include the high energy component, the local absorption and emissivity are calculated with bi-Maxwellian [8]. Calculated radiation temperature spectra are shown in Fig.4 a), assuming the bulk electron temperature $T_{\rm e}(\rho) = T_{\rm e,0}(1-\rho^2)^2$, $T_{\rm e,0}$ =15 keV, and $T_{\rm e,h} = T_{\rm e,h0}(1-\rho^2)^2$, $T_{\rm e,h0}$ =100 keV with 0%-50% fraction. When the high energy fraction increases more than 3%, the situation

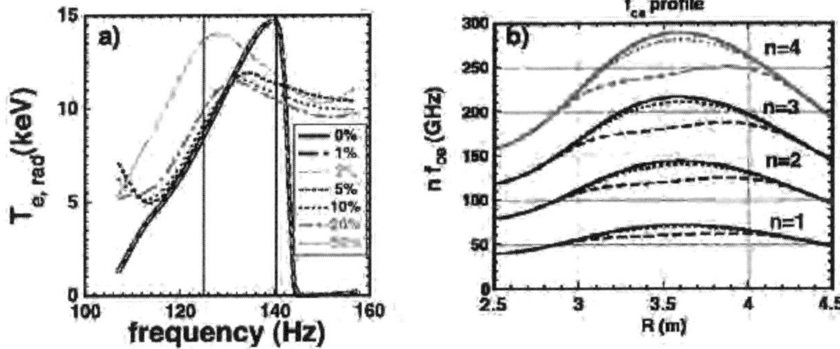

Fig. 4. a) Calculated ECE spectrum assuming $T_{\mathrm{e,bulk}} = 15 * (1-\rho^2)^2$ (keV) and $T_{\mathrm{e,high}} = 100 * (1-\rho^2)^2$ (keV) and $n_e = 2.0 * (1-\rho^8)^2$ ($\times\ 10^{18}\ \mathrm{m}^{-3}$) for various $n_{\mathrm{e,high}}/n_{\mathrm{e}}$. b) The dependence o the electron cyclotron and its harmonic frequencies along line of sight. ECE receiving antenna is located inside and line of sight is along the major radius on the mid-plane in the horizontally elongated cross section in LHD. Frequency down shift effect is shown for the electrons of the energy, E_0=15 keV and 100 keV electrons. Here, electron energy is assumed to depend on the minor radius as $E_0(1-\rho^2)^2$.

changes drastically. The radiation temperature at the frequency corresponding to the cold second harmonics near the center even decreases as the high energy fraction increases. Fig. 4 b) are shown the EC harmonic frequencies along the line of sight. ECE receiver is set inner side of the major radius, R. Each thick line indicates the cold resonance frequency and dotted and broken lines correspond to the relativistically down shifted frequency as-

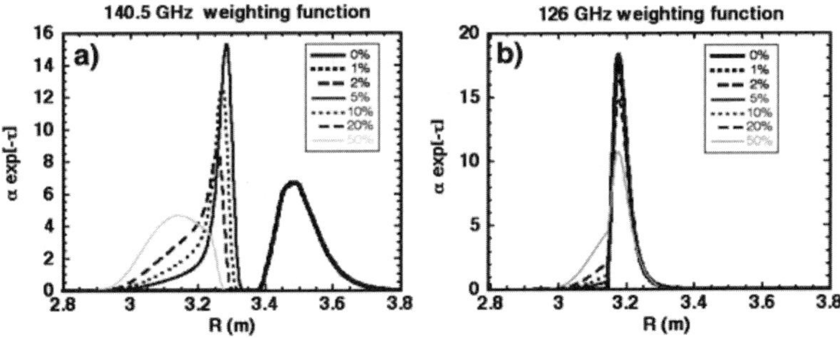

Fig. 5. The weighting function $\alpha \exp[-\tau]$ that relates the observed radiation temperature to the local radiation temperature, for the radiation frequency of a) 140.5 GHz and b) 126.5 GHz

suming the electron energy $E = 15 * (1 - \rho^2)^2$ keV and $E = 100 * (1 - \rho^2)^2$ keV, respectively. The quantity, $\alpha \exp[-\tau]$ works as a weighting function how much the local radiation temperature are weighted in the calculation of the total radiation intensity (equivalent to the radiation temperature). In Fig.5 are shown the the ECE weight function expressed as $\alpha \exp[-\tau]$, for 140.5 and 125.5 GHz respectively. Thus, as far as the absorption at the relativistically down-shifted third harmonic resonance are negligible, radiation temperature gives a good estimate of the local electron bulk temperature near the cold second harmonic resonance position. At the second harmonic cyclotron frequency near the maximum magnetic field strength along the line of sight, if the high energy fraction exceeds a certain value, $\approx 3\%$, in this particular case, the weighting function jumps to the relativistically down-shifted resonance region and gradually shift further out as the high energy fraction increases as shown in Fig. 5a). On the other hand, at the lower frequency, the weighting function gradually spreads to the out and the total radiation temperature is determined by the balance between the increase in the local radiation temperature and the width of the weighting function as the high energy fraction increases.

4. Conclusions

The central bulk electron temperature, $T_{\rm e0}$, exceeded 15 keV at averaged electron density, $n_{\rm e,ave} \leq 3 \times 10^{18}$ m^{-3}. The presence of high energy electrons are confirmed in such low density and high temperature, collision-less plasma. The behavior of ECE radiation spectra are compared with the calculated one. It is indicated that the absorption at the relativistically down-shifted third harmonic electron cyclotron resonance can drastically change the ECE spectra when the high energy electrons of more than 100 keV exist. Quantitive comparison of the radiation temperature and the high energy fraction and its profile are left for the future work.

References

1. H. Takahashi *et al.*, Fusion Science and Technology, Vol. **57**, (2010) 19.
2. M. Yokoyama, K. Ida, H. Sanuki, *et al*, Nucl. Fusion **42** (2002) 143.
3. S. Kubo *et al.*, J. Plasma Fusion Res. SERIES **5** (2002) 584.
4. M. Tamor, Nuclear Fusion **18** (1978) 229.
5. G. Taylor *et al.*, Plasma Phys. and Controlled Fusion **31** (1989) 1957.
6. S. Kubo *et al.*, J.Plasma Fusion Res. Volume **4**, (2009) S1095.
7. M. Bornatici *et al.*, Nuclear Fusion **23** (1984) 1153.
8. G. Bekefi, *Radiation Processes in Plasma* ,John Wiley & Sons Inc., New York (1966).

STUDY OF 70GHZ SECOND HARMONIC X-MODE ECCD IN HELIOTRON J*

K. NAGASAKI[1)], K. MINAMI[2)], H. YOSHINO[2)], K. SAKAMOTO[1)], S. YAMAMOTO[1)], T. MIZUUCHI[1)], H. OKADA[1)], K. HANATANI[1)], T. MINAMI[1)], K. MASUDA[1)], S. KOBAYASHI[1)], S. KONOSHIMA[1)], M. TAKEUCHI[1)], Y. NAKAMURA[2)], S. OHSHIMA[3)], K. MUKAI[2)], S. KISHI[2)], H. Y. LEE[2)], Y. TAKABATAKE[2)], Y. YOSHIMURA[4)], G. MOTOJIMA[4)], Á. CAPPA[5)], B. BLACKWELL[6)], F. SANO[1)]

1) Institute of Advanced Energy, Kyoto University, Gokasho, Uji, Kyoto 611, Japan
2) Graduate School of Energy Science, Kyoto Univ., Uji, Kyoto 611-0011, Japan,
3) Kyoto Univ. Pioneering Res. Unit for Next Generation, Uji, Kyoto 611-0011, Japan,
4) National Institute for Fusion Science, Toki, Gifu 509-5292, Japan
5) Laboratorio Nacional de Fusión, EURATOM-CIEMAT,Madrid, Spain,
6) Plasma Research Laboratory, Australian National Univ., Canberra, Australia

E-mail address: nagasaki@iae.kyoto-u.ac.jp

Second harmonic X-mode electron cyclotron current drive (ECCD) experiments have been made in Heliotron J by using an upgraded EC launching system. A focused Gaussian beam is injected with the parallel refractive index, $N_{||}$, ranging from -0.05 to 0.6. According to ray tracing calculations, the focused Gaussian beam makes the EC power absorption localized, $\Delta\rho<0.2$, modifying the rotational transform profile. The experimental results show that the EC driven current can be controlled by $N_{||}$, and it depends on the local magnetic field structure where the EC power is deposited. A large increase in ECE signals has been observed when the EC current was driven, indicating important role of high-energy electrons on the ECCD.

1. Introduction

Non-inductive current has an important role on realization of high performance plasmas and sustainment of steady state in toroidal fusion devices. Finite plasma pressure drives bootstrap current, and tangential neutral beam injection (NBI) generates a so-called Ohkawa current, both of which modify the rotational transform profile in Stellarator/Heliotron (S/H) systems, thereby affecting equilibrium and stability [1]. The noninductive current also modifies the edge field topology and divertor performance [2]. Furthermore, transition onset to an improved confinement mode in NBI plasmas has been observed in relation to the noninductive current in Heliotron J [3].

Electron cyclotron current drive (ECCD) is recognized as a useful scheme for stabilizing magnetohydrodynamic (MHD) instabilities and analyzing heat

* This work is performed with the support and under the auspices of the Collaboration Program of the Laboratory for Complex Energy Processes, IAE, Kyoto University, and the NIFS Collaborative Research Program (NIFS04KUHL005, NIFS08KOAR010), the NIFS/NINS project of Formation of International Network for Scientific Collaborations, and the Grant-in-Aid for Scientific Research, MEXT.

and particle transport. In S/H systems, ECCD is expected as an effective current drive scheme to suppress the non-inductive current and to tailor the rotational transform profile, particularly in low-shear devices. The ECCD experiment in Heliotron J showed that the EC driven current strongly depended on the magnetic field configuration [4-6], suggesting that the ECCD is determined by the balance between the Fisch-Boozer effect [7] and the Ohkawa effect [8]. Comparative studies have been performed among Heliotron J, TJ-II, CHS and LHD, and common phenomena connected with injection angle dependence and ECCD efficiency has been observed [9].

We so far launched a non-focused Gaussian beam with fixed angle in Heliotron J. We have recently installed an upgraded launcher in order to improve the controllability of ECCD. This paper presents theoretical analysis and recent experimental results on ECCD using this launcher system in Heliotron J.

2. Upgraded launcher system for Heliotron J

Heliotron J is a medium-sized plasma experimental S/H device [10][11]. The device parameters are the plasma major radius R = 1.2 m, the averaged minor radius a = 0.1–0.2 m, the rotational transform $\iota/2\pi$ = 0.3–0.8, and the maximum magnetic field strength on the magnetic axis, B = 1.5 T. The coil system is composed of an L = 1, M = 4 helical coil, two types of toroidal coils A and B, and three pairs of vertical coils. The configuration is scanned widely on Heliotron J by varying the current ratios in each coil, making it possible to investigate the properties of noninductive current. The magnetic field ripple is defined by the ratio of the magnetic field at the straight section (ϕ = 0 deg) to that at the corner section (ϕ = 45 deg), $h = B_{str}/B_{cor}$. It can be changed with keeping the plasma volume and rotational transform almost constant (see Fig. 1).

Plasmas are produced and heated by the 70-GHz second harmonic X-mode ECH, which has a cut-off density of 3.0×10^{19} m^{-3}. The injected power is up to 270 kW, and the maximum pulse length is 140 msec in the experiment reported here. We have recently installed an upgraded EC launching system on Heliotron J in order to extend the controllability of EC driven current. Figure 2 shows the upgraded 70GHz launching system consisting of an ellipsoidal mirror and a steerable flat mirror. A low power test using a

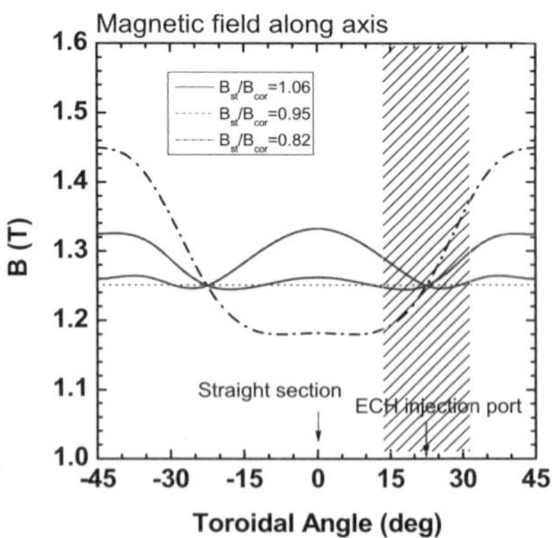

Figure 1. Magnetic field profile along magnetic axis.

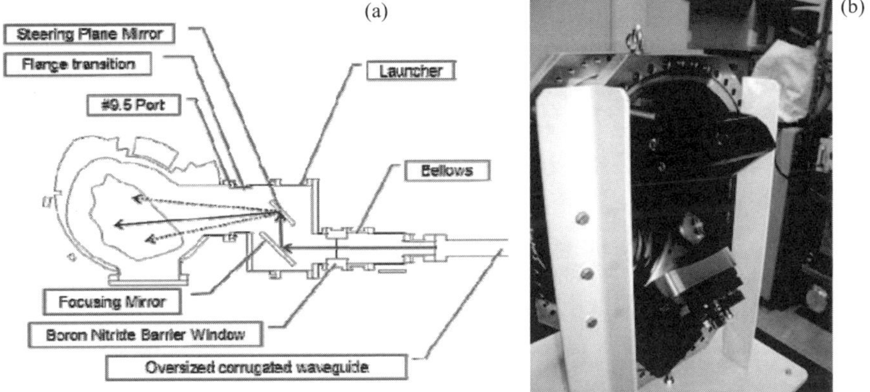

Figure 2. Upgraded70GHz launching system for Heliotron J, (a) schematic view of poloidal cross-section, (b) photo of the launcher.

Gunn Oscillator shows that the beam radius of $1/e^2$ power is 3 cm at the magnetic axis (see Fig. 3), smaller than the minor radius, a~17 cm, and the available $N_{||}$ ranges from -0.05 to 0.6. The new EC launcher is positioned between the straight and corner sections, while it was positioned at the straight section for the previous experiment, meaning that the power is deposited at the different ripple position. The X-mode fraction is more than 80 %, especially more than 90 % for $0.1<N_{||}<0.6$.

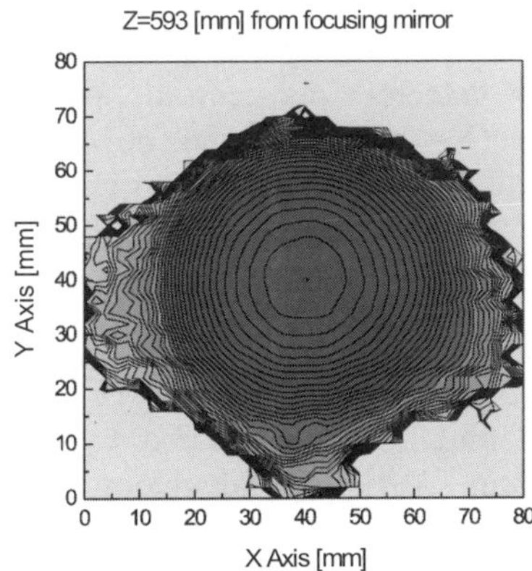

Figure 3. Measured beam profile from launcher in low power test stand

3. Ray tracing calculation for power absorption

Power absorption profiles are calculated by using the TRECE ray tracing code [12] in which the 3-D magnetic field structure is taken into account. Figure 4 shows an example of the ray tracing calculation for standard configuration. The EC power is centrally deposited, $\Delta\rho<0.2$ for the available parallel refractive index by adjusting the magnetic field strength although the profile is relatively broadened at high $N_{||}$. The absorption profile is more localized and peaked than that in the previous launching system where the unfocused Gaussian beam was injected with the injection angle fixed. An plasma experiment scanning the

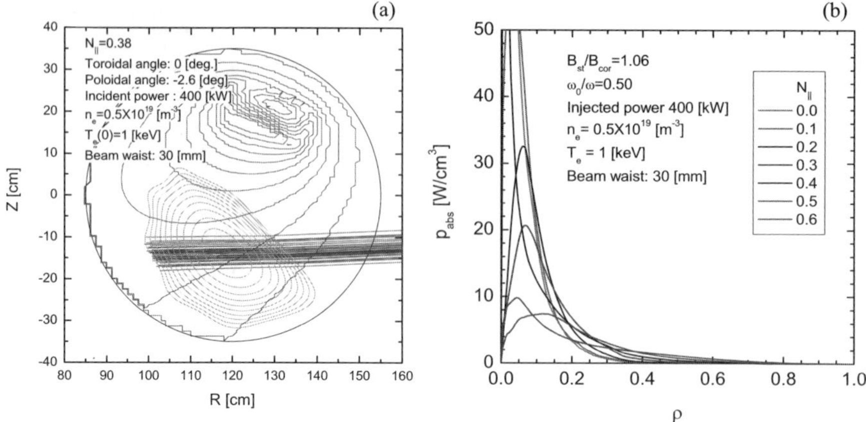

Figure 4. Example of ray tracing calculation, (a) poloidal cross-section of ray trajectory, (b) power absorption profile at some $N_{\parallel}$ conditions.

magnetic field strength confirmed that the SX profile measured with a filtered AXUV detector was centrally peaked at on-axis heating. For perpendicular launch ($N_{\parallel}$=0.0), the total power absorption is weakly dependent on the magnetic field strength due to large magnetic gradient, while for large $N_{\parallel}$, the total absorption rate reaches 100 %, but the magnetic field regime for enough single pass absorption is narrow due to the modest magnetic gradient.

4. Effect of ECCD on rotational transform profile

The poloidal field generated by EC driven current modifies the rotational transform. A simple calculation assuming Gaussian EC current profiles for

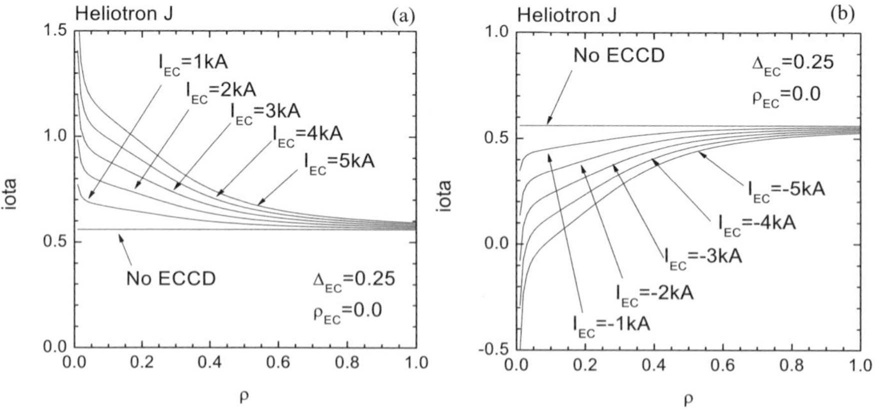

Figure 5. Effect of ECCD on rotational transform profile, (a) positive EC driven current and (b) negative EC driven current.

standard configuration shows that co-ECCD of 5 kA increases the central rotational transform from 0.57 to over unity, forming a negative magnetic shear (see Fig. 5). On the other hand, the ctr-ECCD decreases the central rotational transform to zero, forming a positive magnetic shear. These situations should occur in experiment, since this order of EC driven current has been observed in the Heliotron J experiment. Although the edge rotational transform changes little, even a slight change in rotational transform possibly affects the edge magnetic island structure. The localized ECCD can be a candidate to control the rotational transform profile for the improvement of confinement and transport.

5. ECCD Experiments

ECCD experiments have been conducted using the upgraded launching system in Heliotron J [13]. Figure 6 shows the $N_{\|}$ dependence of the measured toroidal current for two magnetic field configurations. For $N_{\|}$=0.0, it is bootstrap current that mainly contributes to the total current. The dependence of the bootstrap current on the magnetic field configuration agrees with neoclassical theory. Separation of the bootstrap current from the total current at finite $N_{\|}$ will be done in the next experimental campaign by reversing the magnetic field direction. The EC driven current flows in the Fisch-Boozer direction, and it can be controlled by $N_{\|}$, which is maximal around 2 kA around $N_{\|}$=0.4 for the configuration of h=1.06. The experiments at two magnetic field configurations show that the EC current is more driven when the power is deposited at the high field position in magnetic ripple structure. However, this amount of EC driven current is reduced by half compared with a ripple top heating under the same magnetic field configuration. With larger $N_{\|}$, the resonance shift to the edge region makes the single-pass absorption rate lower or the absorption profile

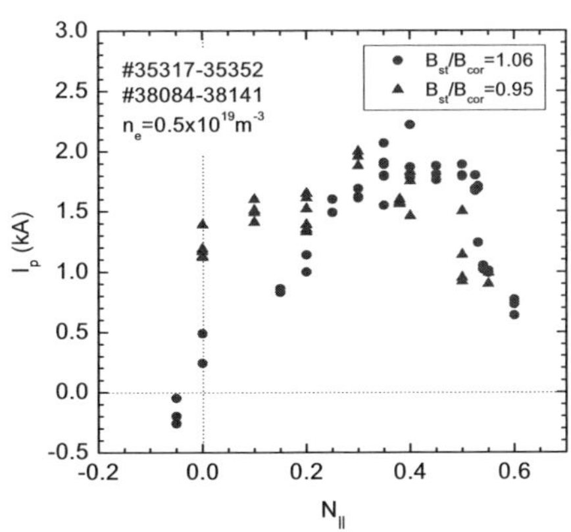

Figure 6. Dependence of measured toroidal current on parallel refractive index.

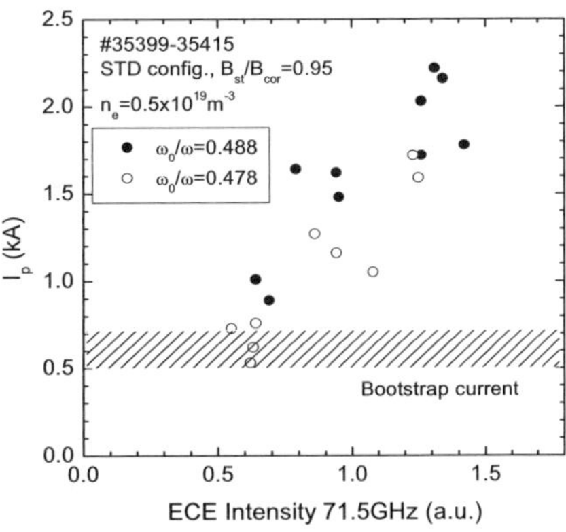

Figure 7. Relationship between ECE intensity and toroidal current.

broad, giving rise to the reduction in the EC driven current. Figure 7 shows the relationship between the ECE signal and the toroidal current. Enhancement of ECE signals by a factor of 3 and high correlation between the toroidal current and the ECE intensity have been observed. Since the optical thickness is gray, $\tau\sim1$, at this low density, the ECE signal reflects not only bulk T_e but also high-energy tail. This suggests that the high-energy electrons may contribute to the ECCD driven by the Fisch-Boozer effect.

6. Summary

Second harmonic X-mode ECCD experiment has been conducted in Heliotron J by introducing the upgraded launching system. The focused Gaussian beam is injected with a wide $N_{||}$ range, making it possible to study the $N_{||}$ dependence. The ray tracing calculation results show that the absorbed EC power is localized, and the absorption position is possible to control. In the initial ECCD experiment using the upgraded launcher, the EC current up to 2 kA flows at $N_{||}\sim0.4$ in the Fisch-Boozer direction for high h configuration. The ECE signal has high correlation with the observed EC current intensity, indicating an important role of high-energy electrons. In the future experiment, we will further study the effect of the magnetic field configuration and $N_{||}$, and rotational transform control for confinement and transport. Comparison with theoretical study is also planned to clarify the Fisch-Boozer and Ohkawa effects.

Acknowledgments

The authors are grateful to Heliotron J staff for conducting the experiments.

References

1. G. Motojima, et al., *Plasma and Fusion Res.* **3** S1067 (2008).
2. T. Mizuuchi, et al., *Nucl. Fusion* **47** (2007) 395.
3. S. Kobayashi, et al., 11th IAEA TM on H-mode Phys. Trans. Barriers (Tsukuba, 2007) (http://www-jt60.naka.jaea.go.jp/h-mode-tm-11/index.html).
4. G. Motojima, et al., *Nucl. Fusion* **47** 1045 (2007).
5. G. Motojima, et al., J. Plasma Fusion Res. SERIES, **8** 1010 (2009).
6. K. Nagasaki, et al., *Nucl. Fusion* **50** 025003 (2010).
7. N. J. Fisch and A. Boozer, *Phys. Rev. Lett.* **45** 720 (1980).
8. T. Ohkawa, General Atomics Report GA-A13847 (1976).
9. K. Nagasaki, et al., *Plasma and Fusion Res.*, **3** S1008 (2008).
10. M. Wakatani, et al., *Nucl. Fusion* **40** 569 (2000).
11. T. Obiki, et al., *Nucl. Fusion* **41** 833 (2001).
12. H. Shidara, et al., J. *Plasma Fusion Res.*, **81** 48 (2005).
13. K. Nagasaki, et al., to be published in Contribution to Plasma Physics.

CONFINEMENT AND PEDESTAL CHARACTERISTICS IN H-MODE WITH ECH HEATING

R. PRATER, J.S. deGRASSIE, P. GOHIL, T.H. OSBORNE
General Atomics, PO Box 85608, San Diego, California 92186-5608, USA

E.J. DOYLE, L. ZENG
University of California-Los Angeles, PO Box 957099,
Los Angeles, California 90095, USA

Access to the H-mode regime of improved confinement and good global confinement when the plasma heating is dominantly of the electrons is critical to the success of ITER, since the proposed heating schemes for ITER, including heating by fusion products, deposit most of their heat in the electron fluid. Most of the world database for the L-H transition and for confinement is derived from discharges with high power positive-ion neutral beam injection (NBI), which primarily heats the ions. Issues for dominant electron heating include the H-mode power threshold and effect on rotation, density profile, global confinement, pedestal height, and edge localized mode characteristics. Experiments have been performed on DIII-D using electron cyclotron heating (ECH) to simulate heating in ITER. These experiments suggest that the H-mode transition power for ECH is significantly lower than that for NBI in deuterium plasmas but about the same for helium plasmas. The global confinement with ECH relative to the H98(y,2) scaling is around 75% under the conditions of these discharges. The pedestal with pure ECH has a higher electron temperature and lower density than for NBI, and the width of the pedestal region is nearly identical.

1. Introduction

The physics of all plasma heating technologies projected for ITER is similar to that of electron cyclotron heating (ECH). MeV neutral beam injection (NBI), ECH, ion cyclotron wave minority heating, lower hybrid current drive, and heating by fusion products all deposit most of their power thermally in the electron fluid. In addition, these heating approaches introduce little or no toroidal angular momentum and little NBI or no fueling. However, most of the world database for confinement and performance in tokamaks [1] was developed using positive-ion neutral beam injection, which on the contrary introduces substantial angular momentum resulting in strong rotation and rotational shear. In order to project to ITER it is necessary to consider tokamak performance in present devices with primarily electron heating. ECH is an excellent approach to developing these projections because it closely simulates the heating in ITER, although the ECH heating profile may be different than the alpha heating profile in ITER.

2. Experimental Considerations

It is widely recognized that confinement at H-mode levels is needed by ITER to achieve its performance goals [1]. This means that the heating power must be sufficient to trigger the transition to the H-mode of improved confinement and that the confinement in the H-mode must be sufficient relative to the H-mode scaling. Recent experiments on DIII-D using a combination of co- and counter-injection of neutral beams have shown strong sensitivity of the H-mode threshold auxiliary heating power to the plasma density, the beam torque, the ion species, and the plasma shape [2]. Hence, comparing performance for ECH dominated plasmas must be done under similar conditions of density and low rotation.

3. Experiments

Experiments in deuterium plasmas show that the H-mode threshold ECH power is smaller than for NBI by about a factor 2. Figure 1 shows measurements from two deuterium DIII-D discharges. On the left side is a discharge in which the ECH power is increased in steps, first by modulating the power and then by stepping it up, and on the right side is a similar discharge with ramped NBI power instead of ECH power. For ECH, the H-mode transition (where the D_α emission drops and the density starts to increase) takes place at a heating power of 1.2 MW, while for the NBI case the transition does not take place until just after a step from 2 MW to 2.8 MW. In this discharge the co- and counter-injection NBI is approximately balanced, so that little net rotation is created. In past studies, low torque NBI has provided lowered threshold power for the H-mode compared to full co-injection [2].

The ECH for Fig. 1 is aimed to deposit power near ρ=0.7, with no current drive. (Here, ρ is the square root of the normalized toroidal flux.) This deposition location is in the region where ECH might be used for control of neoclassical tearing modes. One view of the H-mode transition is that it depends on the power transported from the core to the plasma boundary, so the power deposition profile is immaterial (provided, as in this discharge, that plasma radiation is not a significant loss mechanism). However, it is also possible that the peripheral nature of this ECH heating profile is beneficial to the H-mode transition.

Confinement in these two discharges is better with neutral beam heating than ECH heating. For NBI the H-mode confinement factor $H_{98(y,2)}$ [1] is 1.05 in the steady condition after the transition, while for ECH it is 0.76. It is likely that part of the difference in confinement is due to the far off-axis heating of the ECH case, but it is also likely that heating electrons is subject to greater transport losses in the electron channel. Typically in DIII-D discharges like those shown in Fig. 1, the electron-ion energy coupling is weak in the core of the plasma, so stronger transport processes in the electron fluid are not ameliorated by the lower ion transport rates.

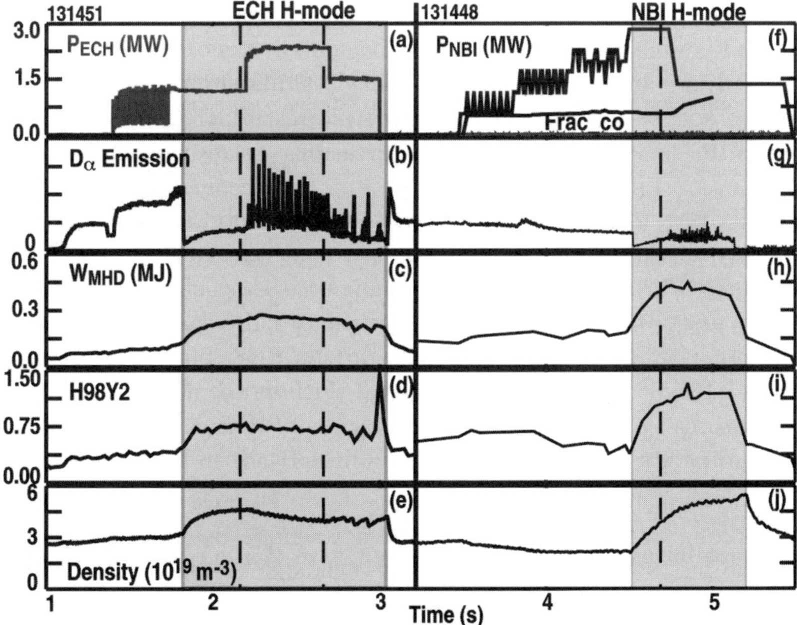

Fig. 1. Two DIII-D discharges in deuterium with toroidal field 2.0 T and plasma current 1.0 MA. The configuration is lower single null. The discharge on the left (a-e) has ECH power deposited near $\rho = 0.7$ and on the right (f-j) NBI power. The NBI power is provided by co- and counter-current sources, so that the net fraction of co-injected power is about 0.5 as shown in (f). (a,f) Incident power, (b,g) D_α emission (au), (c,h) plasma stored energy, (d,i) confinement factor H98(y,2), and (e,j) line-integrated electron density.

After the step up in ECH power at 2.4 s in Fig. 1, the H-factor decreases further to 0.7, and little increase in stored energy is seen, while in the NBI case the H-factor increases to 1.18 after the NBI power is stepped down to 1.45 MW. In addition to changes in transport, a factor affecting the global confinement is the edge localized modes (ELMs), which become very large in the ECH case but remain small and frequent in the NBI case. ELMs cause increased transport from the core, particularly for the Type I ELMs of the ECH case. Also, in the ECH case the density decreases when the ECH power is stepped up, due at least partly to the ELMs but also this may be due to density pumpout effects sometimes observed when electron heating is applied or increased. Another factor is that when the NBI power is stepped down the balance between co- and counter-injection is lost and net torque by the NBI causes plasma rotation to increase. Plasma rotation is known to be beneficial to confinement [6], and hardly any loss in stored energy is seen when the NBI power decreases. An additional factor in both cases is that the total heating power is not much greater than the transition power.

The nature of the pedestal in NBI and ECH H-modes is an important factor for ITER because the plasma performance depends to a large extent on the height of the transport barrier that forms at the plasma boundary in H-mode discharges [7]. In Fig. 2 the boundary profiles are shown for the two discharges (ECH and NBI) of Fig. 1. Figure 2 shows that for ECH only, the density at the top of the pedestal is 20% higher for NBI than for ECH, while the electron temperature is 40% higher in the ECH case. The pressure profile at the edge is therefore higher in the ECH case, assuming that $T_\mathrm{i} = T_\mathrm{e}$ in the ECH case, since the measurement of T_i requires beams which are absent in the ECH discharge. T_i may in fact be significantly smaller than T_e in the pedestal of ECH discharges, so the difference in pedestal pressure is probably larger than that shown in Fig. 2(f).

An important issue for ITER is access to the H-mode in the non-DT phase when heating by the fusion products is not significant so the power will be limited to the available auxiliary heating. Operation in the H-mode during the hydrogen or helium phase is important for such concerns as assessing the strategy for suppressing ELMs and addressing the coupling of the ICRF heating system with an H-mode edge. Because of this relevance, experiments on the H-mode threshold power in helium have been performed. Figure 3 shows that the threshold power in helium is higher than for deuterium by a factor around 1.5. The point for ECH+NBI, which uses the maximum ECH power available plus sufficient balanced NBI power to achieve the transition, is consistent with the NBI-only data points. This is in contradiction to the results from deuterium in which the transition was at much lower in power for ECH. It is noteworthy that both helium and deuterium exhibit the same torque dependence: low or negative torque implies lower threshold power than positive torque.

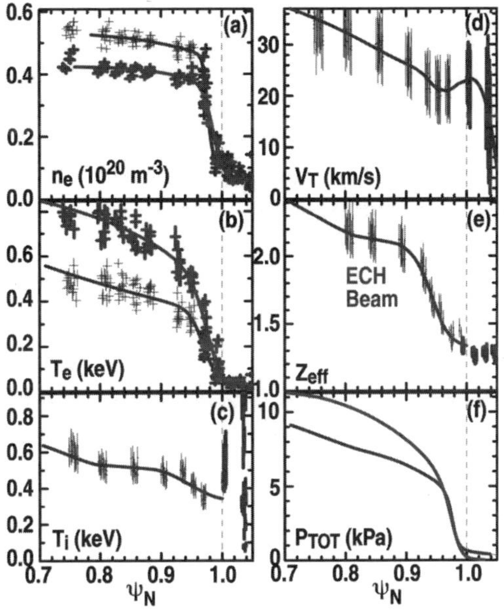

Fig. 2. Measurement of (a) electron density, (b) electron temperature, (c) ion temperature, (d) toroidal rotation velocity, (e) effective ion charge Z_{eff}, and (f) total pressure, as a function of normalized poloidal flux. The profile measurements are made over the last 20% of the period between ELMs. The last closed flux surface is identified by the vertical dashed line.

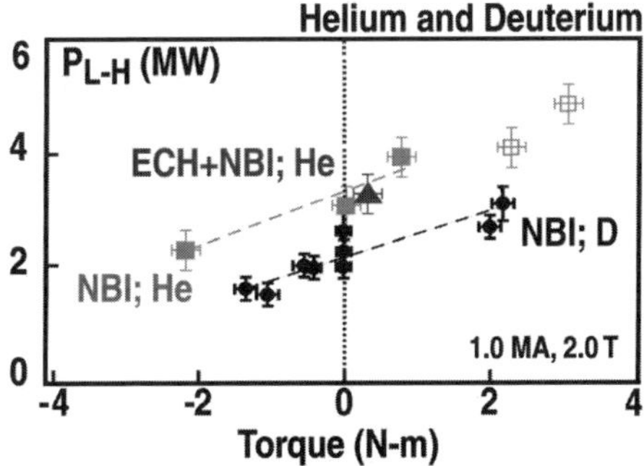

Fig. 3. Threshold power for helium working gas (square points) and deuterium (round points). Most points are for NBI, but the triangle point is for helium with mostly ECH plus a small amount of nearly balanced NBI. The hollow square points are lower limits on the threshold power for helium, as these discharges did not achieve the H-mode.

4. Discussion and Conclusions

Unqualified conclusions are not possible now because the database of ECH discharges is too small. Nevertheless, some tentative conclusions may be drawn. First, the threshold power appears to be significantly lower for ECH than for NBI in deuterium, although the threshold powers are quite similar in helium. Second, confinement as characterized by the value of H98(y,2) for ECH H-mode discharges is smaller than that of NBI H-mode discharges by around 25% in the few shots available with similar parameters. The pedestal widths are nearly the same for ECH and NBI, but as expected for ECH the T_e pedestal height is larger but the density is lower. Distinct differences are seen in the ELM behavior, with ECH more likely to have regular large Type I ELMs.

Acknowledgments

This work was supported by the U.S. Department of Energy under DE-FC02-04ER54698.

References

1. ITER Physics Expert Groups on Confinement and Transport and Confinement Modelling and Database, *Nucl. Fusion* **39**, 2175 (1999).
2. G.R. McKee, P. Gohil, D.J. Schlossberg, et al., *Nucl. Fusion* **49**, 115016 (2009).
3. S. Angioni, A.G. Peeters, X. Garbet, et al., *Nucl. Fusion* **44**, 827 (2004).
4. R. Prater, *Phys. Plasmas* **11**, 2349 (2004).
5. A. Kallenbach, R. Dux, M. Mayer, et al., *Nucl. Fusion* **49**, 045007 (2009).
6. P.A. Politzer, C.C. Petty, R.J. Jayakumar, et al., *Nucl. Fusion* **48**, 075001 (2008).
7. P.B. Snyder, N. Aiba, M. Beurskens, et al., *Nucl. Fusion* **49**, 085035 (2009).

II. Electron Cyclotron Emission

SUMMARY OF ECE PRESENTATIONS AT EC-16

G. TAYLOR
Plasma Physics Laboratory, Princeton University, Princeton, NJ 08543, USA

At the EC-16 workshop there were 17 presentations primarily on ECE and an invited talk on EBE. There was also a discussion session on the ITER ECE diagnostic system design. ECE imaging, correlation ECE, oblique ECE and EBE imaging diagnostic performance has continued to improve since EC-15. There have been some interesting developments in ECE receiver technology and data analysis that show promise for the future. There is still a need to agree on some potentially critical details of the ITER ECE diagnostic design, and it remains unclear how important the Thomson scattering/ECE discrepancy, seen previously on TFTR and JET, will be for ITER.

1. 2-D ECE Imaging and Microwave Imaging Reflectometry

A combination of microwave imaging reflectometry, with a 2-D 7x5 channel receiver array, and ECE imaging (ECEI) is being used to reconstruct 3-D electron density and temperature fluctuations in the Large Helical Device (T. Yoshinaga). Improvements to 2-D ECEI diagnostics on TEXTOR, ASDEX-U and DIII-D were also reported (H. Park). These improvements include the use of mini lens-based arrays and improved video amplifiers with a variable bandwidth between 12.5 and 400 kHz. Recently this improved ECEI diagnostic design has been used to study Alfvén eigenmodes in DIII-D and ELMs in ASDEX-U. A similar ECEI system is now being installed on KSTAR. The frontend optics includes a zoom feature allowing the height of the imaged area in the plasma to be increased from 0.3 to 0.9 m.

On HT-7 an 8x16 channel ECEI system have been used to study sawtoothing plasmas and T_e fluctuation spectra (B. Wan). The ECEI diagnostic has a 100 GHz local oscillator and a DC-18 GHz IF. ECEI results for magnetic reconnection on the low field side of the magnetic axis during compound sawteeth, with both a weak and strong reconnection, were presented. The ECEI system has been upgraded for use on EAST so that it now has an extended frequency range of 96-133 GHz and an optimized optics system that can image a 7x23 cm region near the midplane. The imaged region can be centered anywhere from -30 cm to +30 cm on either side of the magnetic axis. Preliminary results for a sawtooth reconnection in EAST were presented. A new 24x16 channel

ECEI system with a frequency coverage of 105-156 GHz will be installed on EAST in early 2011.

2. 2-D Electron Bernstein Wave Emission Imaging

An electron Bernstein wave emission (EBE) radiometer with a rapidly spinning mirror has been used on MAST to measure the shape of the Bernstein wave to O-mode (B-X-O) mode conversion window (V. Shevchenko). Since the shape of the window depends on the magnetic field pitch at the mode conversion surface this has allowed a measurement of the field pitch versus major radius. The field pitch variation measured by this technique in the edge of H-mode plasmas was found to vary much more strongly with major radius than the pitch measured by motional Stark effect spectroscopy. There are plans to install a phased array EBE imaging system on MAST that has the potential for much better time resolution than the rotating mirror system.

3. Correlation ECE

Correlation ECE (CECE) measurements on TCV show a reduction of correlation length for negative triangularity plasmas, and the dependence of the correlation length on plasma collisionality in negative triangularity plasmas is much weaker (T. Goodman). CECE measurements support the prediction that TEM turbulence is a dominant contribution to energy transport in low collisionality TCV plasmas. Coupled X-mode correlation reflectometry and CECE has allowed electron density (n_e) and temperature (T_e) fluctuation cross-phase angle measurements in DIII-D (A. White). The measured n_e-T_e phase angle in L-mode plasmas, heated by neutral beam injection and electron cyclotron heating (ECH), agrees quantitatively with nonlinear gyrokinetic code predictions.

4. Oblique ECE

Oblique ECE measurements on TCV show evidence of an asymmetry in the electron distribution function (EDF) during on-axis electron cyclotron current drive (ECCD) (T. Goodman). The level of ECCD was increased while the total ECH power was kept constant. Measured oblique ECE co/counter view ratios are consistent with an asymmetric EDF at the location of the ECCD current channel. Modeling of multi-angle oblique ECE spectra with the SPECE code for a lower-hybrid-driven EDF in JET shows good agreement with measured total (X-mode + O-mode) ECE data (L. Figini).

5. Other ECE Diagnostics

On HT-7, a 16-channel, 98.5-125.5 GHz, heterodyne radiometer has been used to study the anomalous Doppler resonance during plasma current ramp down (E. Li). An improved design for the front end of the HT-7 radiometer, which employs two elliptical mirrors to form a Gaussian telescope, was also presented (A. Ti). A 20-channel ECE grating polycromator is now operational on EAST (Y. Liu). Installation of a low-loss corrugated waveguide and improved video ampliers have resulted in increased system signal-to-noise. A periscope arrangement, with a rotateable ellipsoidal mirror, will be installed in late 2010, allowing absolute *in situ* calibration of the instrument. A new 32-channel, 104-152 GHz hetrodyne radiometer is also being developed for EAST (B. Ling).

A 48-channel, 110-162 GHz, heterodyne radiometer has been installed on KSTAR (S. Jeong). The diagnostic employs four highly-integrated microwave circuits for the detector modules. Each module has a 1-12 GHz broadband amplifier, that is followed by a DC-10 GHz low-pass filter, a 2-9 GHz power divider, and eight 0.5 GHz bandwidth bandpass filters. Measurements during sawtoothing plasmas were presented.

The high-end frequency range of the JET ECE hetrodyne radiometer has been exctended from 139 to 207 GHz (E. de la Luna). The upgraded radiometer provides access to higher density plasmas at higher magnetic fields, improved spatial resolution, and access to the plasma edge region on the low-field side.

A 4 GHz Fourier transform (FT) direct IF digitization system has been installed on the in-line ECE system on TEXTOR (W. Bongers). This system directly digitizes the IF bands with no video detection, allowing conservation of the ECE frequency phase. It provides flexible temporal and spatial resolution with the option for a dynamic zoom around specific plasma positions and events. Real time FT ECE systems could be used as an adaptive sensor for MHD stability control by ECH and ECCD.

6. ITER ECE Diagnostic

There were presentations on ITER ECE related issues by H. Pandya, S. Danani, M. Austin and V. Udintsev, and a lively discussion session, led by V. Udintsev, on issues pertinent to the design and operation of the ITER ECE system. Magnetic islands with radial widths $\geq$ 4 cm should be easily detectable by ECE in ITER, but smaller island widths, $\sim$ 1 cm, are expected to be difficult to detect. The front end design of the ITER ECE diagnostic system has been simplified from a Gaussian telescope to an optical system with a single focusing element and the transmission line has been moved closer to the plasma mid-plane. It is not clear whether there is a need to have both a hot and a cold calibration source for each plasma view, or whether a single source that is run hot and then cold would be sufficient. It was noted that microgroove waveguide can only be used up to ~ 300 GHz. Since there is a desire to measure the ECE spectrum up to ~ 1 THz, a dielectric waveguide should be considered.

It has not been decided if there should be a double vacuum window or a window plus a vacuum isolation valve. An oblique view is still being considered for one of the two plasma views, and could be as much as 16 degrees to normal, and still fit inside the envelope of the ITER E9 port plug. This oblique view would provide some additional information if the EDF becomes non-thermal. Modeling predicts that such an oblique view should still measure T_e in thermal plasma. It was noted that some designs being considered had crossed sightlines between the perpendicular and oblique view, providing an opportunity for CECE measurements. However, it was agreed that attempts to provide CECE capability should not determine the choice of oblique angle. There is also a need to measure total EC power loss in ITER, where $T_e(0)$ is expected to reach 25-35 keV, resulting in significant ECE power loss. The ECE spectrum will peak at ~ 1 THz.

There was some discussion of the ECE/Thomson scattering T_e discrepancy, originally observed at $T_e > 7$ keV on TFTR and JET. There have been experiments recently on DIII-D that used ICRF and NBI heating that reached $T_e(0) \geq 7$ keV. Data are still being analyzed from these experiments, but so far there is no evidence for a discrepancy. However, it was noted that in the previous systematic studies of the discrepancy on TFTR and JET, the discrepancy did not become clearly apparent until $T_e(0)$ reached ~ 10 keV.

7. Concluding Remarks

Significant progress has been made in ECEI, CECE and EBEI since EC-15. Some interesting recent improvements in receiver technology and data analysis show promise for the future. There is still a need to agree on potentially critical details of ITER ECE design. There is also a need for dedicated experiments at $T_e \sim 10$ keV to further study of the ECE/TS discrepancy, as well as ECE modeling in support of this work.

EC PROGRAM ON EAST AND HT-7

BAONIAN WAN, BILI LING, TI ANG, LIU YONG, LI ERZHONG, HAN XIANG †
Institute of Plasma Physics, Chinese Academy of Sciences, Hefei, China

CHANGXUAN YU, WANDONG LIU YIZHI WEN， JINLIN XIE, XIAOYUAN XU, JUN WANG， MING XU, BINXI GAO
University of Science and Technology of China Hefei, China

N. C. LUHMANN, C. W. DOMIER, BENJAMIN JOHN TOBIAS, JIAN WANG, ZHENGGANG XIA, ZUOWEI SHEN
UC. Davis, USA

PERRY PHILIPPE , KENNETH GENTLE , WILLIAM ROWAN , HE HUANG
Fusion Research Center, UT at Austin, USA

RON PRATER
General Atomic, San Diego, USA

GARY TAYLOR
Princeton Plasma Physics Laboratory, USA

Program of ECH of 4MW at 140 GHz is launched for pressure and current density profile control on EAST. Several ECE diagnostics are under development as important ingredient of the research program of EAST. HT-7 is equipped with a heterodyne radiometer containing 16 channels and a ECE image system with 8(radial)x16(vertical) channels. Physical issues including fluctuation by electron and ion modes, low frequency Zonal Flow, magnetic reconnection mechanism, anomalous Doppler resonance effect, etc were investigated on HT-7. These two systems have been moved to EAST after some modifications. New ECE systems including a 32-channel ECE system covering 104–168 GHz and a ECEI system of 24(radial)x16(vertical) channels are under developing. These two systems are designed to fit the ECH plasma regimes and synergetic work for long range correlation research of plasma turbulence. A grating polychromator ECE system will be installed soon for Te profile measurement covering whole operation range of toroidal magnetic field on EAST.

---

† Work was supported by the National Natural Science Foundation of China under Grant No. 10725523, No. 10990212 and No. 10721505.

1. Introduction

EAST is a fully superconducting tokamak aimed high performance plasma under steady-state condition. EAST has presently actively cooled plasma facing components (PFCs)[1]. A 2MW lower hybrid current drive (LHCD) system at 2.45GHz and 4.5MW ion cyclotron resonant heating (ICRH) system are in operation. They will be extended to 8~10MW in 2011. A 4MW neutral beam injection system and 4MW LHCD system at 4.6GHz are expected to be available in 3~4 years. Nearly 30 diagnostics are presently equipped, which can provide most of key plasma parameter profiles. These capabilities make EAST a unique platform to address some of advanced tokamak issues for next-step fusion devices such as ITER and beyond.

An attractive tokamak as an energy producing system requires high performance plasmas with high confinement, high power density (which means) and high stability under steady-state condition. The above powers should be enough for high confinement plasma operation with sufficient high β. However, all of these heating and current drive means has less flexibility in control of plasma profiles, such as pressure, current density, which play key role in achieving high performance and controlling stabilities. Electron cyclotron wave, which is capable of well localized heating and current drive, can accomplish reliably plasma profile control. This leads the recent program to launch the ECH/ECCD on EAST. By the way, significant extension of plasma diagnostic capability is another important aspect in the advanced tokamak program on EAST.

2. ECH/ECCD program on EAST

The following may be (and has been) described as 'dangerously irrelevant' physics. The Lorentz-invariant phase space integral for a general n-body decay from a particle with momentum P and mass M is given by: EAST with its full superconducting feature aims at steady-state operation, e.g. fully non-inductive plasma discharges with high confinement performance. Available 10 MW LHCD power is sufficient to achieve discharges up to 1MA of plasma current. But most of the lower hybrid wave driven current is off-axis in such plasmas, while fast wave current drive (FWCD) can provide seed current in plasma center. This implies that controllability of current density profile only by LHCD is not robust, which may affect the plasma performance of confinement and instabilities. ECH and ECCD would be the best choice for these issues.

Based on the mission of EAST and fore-seeing capabilities, first priority of physical objectives of ECH and ECCD is to develop the scenarios of stable high performance fully non-inductive plasma discharges. In this sense, localized electron heating and current drive as well as possibly large current drive is required to sustain a desired current density profile in control of electron transport barrier and pre-empty MHD instabilities. Other applications of EC power, such as MHD control, studies of transport of energy or momentum and plasma startup assistance have to be included to optimization of the conceptual design.

Based on these considerations above, simulation has been carried out to determine two important issues: choice of frequency and choice of launch location [2]. While EAST is capable of operating at fields up to 3.5 T, very powerful heating systems are needed to assess high β_N conditions at high field. Estimation scaled from the DIII-D discharge shows that the power required to obtain the $\beta_N = 3.19$ of the desired EAST equilibrium ($B = 3.0$ T, $I = 0.95$MA, R =1.916 m, $a = 0.462$ m, $q95 = 4.671$, $\kappa = 1.768$, $\beta_N = 3.188$) will be ~20MW. Scaling the toroidal field instead to 2.5 or 2.1 T from 3.0 T, the required power to achieve the same β_N at the same $H98y2$ and N_{GW} is reduced to 14.7 and10.2 MW, respectively, which is a lot more feasible. This estimation makes the choice of frequency at 140GHz better than 170GHz for high β_N plasma discharges. Another consideration is from the commercial availability of the 140 GHz very long pulse gyrotrons.

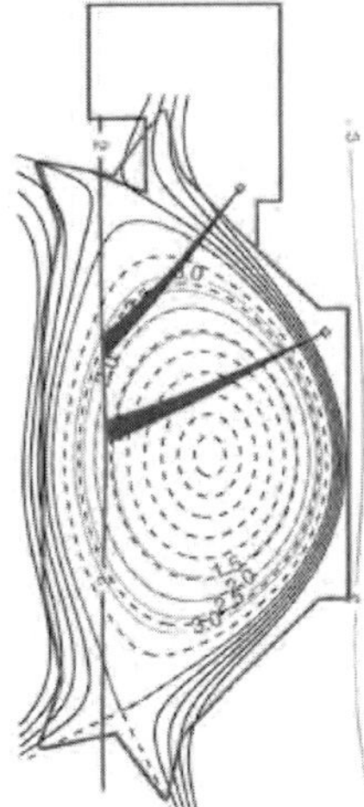

Fig.1 Two possible injection locations

Two possible launch locations (see figure 1), e.g. top and equatorial ports, are investigated by considering physical objectives and based on simulation. Investigation shows that either launch location would be satisfactory driving current at ρ~0.6 for current profile control for negative shear scenarios. However, if the key physics objective is controlling neoclassical tearing modes at the q=3/2 surface or driving current at the q=2 surface, simulation shows that the equatorial port launch is better without great sensitivity to the launch angle. The bottom line is that the top port launch may be satisfactory for some key physics objectives of EAST, but the toroidal field may have to be optimized over a reasonable range of 2.1 to 2.5 T for 140 GHz according to the details of the equilibrium, kinetic profiles, and the physics objective. The top launch has performance similar to that of the equatorial port launcher for ECCD at values of ρ greater than about 0.5, but for high peak heating or current drive objectives the equatorial port launcher is better. Choice of launch location needs further investigation by considering the mechanical constrains for optics.

The engineering design will be further investigated based on the successful experience from worldwide tokamaks, which will be done with joint effort of international collaborations. Presently, 4-5MW source power is planned and system is scheduled in three years for operation on EAST.

3. ECE systems on HT-7

There are a multi-channel ECE heterodyne radiometer and an ECE image system on HT-7. These two systems were installed on two equatorial ports, which are toroidally separated by 135°. The multi-channel ECE heterodyne radiometer use a TPX lens and a circular horn antenna to receive wideband ECE radiation and then transmit the radiation to a 3db power divider by oversized waveguide. Two local oscillators at frequencies of 95GHz and 108.5GHz convert the ECE radiation downshift to 2-18GHz followed by 8 bandpass filters and Schottky barrier diode detectors. Total 16 channels cover the frequency from 98.5 to 125.5GHz. The time resolution is about 50 microseconds and spatial resolution is 2cm.

The ECEI system is developed for HT-7 with joint efforts from USTC and ASIPP in collaboration with UC Davis, USA [3,4]. The collection optics was upgraded after one year of the system operation[5]. This system is quite similar to those applied in other tokamaks [6-9]. The ECE broad bandwidth radiation is collected by a vertically aligned mixer array and separated by frequency band. Using a 16 element array, and an 8 band receiver provides time-resolved 8(radial)×16(vertical) images on HT-7 for investigation of sawtooth activities and temperature turbulence. Resolution of ECEI for each pixel is about 1cm(radial)×1.1cm(vertical). It covers a region of about 5cm(radial)×20cm(vertically symmetric along mid-plane) in a poloidal cross section. The time resolution is about 4 microseconds and the system noise to signal level is about 1%.

2D distribution of electron temperature fluctuations have been investigated by the ECEI system at various plasma conditions. And long distance correlation has been observed by combining the ECEI and multi-channel ECE heterodyne radiometer. Following show some experimental results.

3.1 Electron mode in plasma with low density

The electron temperature fluctuation with a broadband spectrum (figure 2) shows that it propagates in the electron diamagnetic drift direction, and the mean poloidal wave-number is calculated to be about 1.58 cm^{-1}. Figure 3 shows typical poloidal wave number to frequency of electron temperature fluctuation. It indicates that the fluctuation should come from the electron drift wave turbulence. The linear global scaling of the electron temperature fluctuation with the gradient of electron temperature is consistent with the mixing length scale qualitatively. Evolution of spectrum of the fluctuation during the sawtooth oscillation phases is investigated, and the fluctuation is found to increase with the gradient of electron temperature increasing during most phases of the sawtooth oscillation. The results indicate that the electron temperature gradient is probably the driver of the fluctuation enhancement. The steady heat flux driven by electron temperature fluctuation is estimated and compared with the results from power balance estimation.

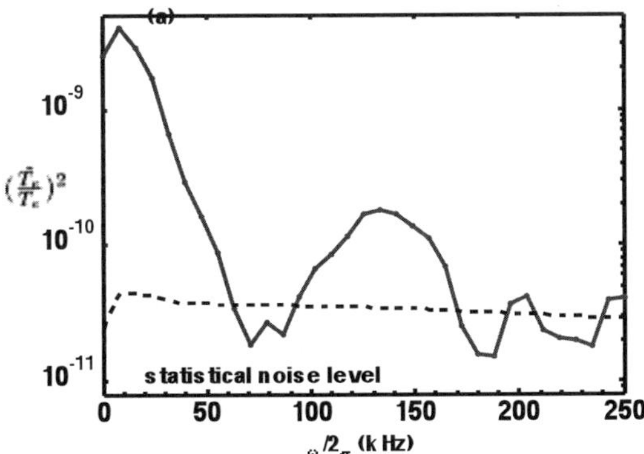

Fig.2 typical spectrum of core electron temperature fluctuation

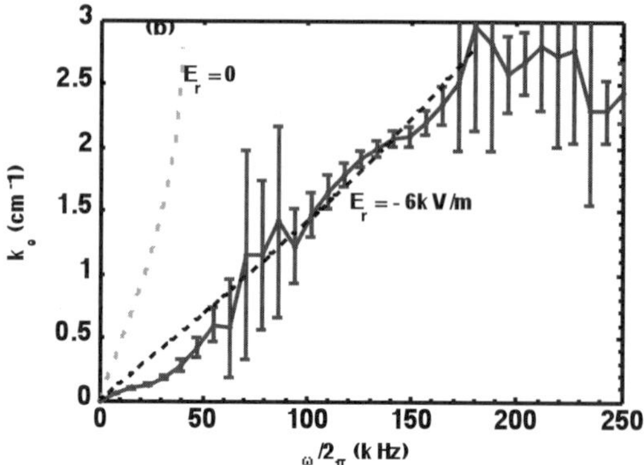

Fig. 3 a typical poloidal dispersion of electron temperature fluctuation.

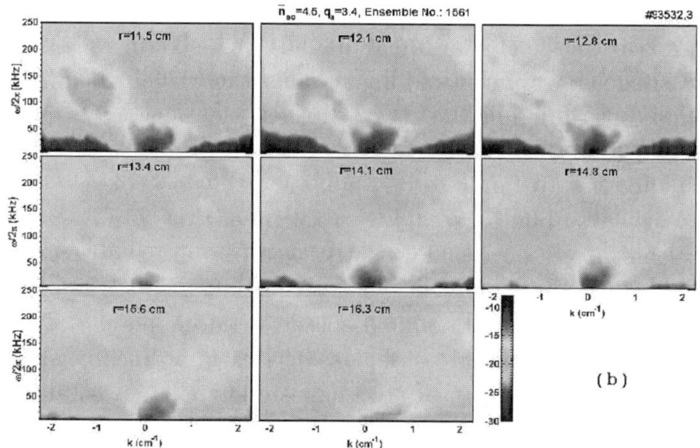

Fig.4 Clear evidence of coexistence of ion and electron mode in high density plasma

3.2 Electron and ion modes coexist in plasma with high density

Electron mode transported in electron diamagnetic direction as well as ion mode transported in ion diamagnetic direction is observed in ohmic plasma (figure 4), when line averaged electron density $\overline{n}_e \gtrsim 4.5\times10^{19}m^{-3}$. The two modes are found coexist at r ~ 11.5 cm in Low field side（LFS） as well as at r ~ 13cm in high field side (HFS). Ion temperature gradient mode in density fluctuation was observed by far-infrared (FIR) scattering and beam emission spectrum (BES), but has not been reported in the temperature fluctuation yet.

3.3 The evidence of the low frequency Zonal Flow

In the low density plasma in HT-7 tokamak, the poloidal dispersion relationship between 0-5 kHz diverts from the linear dispersion relationship, with the wave number about zero. The toroidal dispersion relationship, as the density is about 1.7, the toroidal mode number is near zero at about 0-12 kHz. When the density increases, figure 5 clearly shows that the range of the frequency with the toroidal mode number near zero decreases. This is the first evidence of Zonal Flow in the electron temperature fluctuation. The existence of the Zonal flow are still be confirmed by further measurement of the other main character.

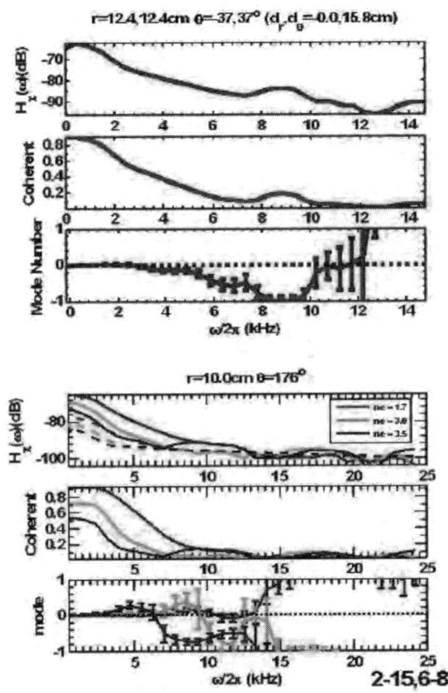

Fig.5 evidence of Zonal Flow in the electron temperature fluctuation

3.4 Magnetic Reconnection mechanism and high-order harmonics

The evolution of the m/n = 1/1 mode and high-order harmonic mode structures in the sawtooth oscillation has been investigated via ECE images on the HT-7 tokamak. Different from the Kadomtsev model which predicts that the m/n = 1/1 mode causes a sole reconnection point at the q = 1 radius, which is responsible for the sawtooth crash, the high-order harmonic modes observed in sawtooth precursor cause a weak magnetic reconnection to happen in more than one place, which is localized and not preferential on the q ~ 1 surface. However, as seen in figure 6, only a small amount of heat and particles escape through the magnetic reconnection. Most of the heat and particles escape out through the strong reconnection later.

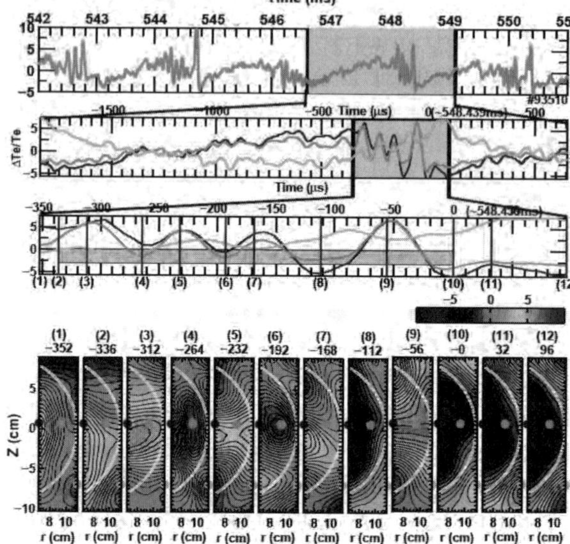

Fig.6 a sawtooth crash process observed by ECEI

4. ECE program on EAST

Mission of the ECE program on EAST is: they should provide electron temperature profiles covering most radial region for normal operational toroidal magnetic field strength; they should provide sufficient information as constrains for kinetic equilibrium reconstruction; some of them should be oriented to specific physics objectives, such as studying of temperature fluctuation, electron thermal transport, instabilities and etc.

Based on these requirements, the program is planned in two phases. In the first phase, the existing systems on HT-7 are modified and moved to EAST. A grating polychromator ECE system from PPPL is included in this phase. In the second phase, two new systems, a 32-channel heterodyne radiometer and a 24(poloidal)*16(radial) ECE image system will be developed. The frequency choice should be applicable to the toroidal magnetic field range for the planned ECH system operation.

The 16 channel heterodyne ECE radiometer[10] is simply modified by changing a new collection lens to fit the longer port extention of EAST, while others remain unchanged. New optics has been designed and installed on EAST for ECEI system shown in figure 7. Three E-plane lens and one H-plane lens are used to cover plasma region of 7cm × 23cm with 8 × 16 channels. This new system has been optimized to keep same sampling volume for large radial extension of -30~30cm. Correspondingly, a new additional local oscillator is used to extend the frequency from 96-133GHz.

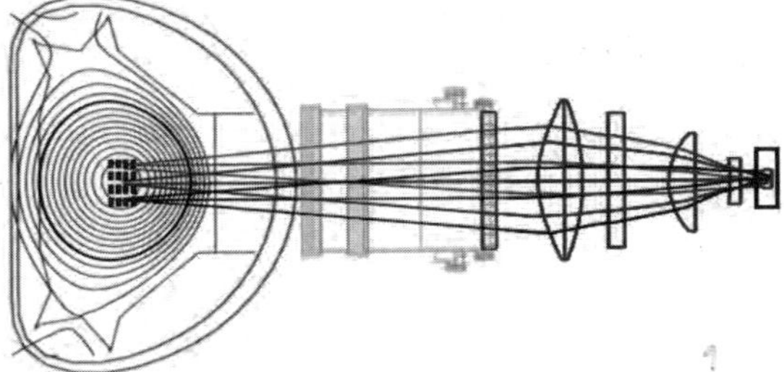

Fig.7 Schematic of new ECEI optics on EAST

A 20-channel grating polychromator transferred from PPPL[11,12], has been modified and applied on EAST. This instrument with two grove gratings of line spacing 1.65mm and 2.75mm, respectively, can applied for toroidal magnetic field of 2-3.5 T on EAST as shown in figure 8. The emission from the plasma is received and introduced to the polychromator by 8-meters long corrugated waveguide. Diffracted emission by the grating is detected by 20 liquid-helium-cooled indium antimonide hot-electron bolometers.

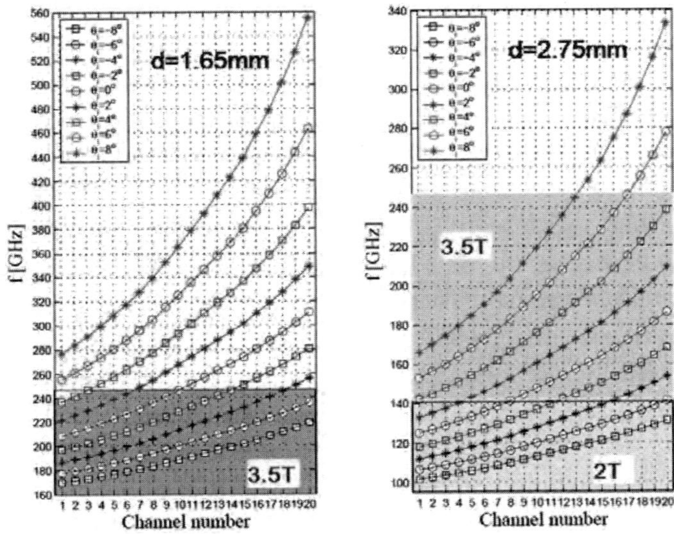

Fig.8 Operation range of PPPL GPC on EAST

These three systems are being operated on EAST and have produced preliminary results. As an example, figure shows time trace of ECE from one channel of each system in one shot shown in figure 9.

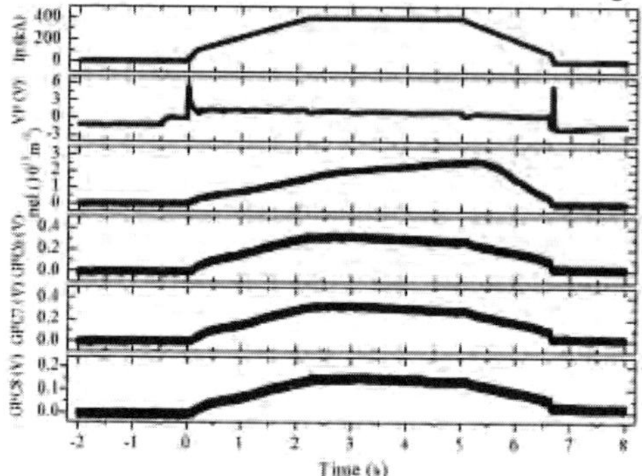

Fig.9 bottom three curves show ECE signals from three systems in one shot

In the second phase, a new multi-channel heterodyne ECE radiometer and a 2D ECE image system will be developed to accommodate some specific physical objectives. By considering toroidal field range of ECH operation, the choice of central frequency for these two systems is selected at one corresponding to the 2.3T.

In collaboration with University of California, DAVIS, the scheme of ECE radiometer system for EAST has been designed[13]. ECE radiation from the plasma is optically divided into four beams using a combination of beam splitters and mirrors as shown in figure 10. The beam splitters are optimized to ensure good coupling to each mixer. Custom designed high density polyethelene (HDPE) lenses focus each of the four beams into harmonic mixers with dedicated Gunn oscillators or dielectric resonator oscillators (DROs) with frequency multiplier together as local oscillators. And the lens will also adjust the beam waist in the vicinity of the ECE radiation layer. A dichroic plates placed before each focusing lens serve as high pass filters. Four beams will cover the frequency from 104-168GHz (figure 11).

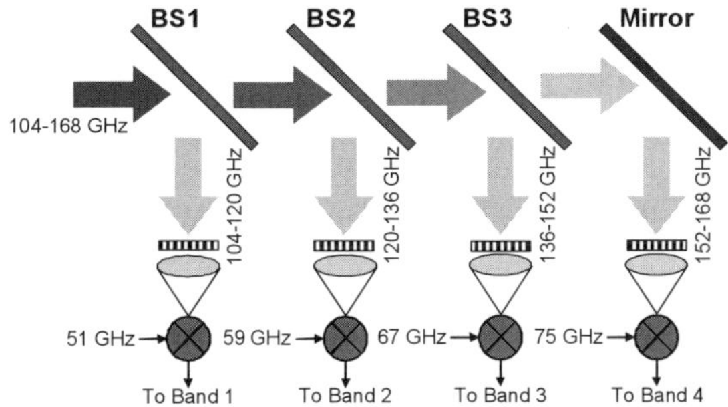

Fig. 10. Schematic layout of a 32 channel ECE radiometer system for EAST

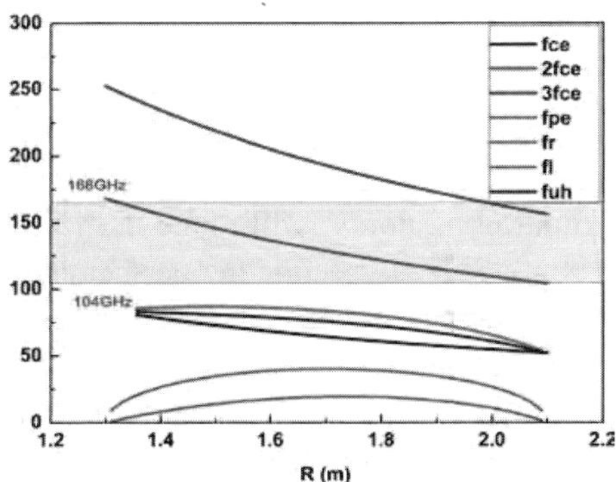

Fig.11 frequency coverage of 32 channel ECE heterodyne radiometer on EAST

The output of each mixer is amplified using low noise 2-18 GHz preamplifiers, and is divided into 8 channels (expandable to 16) and followed by 8 individual bandpass filters and Schottky barrier diode detectors with attenuators placed before each filter to normalize the detector output signals as well as provide improved isolation against out-of-band filter reflections. The bandpass filters are 500 MHz wide at frequencies of 2.5-17.5 GHz in 2 GHz steps.

The ECEI system built for EAST will provide time-resolved 16(radial)×24(vertical) images of electron temperature and temperature turbulence. It covers a region of about ρ~0.5(radial)×24cm(vertically symmetric along mid-plane) in a poloidal cross section. Spatial resolution is about 1cm. The time resolution is about 1 microseconds and the system noise to signal level is about 1%.

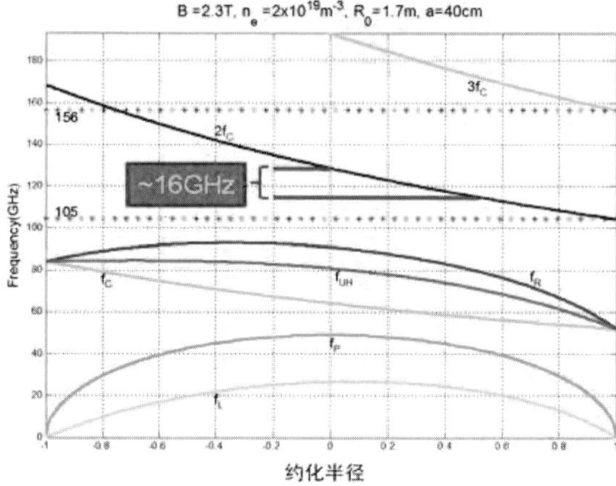

Fig.12 frequency coverage of ECEI system on EAST

Compared with existing ECEI system, new system will adopt some well-developed technologyies to improve its global performance. High power BWO is selected as oscillator instead of Gunn oscillator. It will range the frequency from 105-156GHz, but cover 16GHz defined by IF section in one shot as shown in figure 12. Mini Elliptical lens array will be used with some benefits, such as : no need to trade-off sensitivity and antenna patterns; improved coupling efficiency by combination with front-side pumping; increased gaps between antenna and lens arrays allowing insertion of more ECRH filter for improved S/N ratio etc. Standardized IF modules will be used for easy system extension.

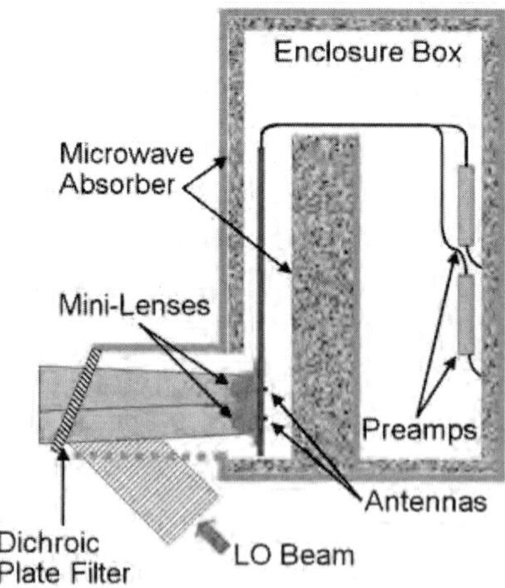

Fig. 13. Schematic layout of ECEI system for EAST

Both new systems have been designed and are building. All components will be fabricated and arrive in Hefei for assembling by end of 2010. They are expected to be ready for the experimental campaign in spring 2011.

5. Summary

EAST is aimed at steady-state high performance plasma and relevant physics. Program of long pulse ECH of 4MW at 140 GHz is launched in accommodation with this requirement. ECE diagnostics as an important ingredient of this program will be developed in next two years. Equipments and experience from HT-7 and wide international collaborations help us quickly to realize overall EC program for EAST. New ECE systems including a 32-channel ECE system covering 104–168 GHz and a ECEI system of 24(radial)x16(vertical) channels are under developing for EAST. They are designed to fit the ECH plasma regimes and synergetic work for long range correlation research of plasma turbulence. It is expected that new ECE systems are ready for 2011 spring experimental campaign.

References

1. Wan B. et al 2009 Nucl. Fusion **49** 104011
2. Ron Prater, Selection of ECH parameters for the EAST tokamak, private communication September 2009
3. J. Wang et al., Rev. Sci. Instrum. **75** 3875 (2004)
4. Wang Jun, et al., Plasma Sci. Technol. **6** 2166 (2004)
5. Wang J, et al., Plasma Science Technol. **8** 76 (2006)
6. B. H. Deng, et al., Rev. Sci. Instrum. **70**, 998 (1999)
7. G. Cima, et al., Plasma Phys. Controlled Fusion **40**, 1149 (1998)
8. B. H. Deng, et al., Phys. Plasmas **5**, 4117 (1998)
9. B. H. Deng, et al., Rev. Sci. Instrum. **72**, 301 (2001).
10. A Ti, et al., International Journal of Infrared and Millimeter Waves **28** 243 (2007)
11. A. Cavallo, R. C. Cutler, and M. P. McCarthy, Rev. Sci. Instrum. **59,** 889 (1988)
12. Y. Liu, et al., "20-channel grating polychromator for electron cyclotron emission measurements on EAST" Poster session I-15, this meeting
13. B. L. Ling et al., "Design of 32-channel heterodyne electron cyclotron emission radiometer for EAST tokamak" Poster session I-14, this meeting

CORRELATION ECE MEASUREMENTS OF TURBULENT ELECTRON TEMPERATURE FLUCTUATIONS IN DIII-D

A.E. WHITE

Massachusetts Institute of Technology,
Plasma Science and Fusion Center, 175 Albany St. Cambridge, MA 02139, USA

W.A. PEEBLES, T.L. RHODES, G. WANG, L. SCHMITZ, T.A. CARTER,
J.C. HILLESHEIM, E.J. DOYLE, L. ZENG

University of California, Los Angeles,
Los Angeles, CA 90095, USA

C.H. HOLLAND

University of California, San Diego,
San Diego, CA 92093, USA

G.R. MCKEE

University of Wisconsin, Madison,
Madison, WI 53706, USA

G.M. STAEBLER, R.E. WALTZ, J. CANDY, J.C. DeBOO,
C.C. PETTY, K.H. BURRELL

General Atomics,
San Diego, CA 92186-5608, USA

This paper describes measurements of long wavelength, turbulent electron temperature fluctuations in the core plasma of the DIII-D tokamak made with a correlation electron cyclotron emission (CECE) radiometer-based diagnostic. Experimental and simulation results indicate that long wavelength electron temperature fluctuations (1) are similar in amplitude and spectrum to density fluctuations, (2) can be associated with both ITG and TEM turbulence, (3) exhibit changes in the relative fluctuation level that correlate with changes in electron thermal transport, and (4) are correlated, but out of phase, with density fluctuations measured simultaneously with reflectometry.

1. Introduction

The study of long wavelength turbulent electron temperature and density fluctuations is important for understanding the physics of turbulent driven transport. Advancements in nonlinear gyrokinetic turbulence simulations [1] and the development of synthetic diagnostics [2] have allowed for direct and quantitative comparisons between theory and experiment.

2. Experimental Set-Up

The correlation electron cyclotron emission (CECE) diagnostic at DIII-D allows for the measurement of long wavelength ($k_\theta\rho_s < 0.3$) electron temperature fluctuations associated with drift-wave type instabilities such as the ITG mode or TEM. The CECE diagnostic is a radiometer-based turbulence diagnostic, shown schematically in Fig. 1 [3]. The CECE diagnostic uses a spectral decorrelation method [4] to measure low amplitude (~1%), broadband ($0 < \Delta f < 800$ kHz) electron temperature fluctuations.

The CECE diagnostic can be used to measure temperature fluctuations at the same radial location as a beam emission spectroscopy (BES) diagnostic [5], although the sample volumes are separated poloidally and toroidally, with the poloidal separation of the CECE sample volumes and the 2-D BES array shown in the inset in Fig. 1.

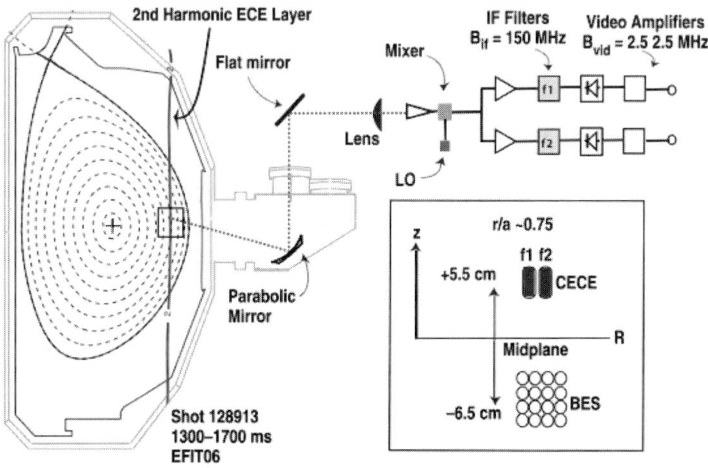

Figure 1. Diagram of the CECE system at DIII-D. Inset shows location of BES and CECE sample volumes.

Recently, the CECE diagnostic has been coupled with a multichannel X-mode heterodyne reflectometer [6] that measures long-wavelength density fluctuations. The CECE and reflectometer diagnostic share the same antenna and the same line of sight to the plasma. This allows for the poloidal and toroidal overlap of the sample volumes. The CECE and reflectometer radial viewing locations can be remotely tuned so that the sample volumes can be made to overlap radially as well. The CECE radiometer signals can then be correlated with the reflectometer signals using standard techniques [7], which allows for the coherency and cross-phase angle between temperature and density fluctuations to be measured.

3. Comparisons Between Electron Temperature Fluctuations, Density Fluctuations and Nonlinear GYRO Simulations

Temperature fluctuations have been measured with the CECE diagnostic in a variety of plasmas at DIII-D: Ohmic, L-mode, H-mode and QH-mode. In Ohmic and L-mode plasmas, electron temperature fluctuations have relative fluctuation levels between 0.5% and 2.0%, increasing with radius. During H-mode and QH-mode plasmas the amplitude of electron temperature fluctuations in the core is reduced by at least a factor of 5 to below the sensitivity level of the diagnostics, concomitant with the improved confinement during H-mode [8]. The profile of electron temperature fluctuations measured simultaneously with density fluctuations using BES during sawtooth-free beam heated L-mode plasma is shown in Fig. 2. The spectra of the two fluctuating fields are very similar, as seen in Fig. 3. These data in Figs 2 and 3 are from a 400 ms time average of the turbulence data during quasi-steady sawtooth-free L-mode plasmas, heated with 2.5 MW of co-injected neutral beam power, described in more detail in Ref. 9. Linear stability analysis indicates that these plasmas are dominantly ITG unstable. Detailed comparisons of these fluctuation profile data with the nonlinear gyrokinetic turbulence transport code, GYRO, have been successful at matching both the fluctuation levels and transport deep in the core ($\rho = 0.5$) but at $\rho > 0.7$ in these plasmas GYRO consistently underpredicts [2,9] the transport and the fluctuation levels when the experimental equilibrium profiles are used as input

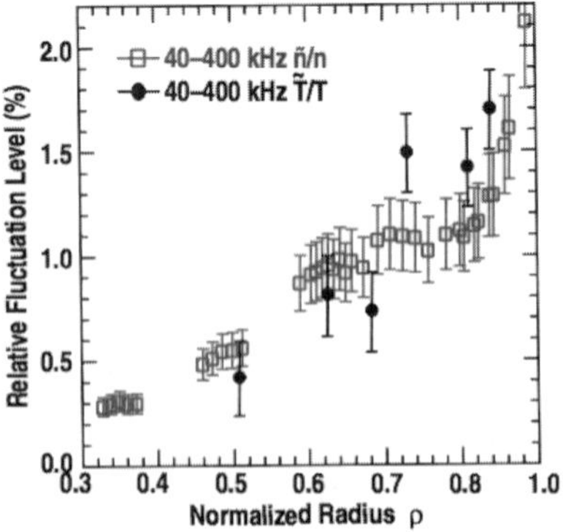

Figure 2. Profiles of turbulent electron temperature (circles) and density (squares) measured simultaneously with CECE and BES in L-mode plasmas.

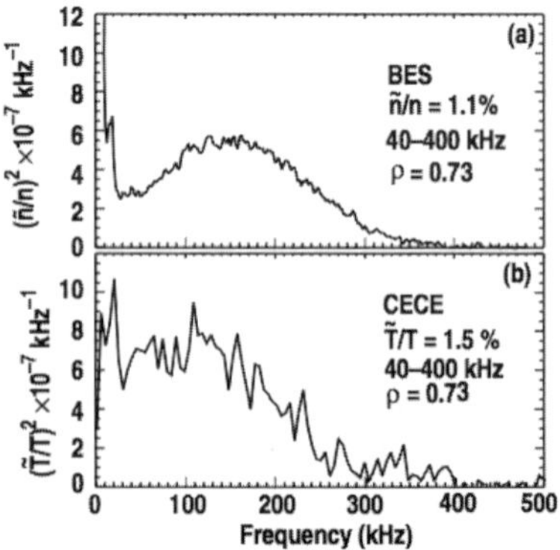

Figure 3. Spectra of density (a) and electron temperature fluctuations (b) at $\rho = 0.73$ in L-mode plasmas, corresponding to Fig. 2 profiles.

in traditional simulation-experiment comparisons [10,11]. New, gyrokinetic transport modeling [12] has shown that the use of power-matching profiles can reproduce the experimentally observed trend that fluctuation level increases with radius. CECE data are playing an important role in these validation efforts [13] by providing a new constraint on the codes.

When electron cyclotron heating (ECH) (X-mode, 110 GHz, radial launch) is used to increase the electron temperature, the ratio T_e/T_i, and decrease the collisionality in these type of L-mode plasmas [14], there is a significant increase in electron thermal diffusivity and a significant increase in electron temperature fluctuations (Fig. 4). There is little to no change in density fluctuations measured simultaneously with BES. Linear stability analysis indicates that in these experiments, before the ECH is added, the plasma is dominantly ITG unstable, but after the ECH is added, the TEM growth rate increases. The increase in the ratio of the fluctuation levels is expected from driftwave theory to increase with the ratio $(\gamma_{TEM}/\gamma_{ITG})$ of the growth rates consistent with experimental observations.

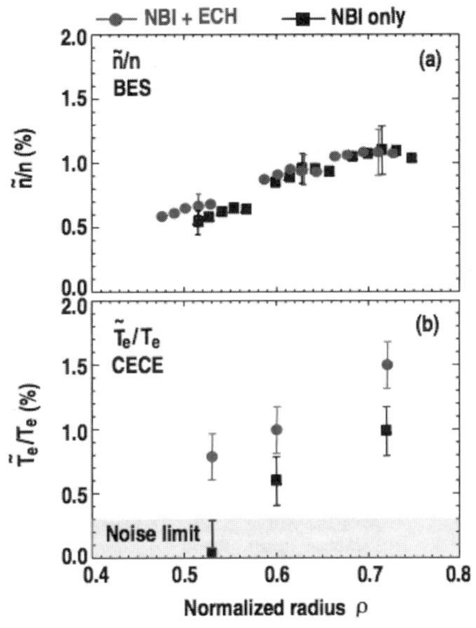

Figure 4. Profiles of density (a) and electron temperature fluctuations (b) measured with BES and CEE showing response to changes in profiles during ECH.

In a different experiment using similar beam-heated L-mode plasmas with and without ECH [15], the coherency and cross-phase angle between the density and electron temperature fluctuations were measured using the coupled CECE and reflectometry diagnostics. Before the experiment, the GYRO code was used to predict what (if any) change in the cross-phase angle would occur when the ECH was added, e.g. when the primary drive was changed from ITG to ITG mixed with strong contributions from the TEM. Linear and nonlinear GYRO results indicated that as the TEM drive increased, the phase angle should decrease towards zero for measurements made near half radius, ρ= 0.5.

Experimentally, a decrease towards zero in the cross-phase angle is observed at two of the core measurement locations in the plasmas with ECH – qualitatively consistent with predictions from GYRO. However, at the outer measurement location, there is no change in the phase angle during ECH (Fig. 5). Nonlinear GYRO simulations for $\rho = 0.65$ have been performed on the plasma with ECH plus neutral beams, and the phase angle is in good quantitative agreement with the measured value (Fig. 6). However, even though the measured turbulent phase angle between density and electron temperature fluctuations agrees well with the GYRO results, GYRO over-predicts the electron thermal flux by 50% and under-predicts the ion thermal flux by nearly a factor of 4 when the equilibrium experimental profiles were used as input to the simulations. The measured electron temperature fluctuation levels are in good agreement with the GYRO results, but the measured density fluctuation levels (BES) are 50% higher than the GYRO predicted levels [15].

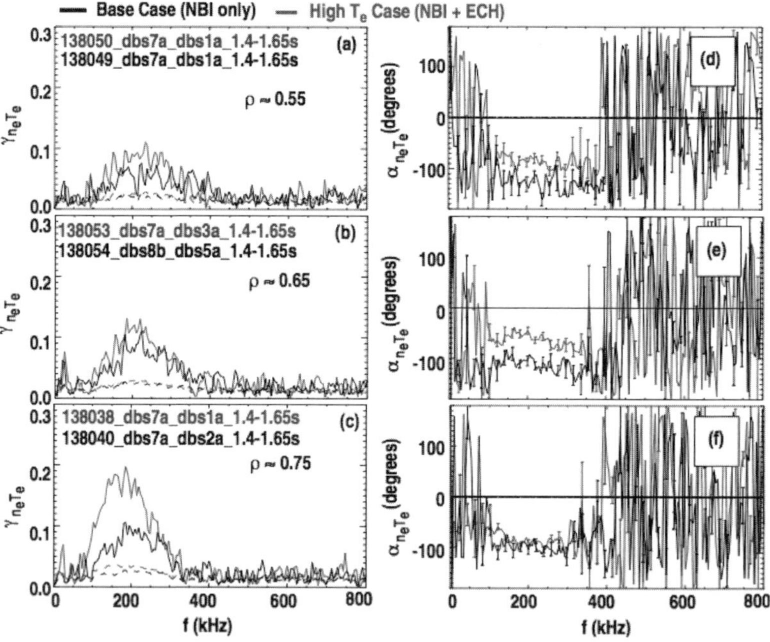

Figure 5. Coherency and cross-phase angle between density and electron temperature fluctuations measured with the coupled multichannel reflectometer and CECE diagnostics.

4. Conclusions

Measurements of long wavelength, turbulent electron temperature fluctuations in the core plasma of the DIII-D tokamak are made with a correlation electron cyclotron emission (CECE) radiometer-based diagnostic. Experimental and simulation results indicate that long wavelength electron temperature fluctuations (1) are similar in amplitude and spectrum to density fluctuations, (2) can be associated with both ITG and TEM turbulence, (3) exhibit changes in the relative fluctuation level that correlate strongly with changes in electron thermal transport, and (4) are correlated, but out of phase, with density fluctuations measured simultaneously with reflectometry. This paper has reviewed some key recent experimental observations and results from quantitative comparisons with the GYRO code using synthetic diagnostics. Comparisons between electron temperature fluctuations and nonlinear GYRO simulations are part of an ongoing validation research thrust at the DIII-D tokamak through Transport Model Validation taskforce. Data from the CECE diagnostic has played an important role in these efforts.

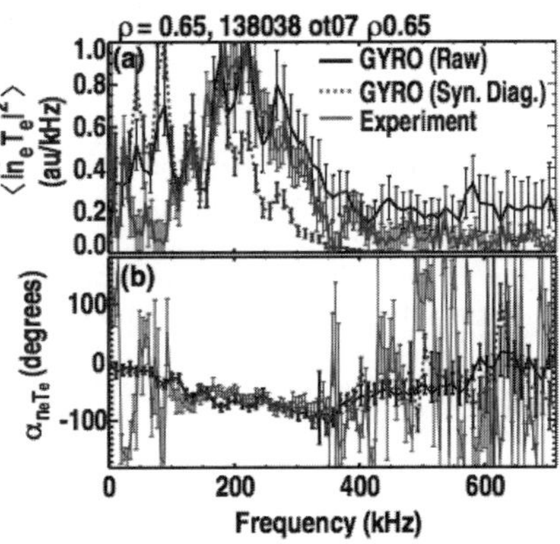

Figure 6. The measured cross-power spectrum (a) and phase angle (b) between density and electron temperature fluctuations (coupled reflectometer and CECE diagnostics) are compared with nonlinear GYRO results.

Acknowledgments

This research was supported by the U.S. Department of Energy under DE-AC05-06OR23100, DE-FG03-08ER54984, DE-FG02-07ER54917, DE-FG03-95ER54309, and DE-FC02-04ER54698. A.E.W is very grateful to the DIII-D team for their support of these experiments.

References

1. X. Garbet, et al., Nucl. Fusion **50**, 043002 (2010)
2. C.H. Holland, et al., Phys. Plasmas **16**, 052301 (2009)
3. A.E. White, et al., Rev. Sci. Instrum. **79**, 103505 (2008)
4. G. Cima, et al., Phys. Plasmas **2**, 720 (1995)
5. G.R. McKee, et al., Rev. Sci. Instrum. **70**, 913 (1999)
6. J.C. Hillesheim, et al. Rev. Sci. Instrum. **80**, 083507 (2009)
7. Bendat and Piersol, Random Data, 3rd Ed. (2000)
8. L. Schmitz, et al. Phys. Rev. Lett. **100**, 035002 (2008)
9. A.E. White, et al. Phys. Plasmas **15**, 056116 (2008)
10. A. Casati, et al. Phys. Rev. Lett. **102**, 165005 (2009)
11. L. Lin, et al. Phys. Plasmas **16**, 012502 (2008)
12. P.W. Terry, et al. Phys. Plasmas **15**, 062503 (2009)
13. P.W. Terry, et al. Phys. Plasmas **15**, 062503 (2009)
14. A.E. White, et al., Phys. Plasmas **17**, 020701 (2010)
15. A.E. White, et al., Phys. Plasmas **17**, 056103 (2010)

OBLIQUE AND CORRELATION ECE IN TCV

T.P. GOODMAN*, V.S. UDINTSEV, E. FABLE, F. FELICI, A. POCHELON, L. PORTE, M. RANCIC, O. SAUTER, CH. SCHLATTER, C. ZUCCA

Centre de Recherches en Physique des plasmas, Association EURATOM-Confédération Suisse, Ecole Polytechnique Fédérale de Lausanne (EPFL), CH-1015 Lausanne, Switzerland
**E-mail: timothy.goodman@epfl.ch*
crppwww.epfl.ch

P.K. CHATTOPADYAY

Institute for Plasma Research, Bhat, Gandhinagar-38428, India

The Tokamak à Configuration Variable, TCV, is equipped with a moveable ECE receiving antenna, identical to the 6 second harmonic (X2) ECH/ECCD launchers, which can track the plasma at any position within the vacuum vessel in real time. In contrast to this poloidal plane viewing, the receiver can be rotated to obtain an oblique view. In this configuration, measurements of the asymmetry of the electron distribution function, EDF, are performed during co/counter ECCD sweeps at constant input power. Direct evidence of the driven current is provided by the ratio of radiation temperatures from co and counter views of identical plasmas. The results are simulated by the NOTECTCV radiation-transfer code in which current profile broadening can be included to reproduce the measurements. Studies of core broadband temperature fluctuations, induced by plasma turbulence, are performed with a high-resolution X2 correlation ECE diagnostic. ECE correlation is measured by two frequency-tunable YIG filters that can be placed between r/a = 0 - 0.9 using both their wide tuning range, and plasma geometry changes from shot to shot. Evidence of broadband (20–150 kHz) fluctuations with peak frequencies ranging from 20 kHz up to 90 KHz at r/a = 0.3 - 0.8 in Ohmic sawtooth-free discharges was obtained. The amplitude of the temperature fluctuations decreases with increasing density and thus, increasing collisionality, which is in qualitative agreement with predictions from quasi-linear gyrokinetic calculations performed with the gyrokinetic code GS2. The mixing length heat diffusivity calculated from GS2 decreases with increasing collisionality, as does that obtained from a power balance analysis. The real frequency of the broadband turbulence stays positive over the range of collisionalities explored, indicating dominant TEM turbulence.

Keywords: Oblique ECE, Correlation ECE, Fluctuations, Turbulence

1. Introduction

Microinstabilities and the larger self-organized turbulent structures that result from them are thought to lead to the observed anomalous heat transport measured in tokamaks [1]. The break-up or modification of the larger structures can lead to so-called transport barriers. As the anomalously large heat and particle losses are directly responsible for the size and therefore cost of future fusion reactors, understanding and reducing this turbulence is a major goal in magnetic fusion research.

Turbulence can result from magnetic or electro-static fluctuations in the plasma. The former breaks up the nested structure of the field lines, while the latter produces cross-field transport via particle drifts generated by the fluctuation electric field and the background magnetic field, under certain conditions. In order to quantify the transport due to the fluctuating fields, it is necessary to measure the temperature, density and electric potential fluctuations and the phase relationship between them. This is a very challenging task. In the absence of such detailed measurements much can still be inferred through a comparison of trends in the fluctuation measurements and those of gyrokinetic simulations.

One experimental tool set used to investigate the nature of turbulence relies on correlation techniques to seek out the underlying temperature fluctuations in signals dominated by thermal noise. Several measurement setups are possible [2] to achieve independence of the thermal noise but correlation of the thermal fluctuations. All techniques rely on long-time averaging-to-zero of the statistically independent noise. An electron cyclotron emission (ECE) diagnostic can be modified to take advantage of these techniques. Such a system is often referred to as correlation ECE or CECE. On TCV the modification is carried out by adding narrow band frequency filters in the radiometer attached to a single-line-of-sight.

In addition to measurements of the fluctuations, another type of modification to a standard ECE diagnostic — inclining the radiometers line-of-sight (antenna) to provide an oblique view of the plasma — may provide a direct measurement of one consequence of enhanced particle transport, as described in the next paragraph. Measurements made with this antenna are referred to as oblique ECE, or ObECE on TCV.

The higher confinement (reduced heat transport) measured in some plasmas can result from modifications of the magnetic shear. On TCV this is accomplished using electron-cyclotron-current-drive (ECCD). The very localized absorption of millimeter waves in the plasma results in a correspondingly narrow driven current channel. If the fast particles that carry the current diffuse across the flux surfaces, the driven current channel is smeared out and may be less effective in modifying the shear. ObECE measurements combined with modeling of the electron distribution function (EDF) have the potential to provide both a direct measurement of the driven current direction, and an indication of the width of the current channel.

This paper discusses results from both of these modified ECE systems on the TCV tokamak, CECE and ObECE. Section 2 outlines, briefly, the ECE system, its correlation ECE branch and moveable ObECE antenna. We then present results of CECE measurements of temperature fluctuations and trends in the fluctuation amplitude with collisionality in Sec. 3. A comparison with the gyrokinetic code GS2 is given in Sec. 4. Measurements made with the ObECE of the driven current channel are described in Sec. 5 and modeling of the current channel width with the emission code NOTECTCV is outlined in Sec. 6. Finally, Sec. 7 summarizes the main results and gives some concluding remarks.

2. ECE setup on TCV

The basic CECE and ObECE equipment is described in Refs. 3 and 4, respectively. A few changes to the setups described in those references have been made and are outlined here.

The CECE is done by splitting off power going to the 24 channel profile ECE system which comes from any one of the 3 low-field-side (LFS) antennas. The CECE branch has a single LO at 63 GHz followed by a power splitter and 2 low-noise amplifiers. The signals are subsequently passed through two, independent, center-frequency-tunable (6-18GHz), narrow-bandwidth (100-160MHz), YIG filters. This allows selection of the frequency spacing between the filters to ensure that the thermal noise is *not* correlated (i.e. the filter bandwidths don't overlap) while maintaining a small spatial separation between the measurement regions in the plasma so that the fluctuations remain correlated. After detection and amplification, the video signals are split into three branches each of which are filtered differently and acquired at $\geq$ 1 MHz. The filtered signals are primarily used for low frequency "MHD" studies (low-pass filtered at 40 kHz), and thermal fluctuation studies (band-pass filtered between 40kHz and 250kHz to avoid aliasing). The final branch is low-pass filtered at 600 kHz. In this paper we refer to the fluctuation branch. The changes to the system are the filters and the amplification.

ObECE is carried out using a receiver identical to the 6 launching antennas of the 3 MW, X2, ECH & ECCD auxiliary heating system: it is the so-called "7th launcher". The antenna consists of 2 focussing and 2 flat mirrors. The flat mirror closest to the plasma can sweep the plasma view at up to $\sim 100^\circ$ s^{-1} in a plane inclined at any angle to the toroidal plane, about the major radius of the tokamak: the two most useful planes being the poloidal plane (standard ECE view with $\mathbf{k} \perp \mathbf{B}$), or the toroidal plane (ObECE). The main changes to the ObECE system are the installation of a 63.5 mm diameter HE_{11} corrugated waveguide at the output of launcher 7, replacing the one-inch diameter, smooth, circular waveguide, and replacement of the fundamental-waveguide circular polarizer. The HE_{11} waveguide mode couples well to a free-space Gaussian beam mode, which passes through a free-space grating-polarizer / convertor section and couples to fundamental-rectangular waveguide. The polarization is selected by choosing a polarizer angle which best couples the elliptically polarized plasma emission, at a given oblique angle, to the linearly polarized TE_{10} mode of the rectangular guide. The rectangular section is followed by a TE_{10} to TE_{11}-circular convertor, an uptaper to overmoded circular waveguide for long distance transmission, then a similar downtaper to fundamental rectangular waveguide at the radiometer. The grating mirror is identical in design to one of the two used to set the X2 gyrotrons' polarizations.

It should be noted that the ObECE antenna is connected to the LFS radiometer and therefore CECE can be done through this antenna, though this is not discussed in this paper.

3. Temperature fluctuation measurements

Plasmas were positioned in front of the LFS lens antenna located on the machine midplane ($z = 0$) so that the view was through the plasma center and the poloidal direction is transverse to the $\mathbf{k}$-vector of the viewing beam. The YIG center frequencies were chosen so that the resonance location was aligned with the expected minimum of the beam power-spot radius (w_p = 11mm) and were separated by 0.24 GHz. The thermal noise in this case is not correlated. By choosing the minimum beam spot we maximize the observable k_{pol} and thus $k_\perp$ of the turbulence (k_r is already much larger than k_{pol} since the relativistic broadening is small at the low temperatures, $\sim$ 350eV, involved here). The diagnostic is thus sensitive to $k_{turbulence} \leq 1.3cm^{-1}$ at $\rho_{vol} \sim 0.55$ (i.e. ρ_{vol} is the square-root of the enclosed normalized plasma volume). The spatial separation is $\Delta\rho_{vol} \sim 0.005$.

Broadband fluctuations are observed between $\sim$ 20–150 kHz in Ohmically heated plasmas, as seen in Fig. 1; which shows the cross-spectral-

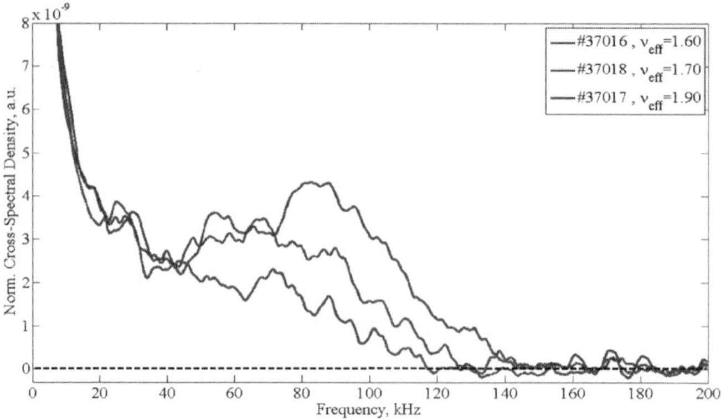

Fig. 1. Broadband fluctuations are visible between 40 kHz and 120 kHz. Their amplitude decreases as the collisionality is increased.

density of the signal fluctuations, normalized to their mean values, as a function of frequency. When the density is increased to increase the collisionality, these fluctuations decrease in amplitude. When ECH is added to similar discharges, the fluctuation level increases significantly and the frequency spectrum broadens, as seen in Fig. 2.

In the plasma frame the wave frequency and the wave-number define the phase velocity of the modes: $v_{phase} = \omega/k$. The plasma $E \times B$ rotation velocity, determined from charge-exchange-spectroscopy measurements of the Carbon impurity in TCV, produces a Doppler-shift in the mode frequency measured by the CECE. DIII-D [5] has reported a broadening of the frequency spectrum due to an increase in the plasma rotation during neutral beam injection. They confirm that the estimated poloidal **k**-vector component being measured is within the resolution limits of the diagnostic set by the beam spot size, assuming a negligible phase velocity of the mode itself. Similarly, on TCV if the measured peak frequency in the spectrum results from plasma rotation alone (i.e. due to a Doppler-shift of $k_{mode}v_{\mathbf{E}\times\mathbf{B}}$) then $k_{mode} \sim 2.1cm^{-1} \pm 50\%$ for shot 37016 of Fig. 1. The temperature fluctuations that we measure are therefore likely to be strongly attenuated. When the collisionality is increased, the frequency of the peak decreases and the deduced k_{mode} decreases, so at fixed power-spot radius, the diagnostic sensitivity would increase. Nevertheless, the radiation temperature fluctuation amplitude decreases, as shown in Fig. 3.

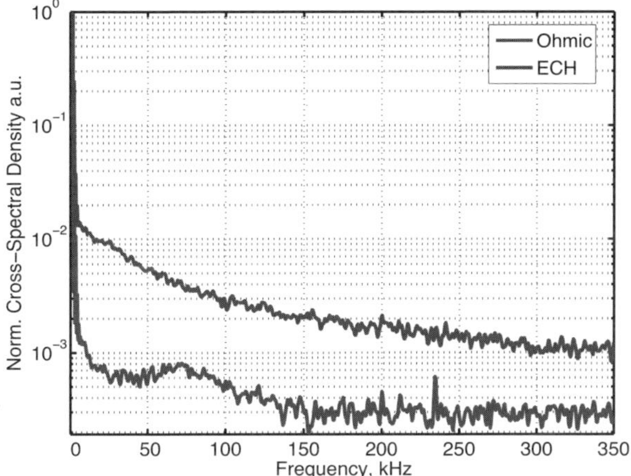

Fig. 2. The broadband fluctuations increase when 500kW of ECH is added to the discharge.

4. Gyrokinetic simulations

The local gyrokinetic code GS2 [6] is used, along with the quasi-linear model of Ref. [7], to investigate the theoretical trend in fluctuations with increasing collisionality. The calculations are carried out at a radial location of $\rho_{vol} = 0.35$ instead of the experimental $\rho_{vol} = 0.55$ because a larger database of simulated shots already exists at this radius. Other relevant plasma parameters are taken directly from the experimental database (except that only deuterons and electrons are considered, i.e. no impurities). In the simulations R/L_n is set to a fixed value of 5, where R is the major radius and L_n is the density gradient scale length (logarithmic derivative of the density profile at $\rho_{vol} = 0.35$). In the experiments this value is 4.5–6 $\pm 20\%$ at $\rho_{vol} = 0.55$. Figure 4 shows the relative density and temperature fluctuation levels (squared) for a mode which is representative of the diagnostic resolution, $k_{pol} = 1.1 cm^{-1}$. The average real frequency of this mode remains roughly constant as collisionality is increased, with a sign consistent with TEM turbulence. The mixing–length heat diffusivity calculated from GS2 decreases with increasing collisionality, as does the measured χ_e from a power balance analysis. Together, the trend of these measurements and simulations is consistent with a decrease, by collisional detrapping, of TEM driven losses as collisionality increases.

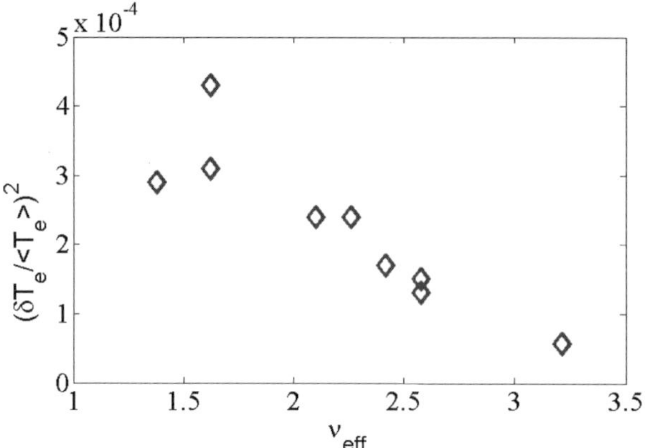

Fig. 3. The measured fluctuation amplitude decreass with increasing collisionality.

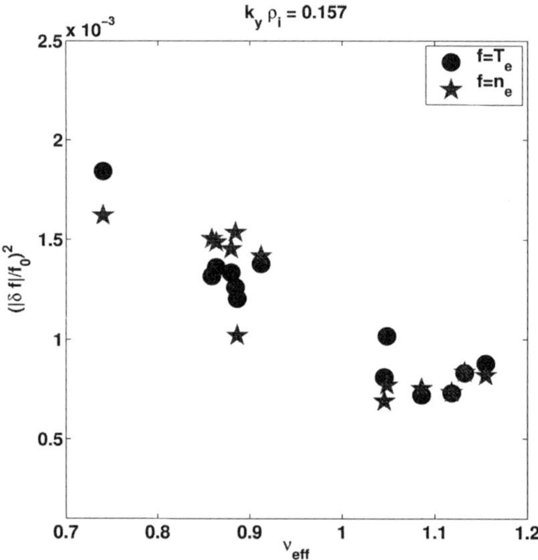

Fig. 4. The GS2 simulated fluctuation amplitudes decrease with increasing collisionality.

5. Current channel measurements

The location and sign of the driven current channel from ECCD is put into evidence in a pair of identical discharges where only the viewing direction of the launcher 7 is changed; once looking with the plasma current (co-viewing) and once against (counter-viewing). The plasma is heated by 2 clusters of 3 gyrotrons each. Two of the gyrotrons of each cluster heat off-axis. One gyrotron of one cluster drives counter-ECCD on axis and one gyrotron of the other cluster drives co-ECCD in the same location. The power of one cluster is stepped up at the same time as the power from the other cluster is stepped down. This produces a constant input power with a net, centrally driven current which increases from counter- to co-ECCD in 6 steps. The local modification of the EDF is evidenced by taking the ratio of the co-viewing to counter-viewing radiometer signals, at each frequency, averaged over each of the 6 steps. These ratios from one such pair of shots are shown in Fig. 5. A ratio of 1 indicates an EDF which is symmetric around $v_{\|} = 0$. The current is seen to step from the counter to the co direction.

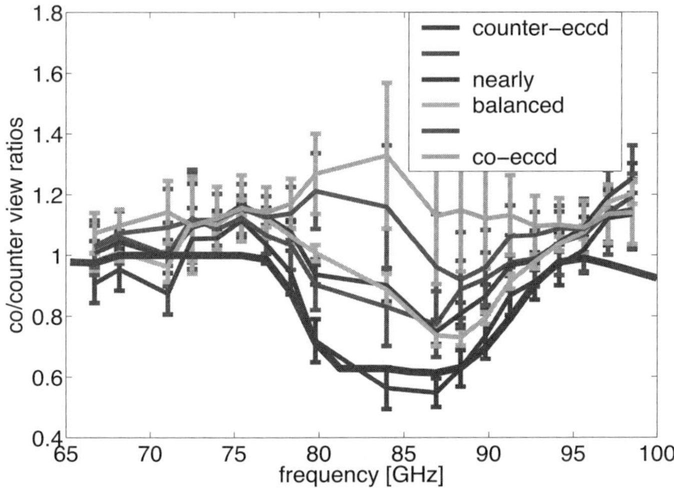

Fig. 5. A central current channel is evident. The solid blue line without errorbars shows the NOTECTCV simluated ratio with current limited within $0 < \rho < 0.2$. (see text)

6. Emission simulations

Note that the radiometer has not been calibrated for this analysis. Only shot-to-shot reproducibility of the plasmas and proper selection of the receiver polarization is required [4]. To provide quantitative information, the ObECE ratios are simulated using the radiation transport code NO-

TECTCV. The plasma emission is simulated from a sum of up to 20 bi-Maxwellian electron populations drifting, independently, parallel to the magnetic field. Each population can have separate parallel and perpendicular temperatures and can be restricted spatially to a given region in ρ. The density of each population is chosen to be proportional to the measured (Thomson Scattering) bulk density profile in that region. Starting with the bulk (essentially non-drifting) population, we keep to a minimum the number of populations needed to reproduce the measured HFS (perpendicular) and LFS (oblique) ECE signals while constraining the drift velocities, densities and regions so as to produce the expected (calculated by the TORAY ray-tracing code) total driven current [8]. The solution is not unique and at present a manual procedure is followed. An example of the match that has been achieved is shown in Fig. 6 for the signals during the counter-ECCD phase of Fig. 5. One bulk, one *hot*, and one drifting population are required.

To address the question of current diffusion, the above populations were taken as a starting point and the region in which the drifting central current was found was increased while keeping the integrated driven current value constant. Figure 5 shows (thick solid line close to the counter-ECCD measurements) the expected ratio of ObECE signals when the drifting population is extended beyond the maximum $\rho < 0.1$ used in Fig. 6 to $\rho < 0.2$.

While these results are encouraging, a more self-consistant, less time-consuming approach is required. Fokker-Planck based ObECE simulations would be best suited to the problem since the wave absorption, particle diffusion and emission can be treated within the same framework.

7. Conclusions

Correlation ECE measurements of temperature fluctuations that decrease with collisionality are shown to be consistent with GS2 simulations of fluctuating temperatures and densites. These are accompanied by heat diffusivity measurements which follow the same trend as values determined from a quasi-linear mixing–length model, also based on linear GS2 calculations, which suggests that the anomalous heat loss in these TCV plasmas is the result of TEM generated turbulence: collisional detrapping of particles being the root of the trend with collisionality for this mode.

Oblique ECE measurements of the co- and counter-ECCD current channel are consistent with NOTECTCV simulations of an asymmetric EDF from one drifting bi-Maxwellian population. Current diffusion is required to match the measure co– to counter– viewing ratios of the experiments. A self-consistent simulation approach based on the EDF from the CQL3D

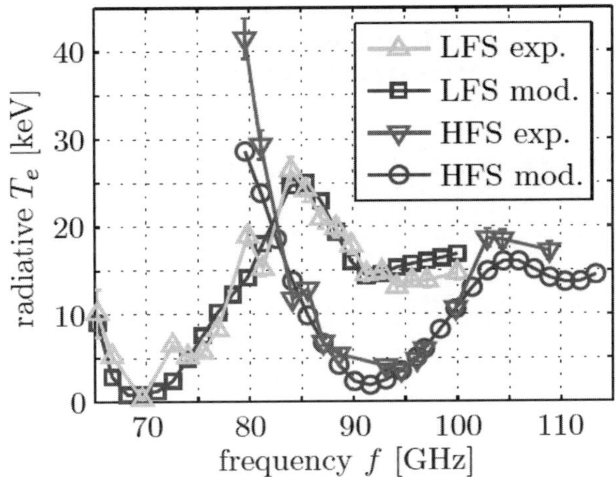

Fig. 6. Simulated and measured HFS and LFS signals during the central counter-ECCD phase of Fig. 5. with the current channel limited to $0 < \rho < 0.1$. Two populations are required. [taken from Ref. 8]

Fokker-Planck quasi-linear interaction code [9] is being pursued.

Acknowledgments

The Authors are grateful to W. Dorland and M. Kotschenreuther for having provided the source of the GS2 code. The work was supported in part by the Swiss National Science Foundation.

References

1. X. Garbet, *et al.*, Nucl. Fusion **50** 043002 (2010)
2. C. Watts, *et al.*, *Proc. of 10th Joint Workshop on Electron Cyclotron Emission and Electron Cyclotron Resonance Heating (Ameland, The Netherlands)*, (World Scientific), 147 (1997)
3. V.S. Udintsev, *et al.*, Fusion Sci. and Technol. **52**, 161–168 (2007)
4. T.P. Goodman, *et al.*, Fusion Sci. and Technol. **53**, 196–207 (2008)
5. A. White, *et al.*, Phys. Plasmas, **15**, 056116 (2008)
6. M. Kotschenreuther, *et al.*, Comput. Phys. Commun. **88**, 128 (1995)
7. E. Fable, *et al.*, Plasma Phys. Control. Fusion **52**, 015007 (2010)
8. Ch. Schlatter, "Turbulent Ion Heating in TCV Tokamak Plasmas" EPFL PhD Thesis No. 4479 (2009)
9. R.W. Harvey and M.G. McCoy, *Proc. IAEA Technical Meeting on Advances in Simulations and Modeling in Thermonuclear Plasmas (Montreal)*, (Vienna: IAEA), 498 (1992)

MULTI-ANGLE MEASUREMENT OF EC EMISSION BY FAST ELECTRONS: SENSITIVITY STUDY

L.FIGINI[1], P.PLATANIA[1], D. FARINA[1], S.GARAVAGLIA[1], G.GROSSETTI[1], S.NOWAK[1], C.SOZZI[1], AND JET-EFDA CONTRIBUTORS*

JET-EFDA, Culham Science Centre, Abingdon OX14 3DB, UK

[1]*IFP-CNR, EURATOM-ENEA-CNR Association, Milan, Italy*

Multiple angle (0, 10, 20 degrees with respect to the radial direction) and polarization (X, O modes) ECE spectra of JET plasmas with significant Lower Hybrid additional power were obtained with the Oblique ECE diagnostics. Such data have been analyzed with the emission code SPECE that encompasses a multi-Maxwellian model of the fast electron tail driven by Lower Hybrid waves. The model has been used to fit, varying the control parameters, the five experimental signals of the ECE diagnostic aiming to characterize the LH power absorption and driven current.

1. Motivation and analysis tools

Electron Cyclotron Emission (ECE) is strongly dependent on the electron distribution function, and even a tiny fraction of the electron population having sufficient energy can introduce a significant deformation in the thermal ECE spectrum when the cold resonance lies outside the plasma volume. Downshifted second harmonic extraordinary (X) mode emission is a distinctive feature of ECE spectra in presence of energetic electrons, and the details of this feature are mainly related to the spatial location, the parallel momentum and the perpendicular temperature of the energetic electrons. The analysis of the suprathermal emission can thus reveal details of the mechanism sustaining the involved non-thermal electron fraction.

The Oblique ECE diagnostics of JET provides five simultaneous spectra (nominally 0° X mode, 10° and 20° X and O modes, where the angle refers to the radial direction) over an extended frequency range (70-350 GHz), with spectral resolution up to 7 GHz and time resolution of 5 ms [1]. These spectra probe the electron distribution function at different electron energies thanks to

* See the Appendix of F. Romanelli et al., Proceedings of the 22nd IAEA Fusion Energy Conference 2008, Geneva, Switzerland

the energy dependence of the resonance condition at different emission angles [2]. The data analyzed in this paper have been relatively calibrated following a careful procedure based on both absolutely calibrated data and on benchmarked simulations [3].

A simple multi-Maxwellian model [4] is included in the emission code SPECE [5] to describe the typical Lower Hybrid (LH) driven plateau in the electron distribution function (e.d.f.):

$$f(\psi,\vec{u}) = (1-\eta(\psi))\cdot f_{M\,T_b}(|\vec{u}|) + \eta(\psi)\sum_{i=1}^{N} f_{M\,T_b}(|\vec{u}-\vec{u}_{//0,i}|), \tag{1}$$

where $u=p/mc$ is the normalized momentum, ψ is the normalized poloidal flux coordinate, and T_b and T_{tail} are the temperatures of the Maxwellian distributions f_M in the bulk and in the tail respectively. The value of T_{tail} determines both the spread of the tail in $u_\perp$ and also the spacing in $u_{//}$ among the Maxwellians:

$$u_{//0,1} = \sqrt{T_b/mc^2}, \qquad \Delta u_{//} = u_{//0,i} - u_{//0,i-1} = 2\sqrt{T_{tail}/mc^2}\,. \tag{2}$$

The number of Maxwellians N can be chosen to control the extent of the suprathermal tail along the $u_{//}$ axis. A plateau extending up to the expected value $u_{//,\max}\sim(N_{//,\min}{}^2-1)^{-1/2}$, where $N_{//,\min}$ is the minimum value of the parallel refraction index component in the launched wave spectrum, can be approximated by choosing N such that

$$N = 1 + \mathrm{int}\left[\sqrt{mc^2/T_{tail}}\,(u_{//,\max} - u_{//0,1})/2\right]. \tag{3}$$

Finally, the density fraction of superthermal electrons η is assumed to be peaked at flux coordinate ψ_0 with peak value η_0 and gaussian decrease with width ψ_c:

$$\eta(\psi) = \eta_0 \cdot \exp\left[-(\psi-\psi_0)^2/\psi_c^2\right]. \tag{4}$$

When the cold resonance for harmonic $n = 2$ is inside the plasma volume the plasma itself is optically thick due to interaction of the radiation with the bulk of low energy electrons, and the radiation temperature T_r corresponds to the bulk electron temperature T_b. The downshifted emission instead is visible when the cold resonance $n = 2$ is outside plasma volume: the number of interacting electrons is low and plasma is optically thin, so that the radiation temperature is $T_r \sim n_{tail}T_{tail}{}^2$. The spectral shape of the downshifted emission is dependent on the distribution in space and energy of the fast electrons, i.e., in the adopted model, it depends on $u_{//,\max}$, T_{tail}, ψ_0, and ψ_c.

2. Experimental data

In order to characterize the dependence and sensitivity of the simulated spectra from the five parameters η_0, ψ_0, ψ_c, $u_{//,\max}$, T_{tail} a database of JET pulses having a rather wide range of macroscopic parameters has been considered (see Table 1), in both transient and steady phases.

Table 1. Range of main parameters for the analyzed shots

Pulse parameter	Min	Max
Vacuum field B_0 (T)	2.3	3.4
Plasma total current I_p (MA)	1.75	2.5
Integrated density n_e ($10^{19}/m^2$)	4.3	11
Electron bulk temperature $T_{e,b}$ (keV)	3	8
Superthermal rad. Temp. $T_{rad,st}$ (keV)	1	8
Additional power P_{add} (MW)	2	25
Lower Hybrid power P_{LH} (MW)	1.9	5
$N_{//}$ at LH's antenna	1.8	2.3

Figure 1 shows the experimental spectra for shot 77874 during the switching-on of the additional power (at the time of the analysis: n_e=7, P_{NBI}=7, P_{LH}=2, I_p=1.8, B_0= 2.66, $N_{//}$=1.8 in the units of Table 1). The experimental data are compared with the SPECE simulations obtained with the tail parameters reported in the caption, and the downshifted emission peak due to the tail of fast electron is visible around 110 GHz.

3. Scan of the tail's parameters

Figure 2 shows the variation of suprathermal peak shape for the perpendicular emission while scanning the tail parameters, compared with the experimental data of JET pulse 74087 at t=60.1s. When varying the spatial localization of the suprathermal population, the LH driven current has been kept unchanged at I_{LH} /I_p=430kA/2MA. In Fig. 2a and 2b the location and the width

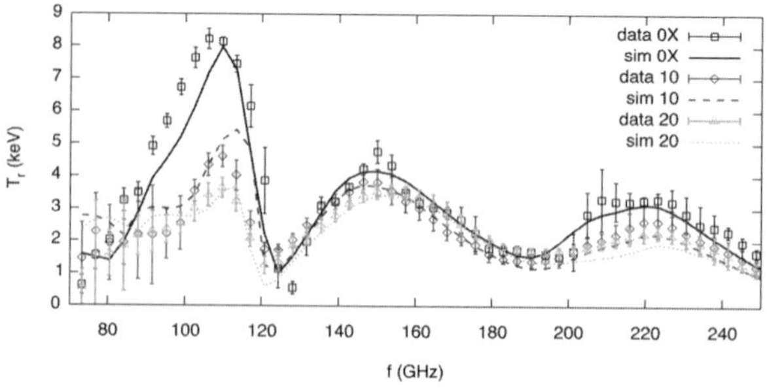

Figure 1: Experimental (open symbols) and simulated (lines) emission spectra at 0° (X mode), 10°, and 20° (O+X mode), for pulse no. 77874 at t = 4.11 s. Spectra are simulated with the following e.d.f. parameters: T_{tail} = 45 keV, $u_{//,max}$ = -0.8, η_0 = 1.4 10^{-3}, ψ_0 = 0.3, ψ_c = 0.12 (I_{LH} = 380 kA). Error bars include calibration uncertainty and signal variance over 30 ms time average.

in space of the suprathermal population are varied respectively. Larger ψ_0 means fast electrons localization at more external radii, therefore the suprathermal emission has a lower peak (lower density) at lower frequency. Fig. 2c clearly shows that, for a given current I_{LH} localized around a given flux surface ψ_0, the ECE spectrum is almost insensitive to the width of the current channel (ψ_c). In Fig. 2c the tail temperature T_{tail} has been varied, keeping $\eta_0 T_{tail}$ = constant. The downshift effect increases with the T_{tail} value, which also affects the value of $u_{//,max}$ through equations (2) and (3). Therefore low T_{tail} also means low $u_{//,max}$. The N parameter (kept constant at N=2 in the previous cases) is varied in Fig. 2d. In this case the scan is step-like, and the plateau extends approximately up to an energy

$$E_{max} = mc^2(\gamma_{max} - 1) = mc^2\left(\sqrt{1 + u_{//,max}^2} - 1\right) \tag{5}$$

that, for T_{tail} = 23 keV and N = 1, 2, 3 ($|u_{//,max}|$ = 0.2, 0.6, 1.0), is about E_{max} = 10, 85, and 210 keV respectively. This latter scan performed for the oblique views is shown in Figure 3. Both in perpendicular and oblique spectra, a large variation in the spectrum shape is associated to the lower N steps (low $u_{//,max}$ values), because at large N values the e.d.f. is modified in a region of the $u_{//}$, $u_{\perp}$ space not

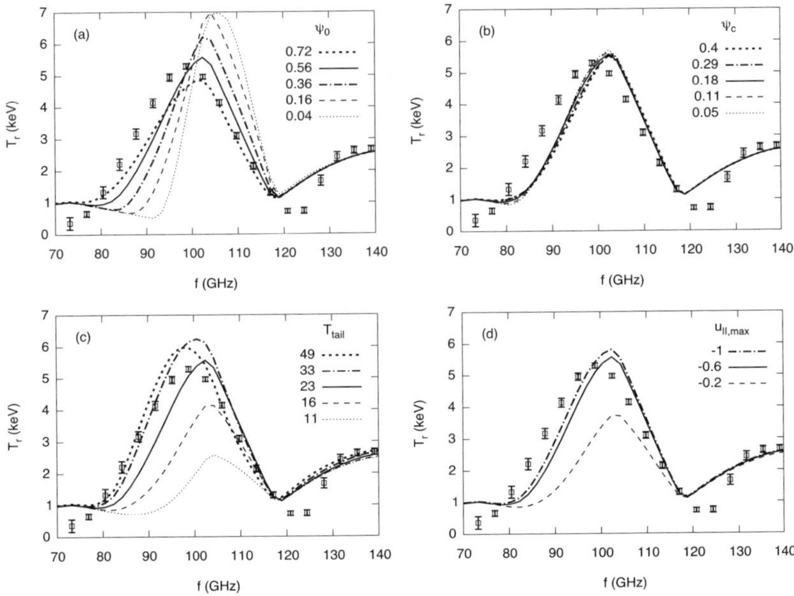

Figure 2: e.d.f. parameters scan around the estimated values for pulse no. 74087 at t = 20.1 s (T_{tail} = 23 keV, $u_{//,max}$ = -0.6, $\eta_0 = 8.5\ 10^{-4}$, ψ_0 = 0.56, ψ_c = 0.18, I_{LH} = 430 kA). Open symbols represent measured data at 0°, lines are simulated spectra. (a) ψ_0 scan, (b) ψ_c scan, (c) T_{tail} scan, (d) $u_{//,max}$ scan. When scanning a parameter, the other values are kept unchanged except: (a) ψ_c adjusted to keep the current channel width constant, η_0 rescaled to keep I_{LH} constant; (b) η_0 rescaled to keep I_{LH} constant; (c) η_0 rescaled to keep $\eta_0 T_{tail}$ constant; (d) η_0 rescaled to keep η_0/N constant.

involved in the resonance process. The energy range corresponding to the three lines of sight is shown in Figure 4: in the case shown, the oblique views are sensitive to lower electron energies with respect to the perpendicular view, and as a consequence the oblique spectra are less affected by changes in the energetic part of the tail.

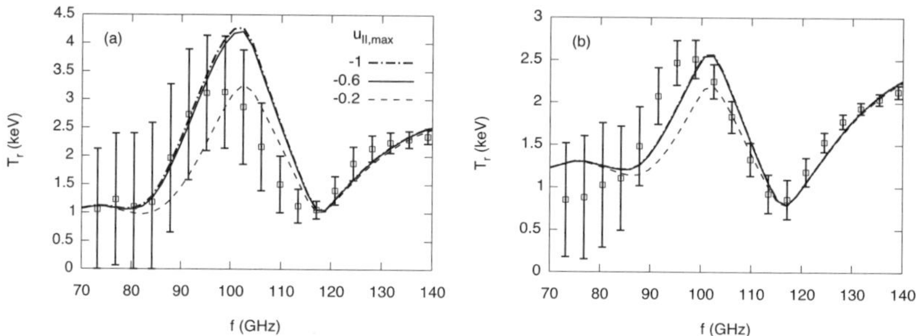

Figure 3: Same as Figure 2d for (a) 10° O+X spectrum, (b) 20° O+X spectrum.

4. Minimal detectable suprathermal fraction

When the cold EC resonance lies in the optically thin plasma layer at the edge and a strong temperature gradient exists towards the hotter inner plasma, downshifted emission might be visible even in a purely Maxwellian plasma. This radiation is emitted by the energetic part of the Maxwellian population, and it might hide or add to the feature generated by a tail of fast electrons. In order to evaluate the minimal fraction of non-Maxwellian electrons that the diagnostic can detect, a case at high temperature where marginal LH wave coupling is expected (JET pulse 77874) was analyzed. Assuming a Maxwellian edf, the simulated spectrum shown good agreement with the experimental data, well reproducing the downshifted feature too, as seen in Figure 5. We then

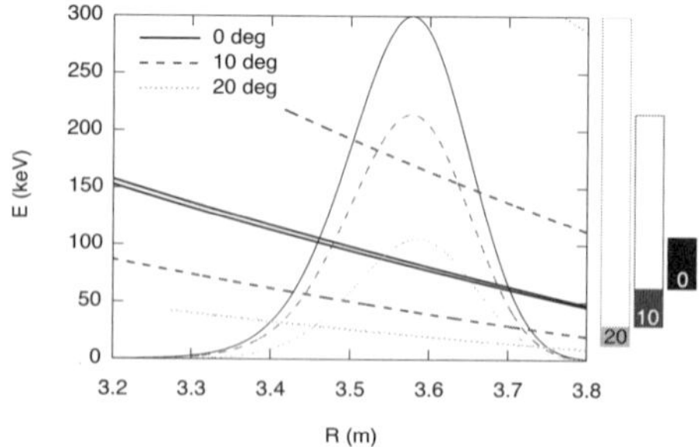

Figure 4: Emission profile $\alpha T_{rad}\exp(-\tau)$ (thin bell-shaped curves) for X mode at 0°, 10°, 20°, f = 102 GHz, and respective resonance energy ranges (thick lines). Lower energies (E < 50 keV) can be "seen" by oblique lines of sight only. Pulse 74087 at t = 20.1 s, same parameters as Figure 2.

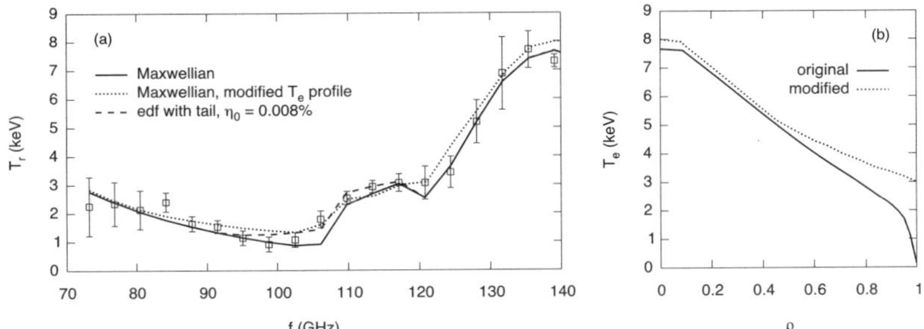

Figure 5: Downshifted emission feature in the perpendicular X mode spectrum (a) for pulse no. 77874 at t = 6.0 s. Measured data (symbols) are compared with spectra simulated either using Maxwellian e.d.f. and different T_e profiles (solid, dotted, profiles shown in (b)), or assuming an e.d.f. with a tail of fast electrons (dashed). e.d.f. parameters are indicated in Figure 1, except η_0 that has been scaled down ($\eta_0 = 8\ 10^{-5}$) to match the Maxwellian simulation within the error bars.

introduced a fictitious suprathermal fraction, large enough to be visible beyond the experimental error bars. For a temperature $T_{\rm tail} \sim 40$ keV, the corresponding density of suprathermal electrons, carrying a current $I_{\rm LH} \sim 50$ kA, was found to be $n_{\rm st} \sim 4\ 10^{15}\ {\rm m}^{-3}$ ($\eta_0 \sim 0.01\%$).

5. Conclusions

Experimental data taken with ObECE diagnostics of JET have been analyzed using the emission code SPECE, in which a model for the LHCD driven emission is included. The parameter space of the model has been explored, showing that it is possible to derive information about the LHCD power localization and its current drive efficiency from the comparison between data and simulations.

References

1. C.Sozzi et al., AIP Conf. Proc., vol. **988**, (2008) 73.
2. E. de la Luna et al., Rev. Sci. Instr. **74**, (2003) 1414 and V. Krivenski, Fus. Eng. Des. **53**, (2001) 23.
3. L.Figini et al, submitted to Rev. Sci. Instr.
4. Brusati et al., Nuclear Fusion **34**, (1994) 23.
5. D.Farina et al. Conf. Proc., vol. **988**, (2008) 128.

Acknowledgments

This work, carried out under the European Fusion Development Agreement, supported by the European Communities and "Istituto di Fisica del Plasma P.Caldirola – CNR", has been carried out within the Contract of Association between EURATOM and IFP. The views and opinions expressed herein do not necessarily reflect those of the European Commission or IFP.

MICROWAVE IMAGING IN LARGE HELICAL DEVICE

T. YOSHINAGA*, Y. NAGAYAMA, H. TSUCHIYA
National Institute for Fusion Science, 322-6 Oroshi, Toki 509-5292, Japan
**E-mail: yoshinaga.tomokazu@lhd.nifs.ac.jp*

D. KUWAHARA, S. TSUJI-IIO
Tokyo Institute of Technology, 2-12-1 Ookayama, Meguro 152-8550, Japan

K. AKAKI, A. MASE
Kyushu University, 6-1 Kasuga-Koen, Kasuga 816-8580, Japan

Y. KOGI
Fukuoka Institute of Technology, 3-30-1 Wajiro-higashi, Fukuoka 811-0295, Japan

S. YAMAGUCHI
Kansai University, 3-3-35 Yamate, Suita 564-8680, Japan

Z. B. SHI
Southwestern Institute of Physics, Box 432 Chengdu, Sichuan 610041, China

H. HOJO
University of Tsukuba, 1-1-1 Tennodai, Tsukuba 305-8577, Japan

Microwave imaging reflectometry (MIR) system and electron cyclotron emission imaging (ECEI) system are under development for the simultaneous reconstruction of the electron density and temperature fluctuation structures in the Large Helical Device (LHD). The MIR observes three-dimensional structure of disturbed cutoff surfaces by using the two-dimensionally distributed horn-antenna mixer array (HMA) of 5×7 channels in combination with the simultaneous projection of microwaves with four different frequency components (60.410, 61.808, 63.008 and 64.610 GHz). The ECEI is designed to observe two-dimensional structure of electron temperature by detecting second-harmonic ECE at 97–107 GHz with the one-dimensional HMA (7 channels) in the common optics with MIR system. Both the MIR and the ECEI are realized by the HMA and the band-pass filter (BPF) arrays, which are fabricated by microstrip-line technique at low-cost.

1. Introduction

Understanding transports in magnetically confined plasmas is one of the important issues to realize fusion reactor. Especially, the anomalous transport, which is considered to be governed by turbulences, has been investigated in many devices. In LHD the MIR system and the ECEI system are under development for the simultaneous observation of two- or three-dimensional (2-D/3-D) structures of the electron density and the electron temperature fluctuations [1,2].

MIR is a multi-channel reflectometry diagnostics [3]. The small antennas are arrayed to form the 2-D receiver. The disturbed cutoff surfaces scatter microwaves which illuminates them in complicated interference patterns. The optics in MIR system focuses the scattered microwave onto the image focal plane where the receiver array is placed. By detecting phases and amplitudes of those scattered waves which are reflected back to each receiver antenna channel, the spatial structure of the fluctuating cutoff surfaces would be reconstructed. It is expected that the cutoff surface positions can be observed in a sensitivity of a fraction of the incident wavelengths along the projection axis of the probe waves. ECEI visualizes the spatial profile of electron temperature by receiving ECE signals with arrayed receiver antennas [2–4]. Since it has high spatial and time resolution, the ECEI systems have been installed in various devices.

Both the MIR and the ECEI system need a number of high-cost microwave (or millimeter-wave) components. Especially, the receiver system, which consists of arrayed receiver antenna, band-pass filters (BPFs) and power or phase detectors, needs many channels. Therefore, the fabricating cost becomes high easily. We have developed the cost-effective micro-stripline circuits for those receiver components [2]. This report describes the characteristics of the important devices in MIR and ECEI as well as the design of common optics.

2. MIR System in LHD

Figure 1 shows the schematics of the MIR and ECEI system in LHD. The optics system consists of the curved mirrors of aluminum (Al) alloy and the dielectric (acrylic) plates as the beam splitters. As the illuminating probe waves, 4 different frequencies at 60.410, 61.808, 63.008 and 64.610 GHz are projected simultaneously onto the plasma in X-mode. This enables simultaneous observation of electron density fluctuations on the 4 different cutoff surfaces at 4 different electron densities, namely, at 4 different minor radii.

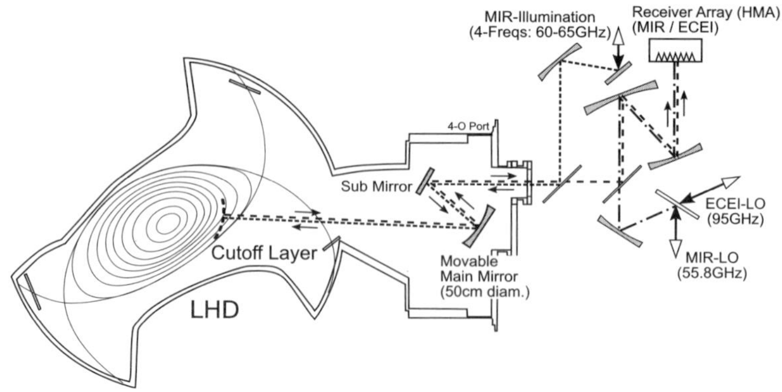

Fig. 1. Schematic view of MIR / ECEI system in LHD.

These probe waves are introduced to the plasma via illuminating optics system so that the waves have the plane equiphase surfaces with the beam widths as wide as the field of view of the receiver array. It is important that the scattered waves from the cutoff surfaces return back toward the receiver array through the same path with the probe waves. Therefore the projection angle of the probe waves is adjusted by changing the angle of the movable main mirror depending on the target LHD plasma with a twisted shape. The acceptable range of the mirror angle was found to be within 1.5 degrees around the optimum angle to obtain meaningful signals with allowable S/N ratio [5].

The scattered waves from the cutoff surfaces are focused onto the image focal plane via the imaging optics. The focused images are received with the 2-D horn-antenna mixer array (HMA) [6,7]. It is designed to receive 2-D image with the arrayed small horn antennas which are aligned in toroidally 5 and poloidally 7 channels. The phases and amplitudes of the scattered waves are detected by using the double superheterodyne scheme. The scattered microwaves at $\sim$ 60–65 GHz are mixed with the first-local oscillator (LO) signal at 55.800 GHz and down-converted to the first-intermediate frequency (IF) signals at 4.610, 6.008, 7.208 and 8.810 GHz inside each small horn apertures of HMA to make the following signal handling convenient. Thus the first-IF signals from each horn channel can be transmitted to the following BPF stages by using the low-cost coaxial cables. The microwave from the first-LO must be projected optically onto the receiver antenna array together with the scattered wave from the plasma. The first-LO beam is required to cover the whole aperture of the HMA with a plane equiphase surface. The whole optics is optimized to satisfy the above requirements by using the finite-difference time-domain (FDTD) method [8,9].

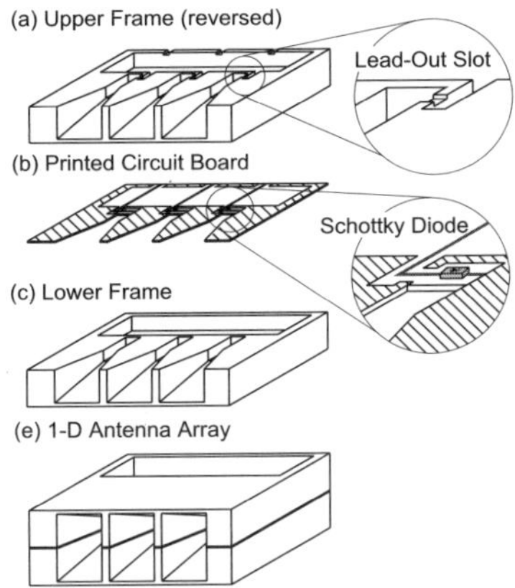

Fig. 2. Illustration of the 1-D HMA assembly. (a) Underside view of upper frame, (b) printed-circuit-board around the horn structure, (c) lower frame and (d) completed form [6].

The 2-D HMA (5×7 channels) is formed by stacking the five 7-channel 1-D HMA, which is made by sandwiching the printed circuit board (PCB) with the upper and lower Al alloy frames (see Fig. 2) [6]. The antenna frames have the half apertures of the pyramidal horn antennas which are connected to the waveguide sections for the down-conversion of received microwave into the first-IF signals which are transmitted by the micro-strip-line on PCB. The Schottky diodes are mounted on PCB inside the waveguide sections for down-conversion. The aperture size of each horn is 13 mm$\times$13 mm. The size of the waveguide section is 1.9 mm$\times$3.8 mm, which is the size of commercially available V-band (WR15) waveguides whose frequency range is 50-75 GHz. Seven small horns' apertures are aligned every 14 mm and thus the 1-D HMA is formed. The obtained first-IF signals are amplified with the 3-stage gallium arsenide (GaAs) amplifiers mounted on the PCB and transmitted to the BPF stages.

The first-IF signals at 4 frequencies are separated to each frequency component signals with the micro-strip-line BPF arrays and mixed with the second-LO signals with the corresponding frequencies at 4.500, 5.898, 7.098 and 8.700 GHz. Finally all the reflected microwaves received from 35 antenna channels with 4 different frequencies at $\sim$ 60–65 GHz can be down-converted to the second-IF signals at 110 MHz. The amplitudes and the phases of these second-IF signals are detected with the amplitude-detection modules and the quadrature-demodulation modules which are used commonly in telecommunication devices.

3. ECEI System

The ECEI system is designed to work simultaneously with the MIR system by using the common optics system. The 1-D HMA for ECEI, which is placed beside the 2-D MIR HMA, is intended to receive the second harmonic ECE from 97 GHz to 107 GHz [7]. The LO wave at 95 GHz is introduced to the same optic axis with the LO wave of MIR system via the acrylic beam combiner plate, whose thickness is optimized for ECEI to give nearly 100 % transmission. The identical 1-D HMA to the MIR system is installed except that the high-pass-filter (HPF) plate is placed at the apertures of the receiver antennas. The HPF plate is 6 mm thick, and has many holes with the diameter at 1.9 mm which are drilled in a hexagonal lattice structure with the neighboring interval at 2 mm. This filter rejects microwaves at lower than 93 GHz by $\sim$ 20 dB to avoid detecting the lower frequency components as well as the strong MIR signals. The IF signals from 2 to 12 GHz are obtained by the down-conversions inside each horn channel as the MIR case. Those IF signals are resolved into 11 frequency components with 1 GHz increments through the BPF arrays, which are fabricated by using the micro-strip-line filter technique [10]. The amplitudes of resolved frequency components are detected and buffered after subtraction of the large DC components which originates from noises.

4. Observed Area

Figure. 3 shows one of the typical configurations of the characteristic frequencies in LHD. In particular plasma configuration as shown in Fig. 3, the electron temperature and the electron density in the same region can be observed simultaneously by using MIR and ECEI.

5. Summary

MIR and ECEI system has been developed for 2-D or 3-D observation of the electron density and temperature fluctuations in LHD. 2-D HMA (5×7 channels) and 1-D array (7 channels) were developed for MIR and for ECEI, respectively. MIR system works at $\sim$ 60–65 GHz, which are down-converted to the first-IF signals at 4.610, 6.008, 7.208 and 8.810 GHz

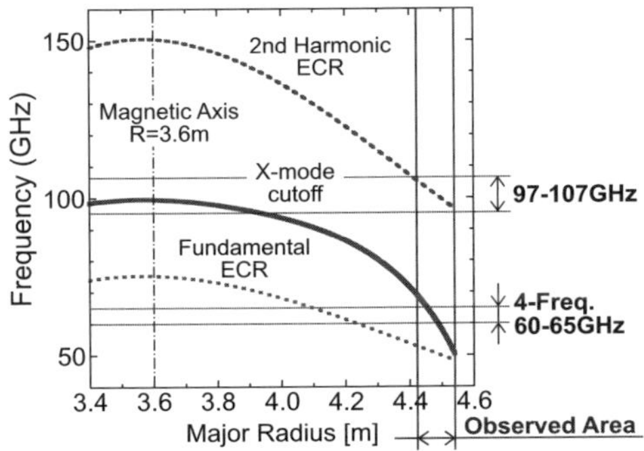

Fig. 3. A typical configuration of LHD. The field strength and the electron density on the magnetic axis located at $R_{ax} = 3.6$ m are 2.75 T and 3×10^{19} cm^{-3}, respectively.

with the HMA. Phases and amplitudes of the received microwave are detected after down-conversion of those first-IF signals into second-IF signals at 110 MHz. ECEI system receives second-harmonic ECE waves from 97 GHz to 107 GHz. The HMA for ECEI is placed beside the HMA for MIR and the common optics with MIR is used. The IF signals from the receiver array at 2–12 GHz are resolved into 11 frequency components by the BPF array. The HMA and BPF arrays can be made by using micro-strip-line technique at low-cost. The common optics system was optimized by using the FDTD calculation.

Acknowledgement

This work was supported by National Institute of Natural Sciences Imaging Science Project (NIFS09KEIN0021), National Institute for Fusion Science (NIFS09ULPP525) and Grant-in-Aid for Scientific Research (21246140).

References

1. S. Yamaguchi *et al.*, *Rev. Sci. Instrum.* **77**, p. 10E930 (2006).
2. Y. Kogi *et al.*, *Plasma Fusion Res.* **2**, p. S1032 (2007).
3. H. Park *et al.*, *Rev. Sci. Instrum.* **74**, p. 4239 (2003).
4. A. mase *et al.*, *Rev. Sci. Instrum.* **74**, p. 1445 (2003).
5. S. Yamaguchi *et al.*, *Plasma Fusion Res.* **2**, p. S1038 (2007).
6. D. Kuwahara *et al.*, *J. Plasma Fusion Res. SERIES* **8**, p. 649 (2009).
7. D. Kuwahara *et al.*, submitted to J. Plasma Fusion Res.
8. M. Ignatenko *et al.*, *Nucl. Fusion* **46**, p. S760 (2006).
9. H. Hojo *et al.*, *Rev. Sci. Instrum.* **70**, p. 983 (1999).
10. Y. Kogi *et al.*, *Rev. Sci. Instrum.* **79**, p. 10F115 (2008).

PROGRESS IN INTERGRATION OF ITER MICROWAVE DIAGNOSTICS

V.S. UDINTSEV, G. VAYAKIS, B. CANTONE, A. ENCHEVA, C. WALKER, M. BENCHIKHOUNE, A.E. COSTLEY, A. DAMMANN, M.A. HENDERSON, I. KUEHN, C.H. LEE, B. LEVESY, A. MARTIN, K.M. PATEL, C.S. PITCHER, A. TESINI, Y. UTIN, M.J. WALSH

ITER Organization, Route de Vinon CS 90 046, 13067 St. Paul-lez-Durance Cedex, France

S. DANANI, H. PANDYA, P. VASU

Institute for Plasma Research, Bhat, Gandhinagar-382 428 India

M. AUSTIN, W. ROWAN

Fusion Research Center, the University of Texas at Austin, Austin, TX, USA

R. FEDER, D. JOHNSON

PPPL, PO Box 451, Princeton University, NJ, USA

D. SHELUKHIN, V. VERSHKOV

RRC Kurchatov Institute, Moscow, Russian Federation

G. COUNSELL, CH. INGESSON, S. ARSHAD

Fusion for Energy, Josep Pla 2, 08019 Barcelona, Spain

This paper reports on the current status of integration of ITER microwave diagnostics, such as ECE, reflectometry systems and Collective Thomson scattering, and gives an outlook on the upcoming technical and design activity. Some open issues are addressed and discussed.

1. Introduction

Integration of ITER microwave diagnostics, such as ECE and reflectometry, within the tokamak requires proper definition of interfaces with many different components located both in-vessel and ex-vessel. Integration of ECE in the port plug containing other diagnostic systems and providing appropriate views for all is a major challenge. This paper explains the present status of integration of

ITER microwave diagnostics and gives an outlook on the upcoming technical and design activity.

The open issues of calibration and stability of ECE systems are discussed in the paper. Proposals for the calibration [1-3], design of the front end and the transmission line are reviewed. The NTM detection and stabilization by ECE and associated challenges are addressed.

Reflectometry on ITER [4] is being designed to measure the electron density profiles, fluctuations, as well as to identify the plasma position by launching the wave below the cut-off frequency into the plasma. The ongoing activity in integration of the reflectometry systems into tokamak complex, issues of alignment and characterisation of the transmission lines after installation are presented and discussed.

For the Collective Thomson scattering diagnostic, a similar approach as for above-mentioned systems is used for the transmission line for the receiver, and a common confinement barrier strategy for all microwave diagnostics in ITER is foreseen. The launching branch of this system, however, can benefit extensively from the work done for the ITER ECH systems [5] with regard to the gyrotron transmission line, especially the location and type of confinement barriers.

2. ECE on ITER

2.1. *NTM detection and stabilization*

From the ITER Measurement Requirements, the electron temperature profile should be resolved with the resolution of a/30 (~70 mm) and accuracy of 10% in the plasma core for the temperature range 0.5 – 40 keV and with a spatial resolution of 5 mm for $r/a > 0.85$. In addition, one important requirement is to detect the Neoclassical Tearing Mode (NTM) when the island size is still not large enough to cause a deleterious effect on the confinement, i.e. before it reaches its saturated width and causes the mode locking, thus triggering the major disruption. NTMs on ITER (2/1, 3/2) will be localised at $r/a = 0.5 – 0.85$. The critical island width is expected to be between 20 – 40 mm, defined as $\Delta r = \Delta T_e/(dT_e/dr)$, whereas the ECH deposition widths for the NTM stabilisation are in the range of 30 – 70 mm. Therefore, the assumption for ECE channel separation of 20 mm or smaller at NTM radii is adequate for the mode detection with $\Delta T_e/T_e = (0.1 – 5.0)x10^{-2}$. The response time of the feedback system once the mode is tracked is in the order of 30-50 ms. The integration time of 100 ms for mode signature detection in the T_e profile is comparable to the growth time of the NTM.

The proposed sequence of operation for feedback is meant to be as follows:

(a) Island is seeded by some event;

(b) Island grows to detection, somewhere between critical island width and its saturated (unrecoverable) width;

(c) ECH feedback is actuated and the island is tracked down to below critical island width to ensure it does not re-appear when ECH is redeployed to other purpose.

Further discussions on Project Requirements are foreseen to ensure that the NTM is detected and stabilized before reaching its saturated width.

2.2. *Calibration and performance issues*

The instruments of ECE on ITER are Michelson interferometry (O- and X-modes) and heterodyne radiometers (O-mode, 122 – 230 GHz; X-mode, 244 – 355 GHz for $B_t(0) = 2 - 5.3$ T). The principal limitations of the system are restricted radial region of observation due to harmonic overlap and degraded spatial resolution due to the relativistic broadening. Therefore, the target resolution can only be met in a limited region, like in the NTM radial range at the LFS discussed in the previous section.

The integration of the system is a challenging task, as various components are located both in-vessel (in the equatorial port plug) and ex-vessel. It is important that the ECE assembly does not obscure partly or fully other diagnostics located in the same port plug. Figure 1 gives one possible option of the diagnostic front end integration in the equatorial port plug 9. The current integration block-diagram is given in Fig. 2. The proper interpretation of the measurements is nearly impossible without the calibration procedure done properly. The current design includes both hot and cold sources in the port plug. The possibilities to use only the hot source operating at different temperatures, or to use the hot source and the vacuum vessel under the assumption that the latter acts as a black body are currently under discussion. The cycling of the hot source needs a reliability estimation, as well as a careful study of how the thermal cycling affects the performance of the facing mirror in the receiver and if additional cooling would be necessary.

Presently, a hot source prototype with non-ITER heating element is being tested under the supervision of the US Domestic Agency [3]. The hot source should withstand high vacuum and high radiation flux environment, have 24-hours stability and provide emissivity of ~0.9 for the frequency range of 100 GHz – 1.5 THz. The most promising material for the hot source surface appears to be SiC: it provides the required emissivity for the desired frequencies; has low activation under typical ITER conditions, and is able to cope with high temperatures required for the diagnostic calibration.

Two lines-of-sight are foreseen [1]. Besides the principal, perpendicular, line-of-sight, a second one exists for additional measurements (and redundancy). For this line, a toroidal angle of about 10 degrees is proposed. The two lines-of-sight cross at the LFS in the plasma. In this way, the second line enables other measurements such as the detection of the presence of non-thermal electron population and radial correlation length measurements for MHD and turbulence (a slight vertical shift between the two lines would also allow poloidal correlation measurements).

The transmission lines of the ECE systems should be designed to operate in the whole required frequency range, 100 GHz – 1 THz. This is a challenging task as the most common solution, corrugated waveguide, typically only operate up to about 350 GHz. At higher frequencies, Bragg reflection and the power absorption by metallic walls degrades the performance. One of the possible solutions for higher frequencies could be the use of evacuated or gas-flushed dielectric waveguide. A possibility to merge both options is currently under study. In addition, the Indian Domestic Agency has proposed to integrate an additional source for checking the stability of the radiometer. It would be similar to the front-end calibration source and installed in the transmission line [2].

3. Reflectometry on ITER

Reflectometry on ITER will allow measurements of core and edge density profiles, MHD, ELM activity, turbulence and the plasma position. Both O- and X-mode reflectometry are planned in ITER covering the frequency range between 15 – 180 GHz. Antennas for different reflectometry systems are placed at the HFS (main plasma HFS and the plasma position reflectometers) and LFS (the plasma position reflectometers) of the vacuum vessel; the main plasma LFS reflectometry system is integrated in the equatorial port 11. Unlike ECE, ITER HFS and plasma position reflectometry have interfaces with the vacuum vessel. The in-vessel waveguides are directly welded on the vacuum vessel surface and are protected by the blanket from the plasma side. The antennas are designed to view the plasma through the cut-outs in the blanket and the first wall. Therefore, the in-vessel waveguide supports and welds should withstand the mechanical stresses caused by the electromagnetic loads during disruptions, and the antennas should be made of a material capable of withstanding thermal and neutron heating loads. Because of the complexity of waveguide routings and introduction of bends near the end of the antenna end and near the entrance to the upper port for HFS and the plasma position reflectometers, the target of the design activity is to ensure the acceptable power losses and low mode conversion loss.

4. Collective Thomson scattering system on ITER: preliminary considerations

Collective Thomson scattering (CTS) system on ITER is planned; the function of this diagnostic, which is located in the equatorial port 12 will be to measure fast ion distribution (100 keV to 3.5 MeV, a/10 ~ 20 cm, 100 ms) near parallel and near perpendicular to the magnetic field at different radii. Unlike ECE and reflectometry, CTS has a launching branch which requires several-MW RF power source around 60 GHz from a gyrotron. The requirements for high-power launching transmission lines and their assembly are yet to be defined, and the engineering design of the front-end quasi-optical components in the port plug are to be finalised. The exact requirements for the gyrotron (output microwave power, power modulation frequency etc) are to be developed.

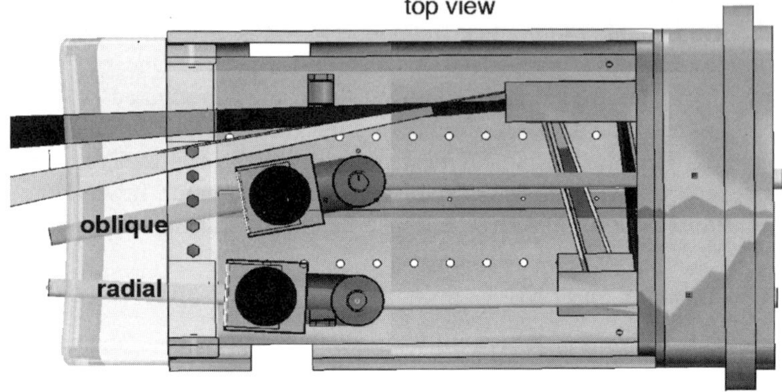

Figure 1. One possible integration option of two ECE lines-of-sight (radial and oblique) into the equatorial port 9 environment (top view). In yellow and grey, the lines-of-sight of core-imaging X-ray spectrometers are shown.

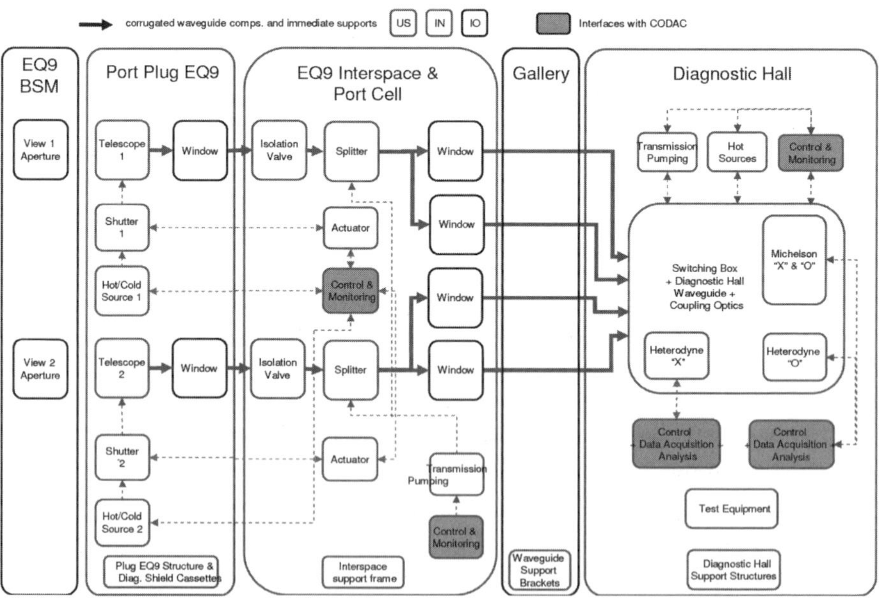

Figure 2. Current integration ECE block-diagram (courtesy of US Domestic Agency). Colors refer to different responsibilities of the Parties.

5. Confinement barrier strategy for ITER microwave systems

On ITER, two confinement systems shall be provided for each principal inventory of radioactive or hazardous material unless formal project approval for a single confinement system is given. Each confinement system shall include one or more static barriers or dynamic components to confine the inventory at risk. The transmission lines of all microwave systems on ITER should provide confinement function in all situations. This means that two windows or a window and an isolation valve should be included in the design. Other requirements, such as boundary monitoring, are also important. The first confinement barrier must withstand 2 bar peak pressure.

The design and choice of material for windows should not only provide the required confinement function, but also not degrade the performance of the diagnostic by introducing reflections and mode conversion. For example, diamond brazed to the copper waveguide section with the profiled quartz plates to enhance the window transmission is proposed as the primary windows for the main plasma HFS reflectometers on ITER. This kind of assembly enables the tolerable (less than 3 dB) power losses for the frequency range of 12 – 120 GHz. The secondary barrier for the same system is less demanding in terms of confinement requirements; 2 cm thick ROHACELL® foam with $\varepsilon < 1.1$ and low RF absorption can be a good candidate.

Acknowledgement

Authors are thankful to Dr. Elena de la Luna from Association Euratom-Ciemat, Spain and to Dr. Timothy Goodman from Association Euratom-CRPP, Switzerland for discussions and useful comments.

References

1. M. Austin *et al.*, these Proceedings.
2. H. Pandya *et al.*, these Proceedings.
3. P. Philips *et al.*, poster presentation at APS Conf. (2009).
4. G. Vayakis *et al.*, Nucl. Fusion **46**, S836 (2006).
5. M.A. Henderson *et al.*, these Proceedings.

FOURIER TRANSFORM BASED ECE SYSTEMS FOR REAL TIME TEARING MODE CONTROL IN TOKAMAKS*

W.A. BONGERS, D.J. THOEN, E. WESTERHOF, A.P.H. GOEDE, M.R. DE BAAR, M.A. VAN DEN BERG, V. VAN BEVEREN AND M.F. GRASWINCKEL
FOM-Institute for Plasma Physics Rijnhuizen, Association EURATOM-FOM, PO Box 1207, 3430 BE Nieuwegein, The Netherlands*

P. NUIJ, J.W. OOSTERBEEK AND B.A. HENNEN,
Eindhoven University of Technology, PO Box 513, NL-5600 MB Eindhoven, The Netherlands

The development of variable-resolution Electron Cyclotron Emission (ECE) diagnostics for Tokamaks based on the digitization and subsequent Fast Fourier Transformation (FFT) of the intermediate frequency signal [1] has enabled the exploration of the space-time domain of ECE diagnosed phenomena. This development depends on advances made in fast multi-Gigahertz sampling ADCs that have recently become available. One important application is the feedback control of neo-classical tearing-modes (NTMs) in Tokamaks by Electron Cyclotron Heating and Current Drive (ECH&CD) [2, 3 and 4]. This requires both a detection of the full mode content in the plasma and an accurate radial localization and poloidal phase of the magnetic islands. An FT ECE system has the possibility to optimize time and frequency resolution (spatial resolution). It could observe the plasma cross-section and then zoom-in on the detailed structure of the island. The scrutiny of the plasma phenomena at variable temporal and spatial scales could be extended by means of wavelet analysis.

A proof of principle on TEXTOR employing an FFT-ECE [1] system in-line with the ECH&CD system [5] enabled detailed observations of scattering of the high power mm-waves on rotating tearing modes [6]. The observed scattering is found to be in phase with the NTM island rotation. The medium and long-term objective is to develop FFT-ECE for NTM control on AUG and ITER.

1. Introduction

In a Tokamak, high temperature plasma is confined by magnetic fields, which form nested toroidal magnetic flux surfaces, characterized by their magnetic winding number q. On surfaces where q has a simple rational value, magnetic islands can develop. These are detrimental for confinement and stability of the tokamak plasma, and a variety of experiments have been carried out for real-time control of these modes (see [5, 6 and 7] and the references therein). One of the approaches to tearing mode control is the integration of the sensing ECE system within the ECRH beam line [8]. During the commissioning of the In-Line-ECE system on TEXTOR, a new physical phenomenon was discovered showing strong anomalous scattering of the high power mm-waves in discharges [6] with a rotating tearing mode. The scattering signal level was at least 6 orders of magnitude higher than the ECE signals. Using fast multi-Giga sample ADC's, a new dedicated FFT spectrometer was developed and mounted on the In-Line-ECE antenna transmission line [1].

This paper describes the design, implementation and first principal test of this new system. Particular emphasis is put on the variable time and frequency resolution that is inherent to such systems, a feature that can be used for optimizing the information content of the signals.

In the following, Section 2 presents a comparison of theoretical characteristics of FFT based ECE systems to classical ECE systems, useful for requirement specification and an exploration of the performance of the current diagnostics. Section 3 describes a test of an ECE measurement by a FFT system on TEXTOR of scattering and preliminary ECE signals. Section 4 concludes the results and proposes future work.

2. Theoretical and practical performance exploration

For classical radiometers, the radiometer formula [9] expresses the minimal detectable power fluctuation of an ECE system looking to the plasma.

$$P_{\min} = k_B B T_{total} \sqrt{\frac{2B_{vid}}{B_{IF}}} \; with \; \sqrt{\frac{2B_{vid}}{B_{IF}}} = \frac{\Delta T_{ECE}}{T_{av\ ECE}} \qquad (1)$$

Here, k_B is Boltzmann's constant, B the input bandwidth, $T_{total} = T_{sys} + T_e$ the sum of the system and plasma electron temperature, B_{vid} the video bandwidth and B_{IF} the IF bandwidth of the radiometer. For a single sideband receiver B and B_{IF} are the same. The ratio of B_{vid} and B_{IF} relates to the ratio of the thermal fluctuations of the ECE temperature, ΔT_{ECE}, and the average ECE temperature, $T_{av\ ECE}$. Preferably, this temperature fluctuation level is at most of the order of a few percent.

2.1. *FT ECE system*

For an FT ECE system, the data acquisition is presently only feasible through a reduction of the data flow by block-wise data acquisition during the sampling time, $\tau_{sampling}$. As a consequence of the finite idling time, τ_{idle}, the number of data points Naq < Nmax , see Figure 1.

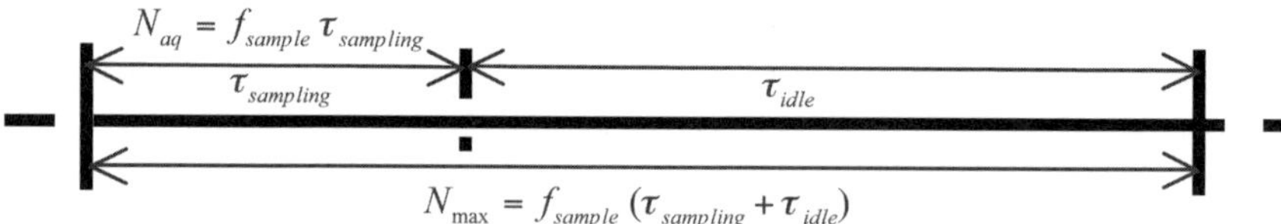

Figure 1 The FT ECE system data acquisition time-domain diagram.

The B_{vid} is determined by $\tau_{sampling}$ (law of Shannon). For an FFT, the spectral resolution, Δf, is a function of the sample frequency of the ADC, $f_{sampling}$, and the number of time domain samples, N_{aq} [10, 11], hence:

$$B_{vid} = \frac{1}{2\tau_{sampling}} \; and \; B_{IF} \geq \Delta f = \frac{f_{sampling}}{N_{aq}} \rightarrow B_{IF} = n\,\Delta f \tag{2}$$

For a given sample rate, increasing the number N_{aq} of samples, increases the frequency resolution (and number of frequency bands) by $N_f = 1 + N_{aq}/2$. This value is similar to the minimal B_{IF} of a classical radiometer [1]. In order to increase the signal to noise ratio and to reduce the plasma inherent fluctuation on the spectra, the intensities in a discrete number of n frequency bands can be summed. Combining equations (1) and (2), yields the minimal detectable power and the relative fluctuation on the ECE temperature:

$$P_{\min FT\,ECE} = \frac{k_B f_{sample} T_{sys}\sqrt{n}}{N_{aq}} \; and \; \frac{\Delta T_{ece}}{T_{av\,ECE}} = \frac{1}{\sqrt{n}} \tag{3}$$

To compare a classical ECE system with our proposed FT-ECE system, it can be derived that:

$$\frac{P_{\min FT\,ECE}}{P_{\min conv\,ECE}} = \sqrt{\frac{\tau_{sampling} + \tau_{idle}}{\tau_{sampling}}} = \sqrt{\frac{N_{max}}{N_{aq}}} \tag{4}$$

The signal to noise ratio, SNR, or dynamic range (full scale sine wave in the power ratio dB) of an ideal digitizer with effective number of bits N_b bits can be calculated [10, 11] by

$$\mathrm{SNR}(f) = 6.02 N_b(f) + 1.76\,\mathrm{dB} + 10\log\frac{N}{2} \tag{5}$$

in which the latter part is the process gain depending on the block size of FFT process. In practice the effective number of bits, N_b, is a frequency dependent value. To calculate the complete frequency dependent SNR behavior of the ADC one can use the equation (6) [11] including additional the dynamical gain. This equation can be fitted for different block sizes to the available frequency dependent values of the SNR obtained from the FFT of the time domain signal, such that the b and σ can be derived.

$$SNR(f) = 10\log\left\{\frac{3\,2^{2b}}{2+3\left(2^{b}2\pi f\sigma\right)^{2}}\frac{N}{2}\right\} \tag{6}$$

Here b is the frequency independent, maximum effective number of bits at DC. The σ is presumed the RMS value sampling window jitter time (assumption is that the jitter has a normal distribution with zero mean and a variance of σ^2).

A practical result of our 8 Gsa/s ADC used in our test FT ECE system is that the effective number of bits is 6.6 of the 10 available and the RMS jitter time is about 0.2 ps. Knowing the SNR range and the maximum power, the minimal detectable power of the ADC can be calculated which can be used to calculate the ADC noise temperature needed for total system noise temperature, T_{total}.

3. Proof of principle: results measured on TEXTOR

The ECE millimeter waves coming from the plasma by additional antennas [9] or by the In-Line principle [5] are heterodyne down-converted, amplified and directly digitized using fast Analog-Digital converters (ADCs) connected by a fast interface to a computer (no hardware video detection and conservation of ECE frequency phase). From the Intermediate Frequency (IF) signal (in the time domain) frequency spectra at discrete time intervals are obtained by Fast Fourier Transform (FFT).

This data processing can be carried out off-line or in real time. Hence, the time interval taken for the FFT determines a flexible frequency resolution in contradiction to a conventional ECE system where the IF frequency signal is splitted to cover various frequency bands corresponding the fixed frequency resolution.

On TEXTOR a line of sight FT ECE system is tested to measure in first instance the scattered ECRH power back into the launcher [6]. In Figure 2, an example is presented of a measurement of the anomalous scattering in TEXTOR. We observe burst of radiation with different frequencies of the ECRH system with a dynamic range of about 40 dB including time-dependent chirps which are related to details of the island location.

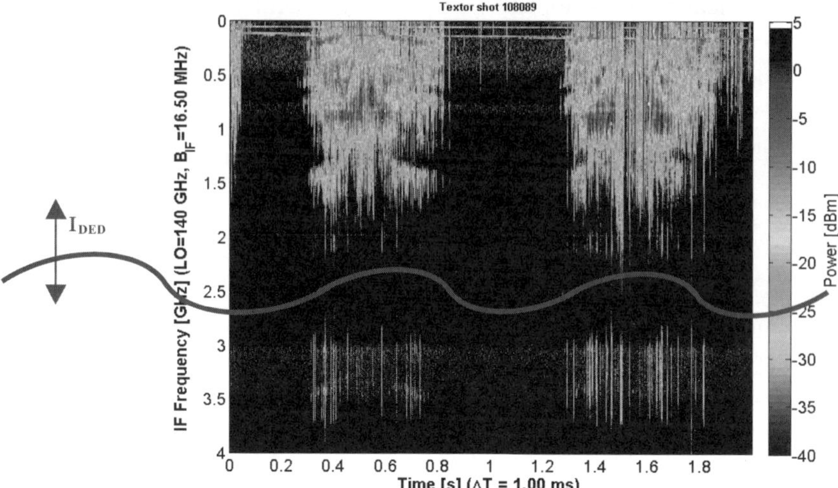

Figure 2 With the Dynamic Ergodic Diverter (DED) an Island was created and moved slowly back and forward across the ECRH beam launched: the symmetrical frequency change of the spectrum in time correlates 100% with MHD. This spectrum contains 8001 frequency domain samples from which 242 groups of 33 summed samples are computed (remaining samples are discarded).

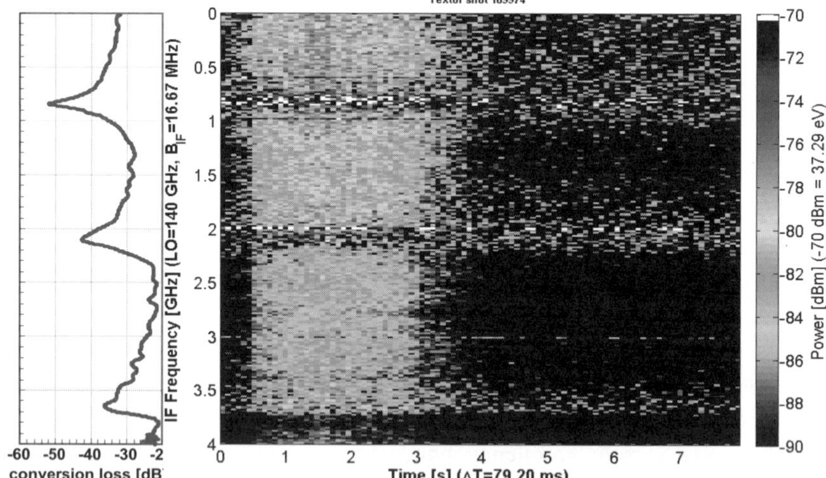

Figure 3 A FFT ECE spectrum of TEXTOR the plasma pulse length was about 2.5s. The marginal signal quality is caused by high conversion loss of the mixer of the FT frontend (left) and TEXTOR operational conditions of the day. This spectrum contains 160001 frequency domain samples from which 239 groups of 667 summed samples are computed to give a theoretical DTECE/Tav ECE of 4%.

After that 15 dB of attenuation was removed (by omitting the quasi optical line of sight diplexer and notch filter [5]), Figure 3 demonstrates measurement of ECE radiation with the FT ECE system in a standard TEXTOR plasma (no ECRH and Bt= 1.9 T).

4. Conclusion and prospect

A novel ECE system with direct IF digitization has been developed: Fourier Transform ECE. By block-wise data acquisition there is a reduced minimal detected power compared width conventional systems (~10 dB)
Limited range (4 GHz) FT system is tested on TEXTOR in line-of-sight concept:

- Good ECRH scattering measured.
- Marginal ECE emission measured (due to frontend and operational conditions).

The next step is to develop a full range FT ECE/scatter system for control with minimalization of idling time and high performance frontend.

As a consequence of the set-up, flexible time/spatial resolution with optional dynamic zoom on certain plasma positions or events comes within reach, with a possible application to adaptive sensing for MHD control. The analysis presented above has been restricted to the characteristics of FFT. It will be investigated if other analysis methods such as Wavelets analysis would yield further capability to optimize the frequency and time resolution.

References

1. D.J. Thoen and W.A. Bongers, et al., Rev. Sci. Inst., **80**, 10, 103504 (2009).
2. A. Isayama, et al., Nuclear Fusion **43**, 1272-1278 (2003).
3. R.J. La Haye, et al., Nuclear Fusion **46**, 451-461 (2006).
4. C.C. Petty, et al., Nuclear Fusion **44**, 243-251 (2004).
5. J. W. Oosterbeek, et al., Rev. Sci. Inst., **79**, 9, 093503 (2008).
6. E. Westerhof, et al., Phys. Rev. Letters, **103**, 12, 125001 (2009).
7. H. Zohm, et al., Nuclear Fusion **39**, 577-580 (1999).
8. B. A. Hennen, et al., Plasma Phys. Control. Fusion, accepted (2010).
9. H. J. Hartfuss, et al., Plasma Phys Control. Fusion **39**, 1693-1769 (1997).
10. S. W. Smith, The Scientist & Engineer's Guide to DSP, (California Technical Publisher. 1997).
11. J. Tsui, Digital Tech. for Wideband Rec. (Artech House Publishers 1995).

Acknowledgments

This work was supported by NWO, ITER-NL and the European Communities under the contract of the Association EURATOM/FOM, and was carried out within the framework of the European

Fusion Programme. The views and opinions expressed herein do not necessarily reflect those of the European Commissions.

Identifying anomalous Doppler resonance effect based on ECE diagnosis in HT-7 tokamak

Erzhong Li*, Liqun Hu, Yong Liu, Ang Ti, Bili Ling and Xiang Gao

Institute of Plasma Physics, Chinese Academy of Science, Hefei, 230031, China
**E-mail: rzhonglee@ipp.ac.cn*
www.ipp.ac.cn

The abrupt steep jump of electron cyclotron emission (ECE) signals during current ramp down has been observed and explained by anomalous Doppler resonance effect (ADR). The identifying process of ADR was presented based on fast Fourier transform (FFT) technique.

Keywords: anomalous Doppler resonance, electron cyclotron emission

1. Introduction

Anomalous Doppler resonance(ADR), due to interaction between runaway electrons and low hybrid waves via electron cyclotron waves, has been regarded as the origin of runaway instability both experimentally [1–3] and theoretically [4]. The instability is usually observed during runaway discharges under the condition of ultra-low density so that the discharge goes into slide-away regime [5,6]. In this type of discharge, a large runaway population implies that the electron distribution function will have a long tail [7,8]. The electron parallel velocity is much larger than the vertical one. The ADR can occur when the resonance condition $v_{\parallel} = (\omega_{ce} + \omega_k)/k_{\parallel}$is satisfied (the low hybrid waves frequency ω_k, the parallel wave number $k_{\parallel}$ and the electron cyclotron frequency ω_{ce}) [9,10]. The ADR increases electron transverse energy so that the electron distribution function tends to isotropy. However, the Cherenkov resonance effect drives the electron distribution function to form a plateau again [11]. The repeat process is usually known as relaxation oscillations which manifest themselves as continuous step-by-step increase on the electron cyclotron emission (ECE) signals during ADR, such as seen in previous experiments in HT-7 [2,3,12]. However, the converting process from parallel energy to transverse energy is not evident experimentally. It is worthwhile to understand the ECE signals variation during ADR. In this paper, we will try to combine the analysis of ECE signals with existing theory to clarify the ADR physical process happened when the plasma current decreases below a critical value in HT-7 tokamak.

2. Observation

A series of discharges were performed in HT-7 tokamak for investigating ADR during plasma current ramp-down. The major diagnostic device for identifying ADR was the ECE system which was introduced in detail in Ref. [13]. It views plasmas horizontally on the low field side (LFS) in HT-7 tokamak. The runaway flux was monitored by runaway electrons diagnosis (RA) whose view field was tangential to the plasma. RA3 is a NaI detector with an energy range $0.5 \sim 7MeV$. RA4 is a BGO detector working in the energy range of $0.5 \sim 12MeV$ [14]. Shown in figure 1 is the arrangement of thesetwo diagnostic devices in HT-7. Shown

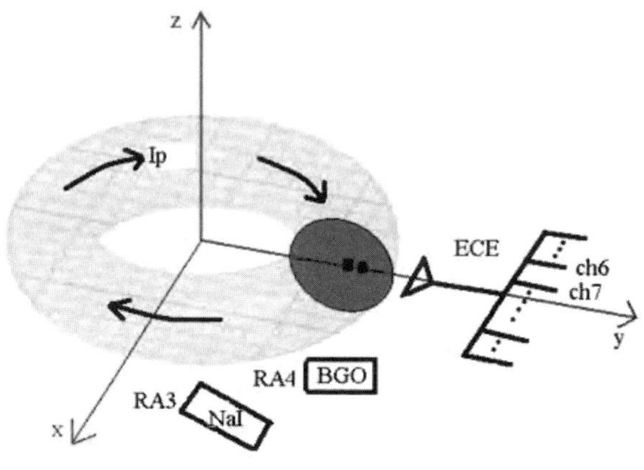

Fig. 1. Schematic view of diagnostics in HT-7 tokamak

in figure 2 is a typical ADR discharge with a long current ramp-down continuation. The black dotted line indicated a possible physical event. From this figure, one can observe distinct distinction between ECE signals and RA signals. At the black dottedline, ECE signals jumped up whilst the RA (both RA3 and RA4) signals began to decrease. However, before the moment indicated by black dotted line, RA3 was almost keeping constant, but RA4 was increasing. It hinted that the higher-energy population of runaway electrons increased. After the moment, the RA signals started to decrease, which indicated a reduced runaway population.

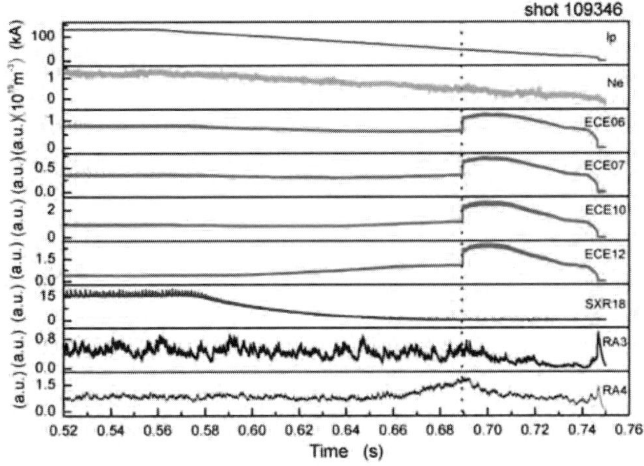

Fig. 2. Waveforms of typical ADR discharge.

On the contrary, Steep jump of ECE signals implied enhance of transverse emission at the black dotted line. SXR signals almost decreased to zero just as normal discharge due to the decreased plasma current and low density. For a low density runaway discharge, the runaway population should be increased acutely during current ramp-down, as a previous experimental study has shown in Ref. [15]. Here, we should keep in mind that a physical event probably happened at the time point indicated by the black dotted line in figure 2.

3. Identification method of anomalous Doppler resonance effect

Shown in figure 3(a) is the waveform of step-by-step increased ECE07 signals just after a steep jump has been seen in figure 2. One observes that the periods of modulated signals change as a function of time. So it is easy for us to use Fast Fourier Transform technology (FFT) to examine the spectrum which is shown in figure 3(b).The frequency of ECE07 perturbation increases with time evolution, from 1 kHz at the beginning to 2 kHz at the last. At the frequency of 2 kHz, the step-by-step increased signals arrive at maximum, such as seen in figure 3(a). This kind of ECE perturbation has been also observed in ASDEX [16].From the knowledge of wave-particle interaction [17,18], one can obtain the relationship, $\frac{dp_{\|}/dt}{dp_{\perp}/dt} = -\frac{E_{\|}\rho_L}{E_{\perp}}k_{\perp}$, to describe the electron motion near a single cyclotron resonance. There, the major parameters are: electron Larmor radius ρ_L, amplitudes of wave electric fields $E_{\|}$ and$E_{\perp}$ respectively, and the transverse wave number $k_{\perp}$. If the parallel momentum increases with respect to time, the transverse momentum will decrease or vice versa. It is well known that the spatial propagation of wave would change the wave phase angle, this holds also for changes of the wave frequency. For comparison, spectrums of ECE06 and ECE07 are shown in figure 4.

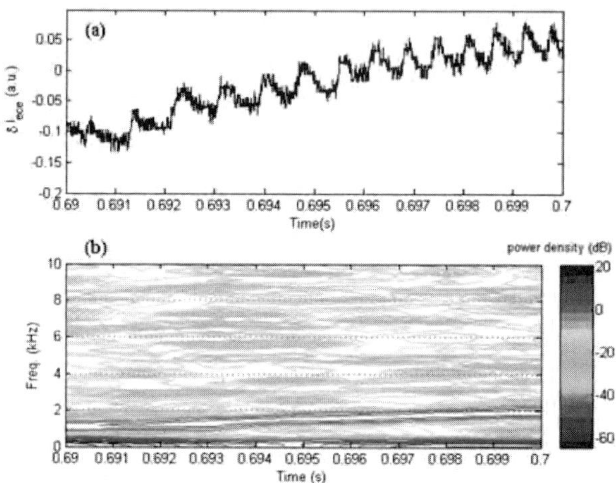

Fig. 3. Time slice of normalized perturbation of ECE07 just after onset of steep jump (a) in the shot 109346 and its spectrum (b).

One can see that their frequency has the same evolution from 1 kHz just at the onset to about 2 kHz. When the frequency reaches 2 kHz, the stepped ECE signals arrive at a saturated state as seen in figure 1. This indicates the interaction between wave and electrons reaches steady state at the frequency of 2 kHz.For further analysis, we examine the phase difference at the frequency 2 kHz between ECE06 and ECE07 which is shown in figure 5. We choose scale of 20 ms for the FFT analysis. The sampling frequency of 250 kHz was used for digitization. Thus, the data length of 20ms is enough to perform FFT. On the other hand, the scale of 20 ms is sufficiently precise to monitor phase difference evolution. It is obvious that the phase difference is a decreasing function of time. The critical value of phase difference is about$\Delta\phi = 0.15rad$ when the steep jump of ECE signals happens.From figure 4, one can find that the frequencies of ECE signals (ECE06 and ECE07) have the same evolution tendency. Thus, the contribution of frequency to phase difference can be excluded. So the variation of phase difference is probably explained by the change of transverse wave number $k_\perp$.When the phase difference arrives at minimum value $\Delta\phi = 0.15rad$ in figure 5, the transverse wave number

$k_{\perp}$ also reaches minimum.

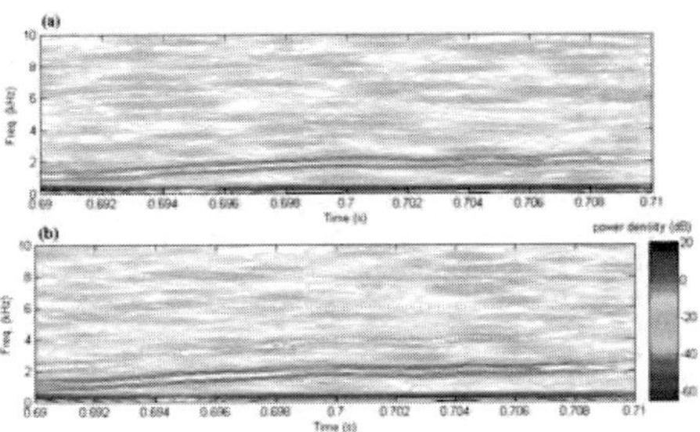

Fig. 4. Spectrums of ECE07 (a) and ECE06 (b) in the shot 109346 with the same scale of power density.

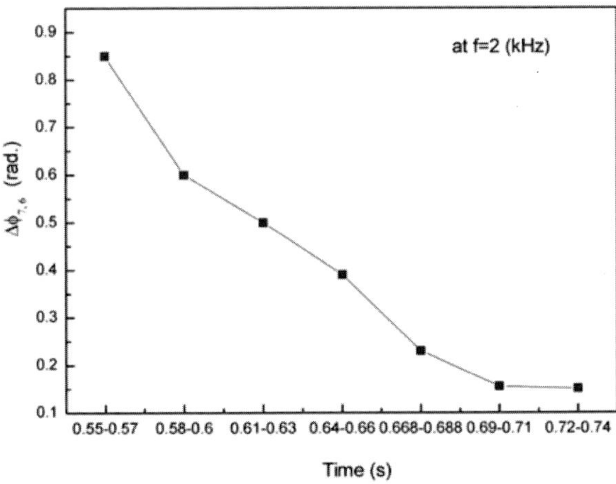

Fig. 5. Phase difference at the frequency of 2 kHz between ECE07 and ECE06 versus several time slices during shot 109346.

4. Conclusion

In conclusion, the steep jump event of ECE signals accompanied by decrease of RA signals had been identified as the interaction of wave with electrons via ADR. ADR effectively converts parallel to transverse energy which is responsible for the steep jump event of ECE signals. The spectrums analysis of ECE06 and ECE07 indicated that ADR happened when the phase difference decreased below the critical value$\Delta\phi = 0.15rad$.

5. Acknowledgments

The authors would like to thank Dr. Gary Taylor (Princeton Plasma Physics Laboratory (PPPL) in USA) for some beneficial discussions, Dr. Perry Phillips and Dr. Huang He (Fusion Research Center (FRC) in USA) for their support. This work is supported by the National Natural Science Foundation of China under Grant No. 10935004.

References

1. Santini F, *Phys. Rev. Lett* **52**,1300 (1984).
2. Chen Z Y, *Chin. Phys. Lett.* **24**,3195 (2007).
3. Sajjad S, Gao X, Ling B, Bhatti S, Ti A, *Phys. Lett. A* **373**,1133 (2009).
4. Luckhardt S C and Bers A, *Phys. Fluids* **30**,2110 (1987).
5. Knoepfel H, *Nucl. Fusion* **19**,785 (1979).
6. Wan B N, *Nucl. Fusion* **44**,400 (2004).
7. Molvig K, Tekula M S and Bers A, *Nucl. Fusion* **38**,1404 (1977).
8. Molvig K, Tekula M S and Bers A (*A steady state solution for the runaway electron distribution function),1984.*
9. *Muschietti L, Appert K and Vaclavik J (Cerenkov and anomalous doppler effects in the relaxation of an electron beam),1979.*
10. *Fuchs V, Shoucri, Teichmann J and Bers A (Runaway electrons distributions and their stability with respect to the anomalous Doppler resonance),1986.*
11. *Parail V V, Pogutse O P,* Nucl. Fusion ***18**,303 (1978).*
12. *Chen Z Y,* J. Plasma Physics. ***75**,661 (2009).*
13. *Ti A, Ling B, Fei Q , Gao X,* Int. J. Infrared Milli Waves ***28**,243 (2007).*
14. *Lu H W, Hu L Q, Chen Z Y, Jang Y, Lin S Y,* J. Plasma Physics ***74**,3(1-9) (2007).*
15. *Chen Z Y,* Plasma Phys. Contrl. Fusion ***48**,1489 (2006).*
16. *Fussmann G,* Phys. Rev. Lett. ***47**,1004 (1981).*
17. *Martin-Solis J R, Sanchez R and Esposito B,* Phys. Plasmas ***9**,1667 (2002).*
18. *Martin-Solis J R and Sanchez R,* Phys. Plasmas ***15**,112505 (2008).*

EC RADIATIVE TRANSPORT IN FUSION PLASMAS WITH AN ANISOTROPIC DISTRIBUTION OF SUPRATHERMAL ELECTRONS

FERRAN ALBAJAR

Fusion for Energy[*†], *Barcelona, 08019, Spain*

MARINO BORNATICI

Physics Department "A. Volta" University of Pavia, Pavia, 27100, Italy

FOLKER ENGELMANN

Max-Planck-Institut für Plasmaphysik, Garching, 85748, Germany

The RAYTEC code for electron cyclotron (EC) wave radiative transport modelling of fusion plasmas has been generalised to cover EC radiative transport in the presence of an anisotropic distribution of suprathermal electrons. More specifically, superposed on a Maxwellian bulk, the population of suprathermals is taken to be the sum of a Maxwellian and a distribution which is Maxwellian in energy and has a distribution in the form of a one-sided loss- or anti-loss-cone in pitch angle, such that suprathermals with an excess of momentum either perpendicular or parallel to the magnetic field (as characteristic, respectively, for EC cyclotron wave heating and current drive) can be dealt with. To display the impact of anisotropy, the profile of the net EC wave power density lost from an ITER-like plasma is evaluated as a function of the cone angle for suprathermals localized either at or off the plasma centre. The results thus obtained are discussed in respect to those for an isotropic distribution of the suprathermals as well as those for solely the bulk plasma..

1. Introduction

In recent years electron cyclotron (EC) radiative transport in fusion plasmas has been investigated, both analytically and numerically, mainly on assuming the electron distribution function to be locally Maxwellian ([1] and references therein). Comparatively less attention has been given to EC radiative transport in

---

[*] The views expressed in this publication are the sole responsibility of the author and do not necessarily reflect the views of Fusion for Energy. Neither Fusion for Energy nor any person acting on behalf of Fusion for Energy is responsible for the use which might be made of the information in this publication.

the presence of a population of suprathermal electrons. Earlier work has analyzed the effect of an isotropic enhanced electron tail on EC radiation loss in a tokamak without accounting for a spatial localization, in the plasma, of the suprathermals [2]. More recently, EC radiative transport has been investigated for an isotropic bi-Maxwellian electron distribution, the radial profile of the density and temperature of the suprathermal component being taken in the form of a Gaussian distribution [3].

Here the EC radiative transport for an anisotropic distribution of suprathermal electrons is investigated by means of the RAYTEC code recently developed for EC wave radiative transport modeling of fusion plasmas [1]. More specifically, superposed on a Maxwellian bulk, the population of suprathermals is taken to be the sum of a Maxwellian and a distribution which is Maxwellian in energy and has a distribution in the form of a one-sided loss or anti-loss-cone in pitch angle [4]. Both an on-axis and off-axis spatial distribution of suprathermals, with a Gaussian radial profile, is considered.

2. The model electron distribution

For an electron population composed of a (relativistic) Maxwellian bulk (labeled "b") and a suprathermal component (labeled "h") that is the sum of a (relativistic) Maxwellian and an anisotropic distribution which is Maxwellian in energy and corresponds to a one-sided loss-or anti-loss-cone in pitch angle [4], the momentum distribution function $F(\mathbf{p})$ can be written

$$F(\mathbf{p}) = (1-\eta) f(p, T_b) + \eta F_h(\mathbf{p}), \tag{1}$$

with $\eta F_h(\mathbf{p}) = \eta_1 f(p, T_{h1}) + (\eta - \eta_1) f(p, T_{h2}) \varphi(\cos\psi)$, where $f(p,T)$ is the relativistic Maxwellian, $\eta \equiv n_h/(n_b + n_h)$ denotes the total density of the suprathermals, $n_h = n_{h1} + n_{h2}$, normalised to the plasma (bulk + suprathermal) electron density $n_t = n_b + n_h$, $\eta_1 \equiv n_{h1}/(n_b + n_h)$ being the (relative) density of suprathermals with Maxwellian distribution (temperature $T_{h1} > T_b$); the pitch angle (ψ) distribution of the anisotropic component (relative density $n_{h2}/(n_b + n_h) = \eta - \eta_1$; temperature $T_{h2} > T_b$) is taken to correspond to a one-sided erf-cone [4]

$$\varphi(\cos\psi) = \frac{1}{1+\mu_c}\left[1 - \mathrm{erf}\left(\frac{\cos\psi - \mu_c}{\sigma\sqrt{2}}\right)\right], \quad |1 \pm \mu_c| \gtrsim 3\sigma, \tag{2}$$

where $\cos^{-1}\mu_c$ is the cone angle and σ characterises the width of the transition. The distribution (2) describes a one-sided loss-cone for $\mu_c > 0$ and an anti-loss-cone for $\mu_c < 0$, thus covering the case of an excess of momentum either

perpendicular or parallel to the magnetic field (as characteristic, respectively, for EC cyclotron wave heating and for current drive).
With the distribution (1) (for simplicity we here take the same temperature, T_h, for the two suprathermal populations),

- the absorption coefficient $\alpha^{(i)}$ of the mode i can be expressed as

$$\alpha^{(i)} = \alpha_b^{(i)} + \alpha_{h1}^{(i)}\left(1+\frac{\eta-\eta_1}{\eta_1}A^{(1)}\left(1+R^{(i)}\right)\right), \tag{3}$$

where $\alpha_b^{(i)}$ and $\alpha_{h1}^{(i)}$ are, respectively, the absorption coefficient of the bulk and of the isotropic suprathermal population (of density n_{h1}); the 2nd term within the square brackets accounts for the contribution of the anisotropic suprathermal population (of density n_{h2}), with $A^{(1)} \equiv \left(1+\mu_c\right)^{-1}\left(1+erfx\right)$, the argument x of the erf-function being given by Eq. (29) along with Eqs. (21) and (41a,b) of [4], and $R^{(i)}$ is given by Eqs. (44)-(47) of the same reference. Note that, $A^{(1)} \to 1$ and $R^{(i)} \to 0$, in the isotropic limit, i.e., for $\mu_c \to 1$ and $\sigma \to 0$, subject to the condition $1-\mu_c \gtrsim 3\sigma$, cf. [4].

- the absorption coefficient $\alpha^{(i)}$ of the mode i can be expressed as

$$I_{bb}^{(i)}\alpha^{(i)} = \frac{\omega^2}{8\pi^3 c^2}\left[T_b\alpha_b^{(i)} + T_h\alpha_{h1}^{(i)}\left(1+\frac{\eta-\eta_1}{\eta_1}A^{(1)}\right)\right], \tag{4}$$

with $I_{bb}^{(i)}$ the (effective) blackbody intensity and $\alpha^{(i)}$ the absorption coefficient (3).
In a form valid also for a suprathermal population with ($T_{h1} \neq T_{h2}$), the average perpendicular (to the magnetic field) energy density of the suprathermals,
$< E > \equiv < m\mathrm{v}_\perp^2/2 > \equiv \int d^3\mathrm{v}\left(m\mathrm{v}_\perp^2/2\right)F_h$, is in the non-relativistic limit

$$< E > = \left[\frac{\eta_1}{\eta} + \left(1+\frac{\mu_c\left(1-\mu_c\right)}{2}\right)\left(1-\frac{\eta_1}{\eta}\right)\frac{T_{h2}}{T_{h1}}\right] n_h T_{h1} \tag{5}$$

and the parallel (to the magnetic field) current density is,

$$j_\parallel\left(\equiv -e\int d^3\mathbf{p}\,\mathrm{v}_\parallel F_h\right) = en_{h2}\left(\frac{T_{h2}}{m}\right)^{1/2}\frac{K_{3/2}\left(\mu_h\right)}{K_2\left(\mu_h\right)}\sqrt{\frac{2}{\pi}}\frac{1-\mu_c^2-\sigma^2}{1+\mu_c}, \tag{6}$$

with $\mu_h \equiv mc^2/T_{h2}$ and valid for $\sigma^2 \ll \left(1-\mu_c^2\right)$.

3. The radiative power density

For a radiating system for which the generalised Kirchhoff law (4) is valid, the net power per unit volume effectively radiated is (Eq.(5) of [5])

$$\frac{dP(\mathbf{r})}{dV} = \sum_{i=X,O} \int_{\omega_{\min}^{i}}^{\infty} d\omega \int d^2\Omega_s \alpha^{(i)}(\mathbf{r},\omega,\mathbf{s}) \left[I_{bb}^{(i)}(\mathbf{r},\omega) - I^{(i)}(\mathbf{r},\omega,\mathbf{s}) \right], \qquad (7)$$

with $I^{(i)}(\mathbf{r},\omega,\mathbf{s})$ the specific intensity of the radiation that is the solution of the radiative transfer equation (which can be obtained numerically following the lines of [5]) and $\omega_{\min}^{i}$ an appropriate lower cut-off frequency.

To show the impact of anisotropy effects, an evaluation of (7) has been carried out for ITER-like parameters (plasma major and minor radius 6.2 m and 2.0 m, respectively; toroidal magnetic field 5.3 T; temperature and density profiles (cf. [3]) specified in Figs. 1), the effective wall reflection coefficient being equal to 0.6.

In the left-hand side of Figs. 2 the power density (7) radiated at $\rho = 0.005$ is shown as a function of the cone angle parameter μ_c of the one-sided erf-cone distribution (2), the transition width being $\sigma = 0.01$, for the on-axis population of the suprathermals. The normalised density of the suprathermals at the plasma centre ($\rho_h = 0$) is $\eta = 0.1$, with a varying fraction η_1/η of "Maxwellian" suprathermals. The case of an off-axis population of suprathermals around $\rho_h = 0.5$, for a local density of the suprathermals of $\eta = 0.1$ at this radial position, is shown on the right-hand side of Figs.2.

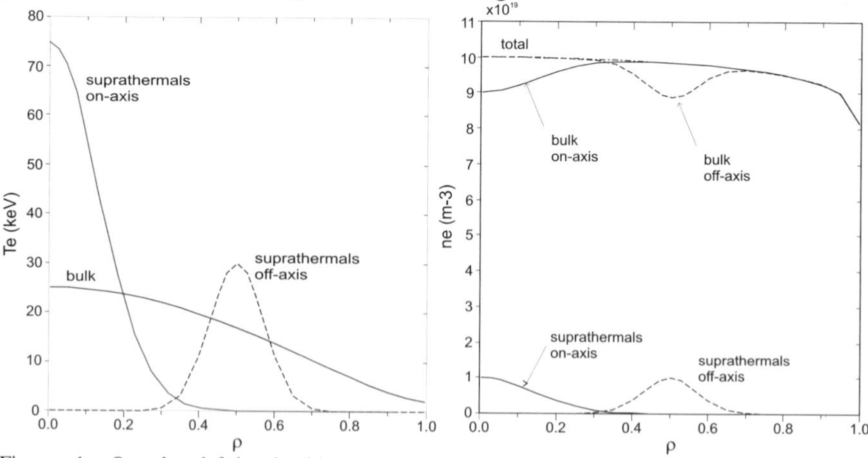

Figure 1. On the left-hand side, electron temperature profiles of the bulk electrons $T_b(\rho) = (T_b(0) - T_b(1))(1-\rho^2)^{1.5} + T_b(1)$, with $T_b(0) = 25$ keV, $T_b(1) = 2$ keV and of the

suprathermals, $T_h(\rho)=\left(T_h(\rho_h)-T_{\min,h}\right)\exp^{-(\rho-\rho_h)^2/(\Delta\rho)^2}+T_{\min,h}$, with $T_h(\rho_h)$ = 75 keV and 30 keV, respectively, for the on-axis ($\rho_h = 0$, $\Delta\rho = 0.2$) and the off-axis ($\rho_h = 0.5$, $\Delta\rho = 0.1$) suprathermals, and $T_{\min,h} = 0.01$ keV. On the right-hand side, radial profiles of i) the total electron density $n_t(\rho)=\left(n_t(0)-n_t(1)\right)\left(1-\rho^2\right)^{0.1}+n_t(1)$ with $n_t(0) = 10^{20}$ m^{-3}, $n_t(1) = 5\times10^{19}$ m^{-3}; ii) the density of the suprathermals $\left(n_{h1}(\rho),n_{h2}(\rho)\right)=\left(n_{h1}(\rho_h),n_{h2}(\rho_h)\right)e^{-(\rho-\rho_h)^2/(\Delta\rho)^2}$ with, respectively, $\rho_h = 0$, $\Delta\rho = 0.2$ for the on-axis and $\rho_h = 0.5$, $\Delta\rho = 0.1$ for the off-axis case; shown is the total density of the suprathermals for $(n_{h1}(\rho_h) + n_{h2}(\rho_h))/n_t(\rho_h) = \eta = 0.1$; iii) density of bulk electrons $n_b(\rho) = n_t(\rho) - (n_{h1}(\rho) + n_{h2}(\rho))$.

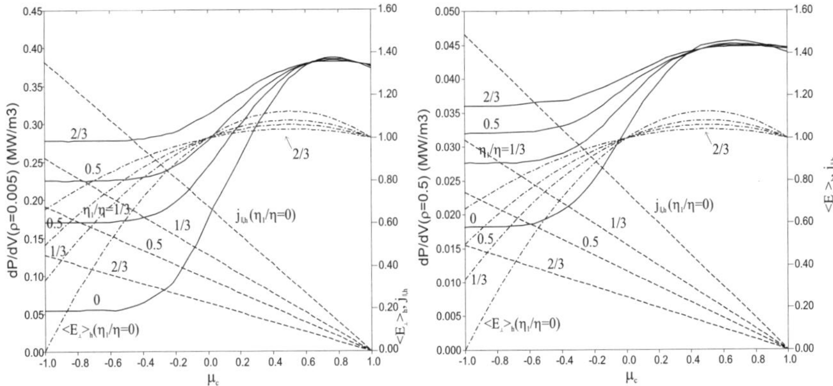

Figure 2. On the left-hand side, dP/dV at $\rho = 0.005$ as a function of the cone angle parameter μ_c for the on-axis, two-component population of suprathermals with $\eta = 0.1$ and $\eta_1/\eta = 0$, 1/3, 1/2, 2/3 ($T_h(0)$ = 75 keV, $\Delta\rho = 0.2$). Also shown is the corresponding average perpendicular energy density <E> of the suprathermals, normalized to $n_h T_h$, cf. Eq.(5), with $T_{h1} = T_{h2} = T_h$, dotted-dashed curves, along with the parallel current density $j_{//}$ (normalized to $e\, n_{h2}(T_{h2}/m)^{1/2}$), cf. Eq.(6), dashed lines. On the right-hand side, the same at ρ = 0.5 for the off-axis population of the suprathermals with $T_h(\rho = 0.5)$ = 30 keV and $\Delta\rho = 0.1$.

From Figs. 2 it appears that i) as to be expected, the net radiated power density dP/dV increases with the density of the Maxwellian suprathermal component, as measured by η_1, over large part of the μ_c-range, namely, for $-1\le\mu_c\le\bar{\mu}_c$ ($\simeq 0.6$ and 0.4, respectively, for $\rho = 0.005$, left-side plot, and $\rho = 0.5$, right-side plot), dP/dV tending to be independent of η_1 when the distribution of the suprathermals as a whole tends to become isotropic ($\mu_c \to 1$); a plateau-like feature extending from $\mu_c = -1$ due to the contribution of the beam-like suprathermals being small, is also apparent; ii) the average perpendicular energy density <E> exhibits a non-monotonous dependence on μ_c with a maximum at $\mu_c = 0.5$ since the formation of a loss-cone at fixed density implies an increase of electrons with prevalently perpendicular energy, while <E> increases with η_1 for $-1\le\mu_c<0$, i.e., for suprathermals having an excess of parallel momentum, and slowly decreases with η_1 for $0 \leq \mu_c \leq 1$, i.e., for a loss-cone distribution as is intuitive; this behaviour of <*E*> is reflected in that of *dP/dV*; iii) for the current density $j_{//}$, the linear dependence on $\eta_2 = \eta - \eta_1$ and almost linear dependence on μ_c (see (6)) is displayed.

The radial profile of the net radiated power density *dP/dV*, for the on-axis and off-axis anti-loss-cone distributions of the suprathermals, is shown in Figs. 3 which display i) the sharp local enhancement of *dP/dV* with respect to the case without suprathermals over the spatial range where suprathermals are present; ii) that a slight reduction of *dP/dV* occurs adjacent to the range of enhanced emission [5]. In Fig. 3a, the case of a distribution of suprathermals extending over a wider radial range ($T_h(\rho) = 3\,T_b(\rho)$, $n_h(\rho) = \eta\, n_t(\rho)$ with $\eta = 0.1$) and leading to a sizeable enhancement of the total emitted EC power P_{EC} is also shown. The relative increase of the total emitted EC power relative to the Maxellian bulk only case, $\Delta(\%) = \left(P_{EC} - P_{EC}^{Max}\right) / P_{EC}^{Max}$, is of about 10% and 5% for, respectively, the on-axis and off-axis cases with $\mu_c = -0.2$ and $\eta_1 = (2/3)\eta$, and of about 200% for the case of a suprathermal population extending over the wider radial range.

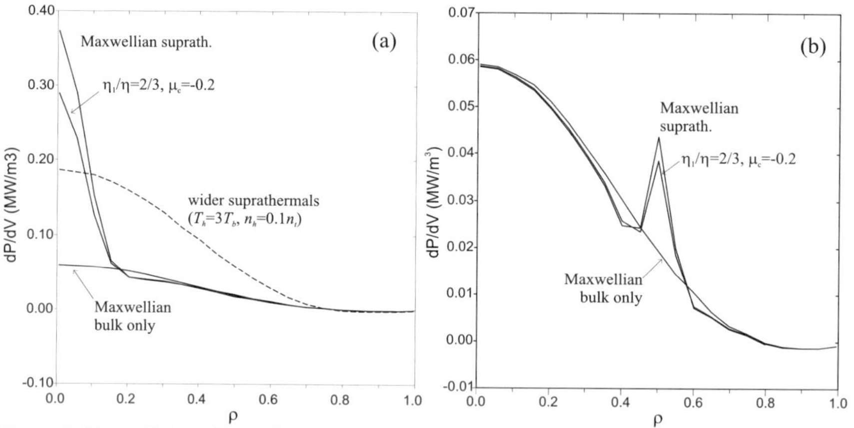

Figure 3. The radial profile of the net EC radiated power density *dP/dV* for an on-axis (plot a) and off-axis (plot b) anti-loss-cone ($\mu_c = -0.2$) population of the suprathermals with the parameters of Fig. 2 and $\eta_1 = (2/3)\eta$ either localized around ρ_h (cf. Figs. 1) or extending over the wider radial range (dashed curve in plot a). The cases of Maxwellian suprathermals, $\mu_c = 1$, and of the Maxwellian bulk only are also shown.

4. Discussions and conclusions

The RAYTEC code was extended to include EC emission of a population of suprathermal electrons having a combination of a Maxwellian and /or a loss-cone or anti-loss-cone distribution. The code has been used to study the impact of anisotropy and localization effects on the net emitted EC power density *dP/dV* for suprathermals present in a narrow range around the plasma centre and half-way to the wall. For comparison, a population extending over a wider radial range was also considered.

The strong local enhancement of *dP/dV* thus obtained, for a well localised population, as to be expected, only amounts to a slight increase of the total EC radiated power which, as noted earlier, is no longer the case for a suprathermal population extending over a wide radial range.

References

1. F. Albajar, M. Bornatici, F. Engelmann, *Nucl. Fusion* **49**, 115017 (2009)
2. I. Fidone et al, *Nucl. Fusion* **31**, 2167 (1991)
3. K.V. Cherepanov, A.B. Kukushkin, Proceedings of the 31th EPS Conf. on Plasma. Physics (London, June 2004)
4. P.A. Robinson, *Plasma Phys. Control. Fusion* **27**, 1037 (1985)
5. F. Albajar, M. Bornatici, F. Engelmann, *Nucl. Fusion* **42**, 670 (2002)

TESTING OF ECESIM CODE IN MATLAB FOR EC EMISSION ESTIMATES FROM ITER PLASMA IN THE PRESENCE OF POLARIZATION SCRAMBLING IN WALL REFLECTIONS

SUMAN DANANI[1], HITESH KUMAR B. PANDYA[1], M.E.AUSTIN[2], P.VASU[1]

[1]*ITER-India, Institute for Plasma Research, A-29, GIDC, Sector-25, Gandhinagar-380025, India*
[2]*Fusion Research Center, the University of Texas at Austin, Austin, TX, USA*

Abstract. We have tested a MATLAB version of the ECESIM code [1] The ECESIM code, originally in IDL, calculates the Electron Cyclotron Emission (Radiation Temperature) given the electron temperature, density and magnetic field radial profiles of a tokamak plasma. This was developed for the analysis of DIII-D ECE and has been shown to be appropriate for the mildly relativistic finite density regime of ITER [2]. The physics of code is based on the theoretical formulation by Bornatici [3]. The MATLAB version of the ECESIM code has been used to calculate the X-and O-mode ECE emission for ITER Scenario-2. From the X and O mode emissivities for ITER, we have identified the frequencies which may be particularly vulnerable to any polarization scrambling which might take place prior to the polarizations being separated.

Email of Suman Danani: sumanipr@gmail.com

INTRODUCTION

The Electron Cyclotron Emission (ECE) is a well established means of measuring electron temperature in tokamak plasma devices. Typically, emission is measured perpendicular to the toroidal magnetic field, such that the radiation is along the major radius R of the tokamak, and the radial temperature profile $T_e(r)$ is inferred. Most measurements use first harmonic Ordinary (O) mode and the second harmonic Extraordinary (X) mode which are optically thick and provide a simple relation between measured radiation and electron temperature. In case of ITER, the first harmonic extraordinary mode is also optically thick but is not available for measurement as it cannot propagate out as can be seen from Fig. 2. It can also be seen from Fig. 2 that at a particular frequency more than one cyclotron harmonic may exist in the plasma along the viewing path. For the case of harmonic overlap, the measured emission accounts for emission at resonances and also any absorption, as well as the possibility of wall reflections of the radiation, and its subsequent absorption. Therefore a radiation transport model has to be employed that correctly calculates the absorption and emission at the various harmonics.

In this paper we first discuss the theoretical aspects of electron cyclotron emission and absorption. We present results of the MATLAB version of ECESIM code which calculates the Radiation Temperature for X and O-mode ECE for ITER Scenario-2. The results are compared with those from the original code. Also a comparison between the calculated and experimental data for FTU second harmonic X-mode ECE is presented. In the next section we present the effect of mode scrambling of O-mode in to X-mode at the frequencies where the intensity of second harmonic O-mode is large and comparable to the second harmonic X-mode intensity. In the final section summary of the paper is given.

THEORETICAL CONSIDERATIONS

The transport of radiation is described by the well known equation of radiation transport:

$$N_r^2 \frac{d}{ds}\left(\frac{I(\omega)}{N_r^2}\right) = j(\omega) - \alpha(\omega) I(\omega) \tag{1}$$

where N_r is the ray refractive index of the mode under consideration, $I(\omega)$ is the specific intensity of the radiation i.e. the power flux density per unit of solid angle and frequency, s denotes the trajectory of the ray traversing the plasma. The first term on the right hand side of eq.1 describes the wave emission from the plasma with $j(\omega)$ being the emission coefficient i.e. the power radiated per unit volume per frequency interval $d\omega$ per steradian in the direction of the ray s. The second term on the right hand side accounts for absorption with $\alpha(\omega)$ being the absorption coefficient, and defined as the fractional rate of absorption per unit path length. The solution of Eq. (1) for the intensity $I(\omega)$ emerging from the plasma is given by:

$$I(\omega) = I_{inc}(\omega) e^{-\tau_0} + \int_0^{\tau_0} S(\omega) e^{-\tau} d\tau \tag{2}$$

Here $I_{inc}(\omega)$ is the intensity of the incident radiation, $S(\omega)$ is a source function defined as

$$S(\omega) = \frac{1}{N_r^2} \cdot \frac{j(\omega)}{\alpha(\omega)} = \frac{\omega^2}{8\pi^3 c^2} T_r \tag{3}$$

where T_r is the radiation temperature, $T_r = T_e$ for maxwellian plasma

and τ_0 is the total optical depth, $\tau_0 = \int_{in}^{out} \alpha(s) ds$ (4)

and the optical depth τ is given by $\tau = \int_{s}^{out} \alpha(s')ds'$ (5)

The emission coefficient $j(\omega)$ is defined as

$$j(\omega) = \int \eta(\omega)ds \tag{6}$$

where $\eta(\omega)$ is the differential rate at which energy is emitted spontaneously per unit solid angle per $d\omega$ by an electron given by

$$\eta(\omega) = \alpha(\omega)T_e \exp(-\tau(\omega)) \tag{7}$$

The radiation temperature is determined from the emission coefficient. For a single frequency, the contribution to emission from various harmonics is considered for the wave travelling from inboard side to outboard side. The mathematical expression can be expressed as:

$I_X = j(\omega 1,1)exp(-\tau_1) + j(\omega 1,2)exp(-\tau_1-\tau_2) + j(\omega 1,3)exp(-\tau_1-\tau_2-\tau_3) + \ldots$ *up to 10 harmonics*

This is calculated for both the X and O-mode. Similarly the emission for the wave travelling from outboard side to inboard side is calculated. Total emission is then a sum over all the harmonics.

The effect of polarization scrambling due to wall reflections [4] is considered and the intensity I' after a single reflection is given as follows:

$$I'_X = r(1-p)I_X + rpI_O \qquad I'_O = r(1-p)I_O + rpI_X \tag{8}$$

The number of reflections can be specified in the code. In practice, it is found that after 7-11 reflections the calculated emission intensity reaches a steady state.

ECE SIMULATION

For all the calculations we have used the Temperature, Density and Magnetic field profiles for ITER Scenario-2 as shown in the figures below.

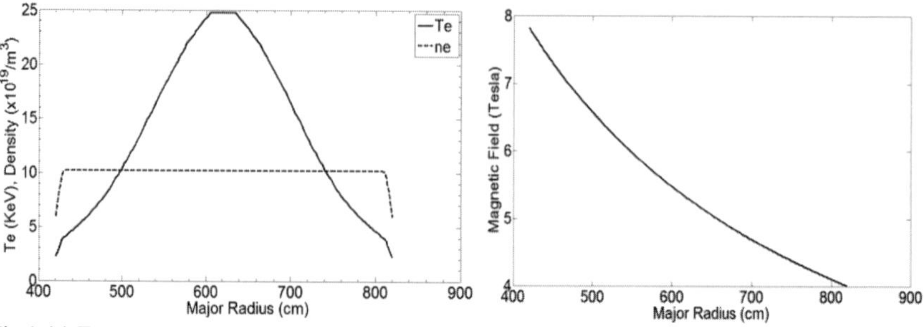

Fig.1 (a) Temperature (T_e) and density (n_e) profile   Fig.1 (b) Magnetic Field (B) profile

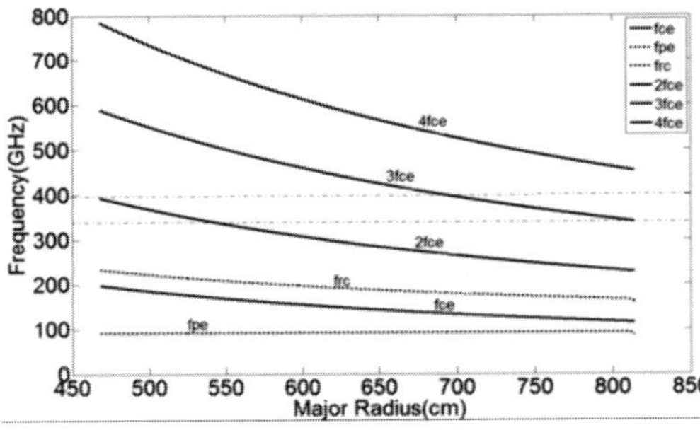

Fig 2 shows the characteristic frequencies for ITER Scenario-2. It can be seen from fig (2) that second and third harmonics are overlapping. There is frequency downshifting due to relativistic effect, which results in an overlapping of the first harmonic and second harmonic also.

Computer codes have to be used that calculate absorption and emission coefficients for multiple harmonics in high temperature plasmas. We developed the MATLAB version of ECESIM code (ECESIM-M) and used it to calculate the Radiation Temperature for X and O-mode ECE for ITER. These results are compared with the original IDL version, which matched reasonably well, as shown in Fig.3 (a). This code was also used for radiation temperature calculation for FTU and the results are compared with the available experimental data for FTU as shown in Fig.3 (b).

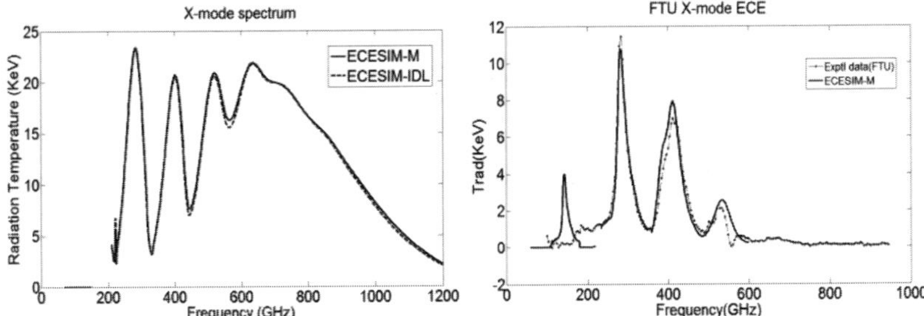

Fig. 3 shows the comparison of T_{rad} calculated using ECESIM-M (a) with the original IDL version for ITER X-mode ECE radiation temperature (b) with the available X-mode radiation temperature spectrum from thermal FTU plasma (with ECRH), for FTU X-mode radiation temperature

Effect of Harmonic Overlap on Temperature Measurement

The effect of harmonic overlap on temperature measurement has been studied. A comparison of input T_e profile and the calculated radiation temperature profile for the first harmonic O-mode and the second harmonic X-mode shows that there is harmonic overlapping due to relativistic downshift of higher harmonics. It is clear that the data needs to be properly analyzed in order to derive the correct temperature T_e information in the regions inboard side of centre.

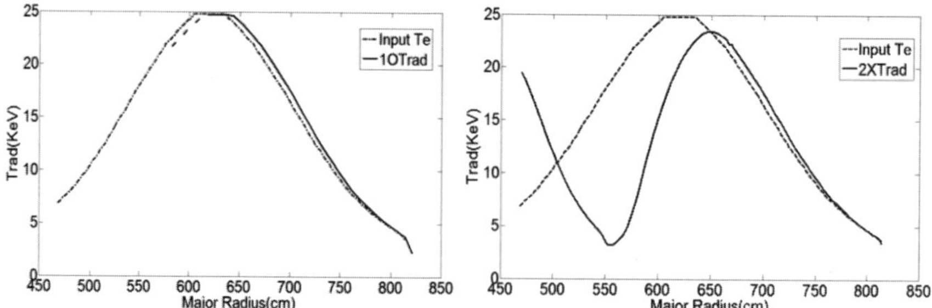

Fig. 4 (a) & (b) The above figures show a comparison of the input profile Te and the simulated profile Trad for the 1st harmonic O-mode and the 2nd harmonic X-mode. As can be seen from the figures, there is a good match between the input and calculated temperature from outboard side upto the centre.

Effect of Polarization Scrambling

There exists a possibility of mode conversion from O-mode to X-mode during propagation of the wave through the waveguide before it gets separated into the two modes via a beam splitter. This effect may become dominant at frequencies where the O-mode intensity exceeds the X-mode intensity in case of X-mode measurement. An example of such a situation is shown in Fig.5 which shows the second harmonic X and O-mode intensities. As can be seen, in the region of frequency from 310-360 GHz, the O-mode intensity is much higher than the X-mode intensity. Hence the risk of mode mixing of O-mode into X-mode has to be considered. This effect is quantified by mixing 5% and 10% O-mode intensity into X-mode intensity as shown in fig.6. For frequencies between 316-357 GHz, the effect of mode mixing varies from 10% to 23% even in case of 5% O-mode mixing, which exceeds the temperature measurement accuracy criterion of 10%.

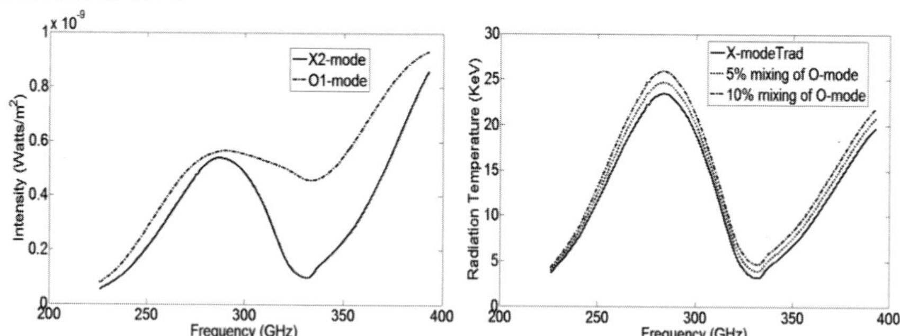

Fig. 5 shows the calculated intensity for the X and O- mode.

Fig. 6 shows the effect of polarization mixing of 5% and 10% of O-mode radiation respectively into the X-mode.

There is also a possibility of mixing of the ordinary and extraordinary modes due to wall reflections which is commonly referred to as polarization scrambling. The effect of polarization scrambling is known to be mainly on the optically thin harmonics [2, 5]. This is clearly seen in Fig. 7 which has been calculated using Eq. 8 with *r=0.76*, for different values of *p*.

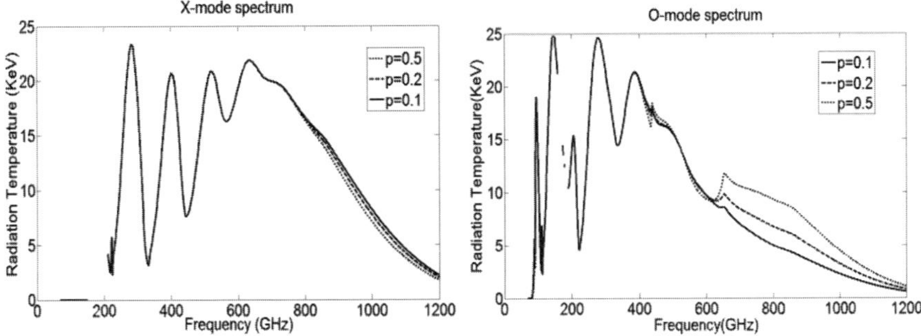

Fig. 7 (a) & (b) represent the effect of polarization scrambling due to wall reflections on X and O-mode ECE, for a wall reflection coefficient, *r=0.76*, and different values of the polarization scrambling parameter, *p*.

CONCLUSIONS

A MATLAB version of ECESIM code has been tested against the original IDL version. As a result of harmonic overlap arising due to frequency downshift, neither X or O-mode radiation temperature can directly yield T_e in the inboard side. In the range of 310-360 GHz, the second harmonic O-mode intensity is considerably in excess of the X-mode intensity, so even a 5-10% conversion of O-mode into X-mode prior to polarization separation, could distort the temperature profile even in the outboard side by more than 10%. In the inboard region, the distortion in the apparent temperature can be more than 40%. Its effect on the derived T_e is to be ascertained by more appropriate error analysis.

REFERENCES

[1] M.E.Austin, private communication.

[2] M.E. Austin, R.F. Ellis, A.E. Hubbard, P.E. Phillips, W.L. Rowan, "*Receiver front-end optics and performance assessment for ITER ECE system,*" S006937-F ITER ECE Report (U. Texas).pdf, February 2007, http://www.pppl.gov/usiter-diagnostics/Instrumentation-Packages/Electron-Cyclotron-Emission/

[3] M. Bornatici, *Review Paper, Electron Cyclotron Emission and Absorption in Fusion Plasmas, Nuclear Fusion, Vol.23, No.9, 1983*

[4] A.E.Costeley, R.J.Hastie, J.W.m.Paul, J.Chamberlain, Physics Review Letters 33, 758 (1974)

[5] F.Albajar, M.Bornatici, F.Engelman Electron Cyclotron Radiative transfer in the presence of polarization scrambling in wall reflections, Nuclear Fusion 45(2005) L9-L13

A 105 GHz Notch Filter for mm-Wave Plasma Diagnostics

V. Furtula[1], P. K. Michelsen[1], F. Leipold[1], M. Salewski[1], T. Johansen[2], S. B. Korsholm[1], F. Meo[1], D. Moseev[1], S. K. Nielsen[1], M. Stejner[1]

[1] *Association Euratom - Risø National Laboratory for Sustainable Energy, Technical University of Denmark, DK-4000 Roskilde, Denmark*

[2] *DTU Elektro, Technical University of Denmark, DK-2800 Lyngby, Denmark*

vefu@risoe.dtu.dk

Notch filters are the most important mm-wave components to protect front-end receivers from intensive gyrotron stray radiation in fusion plasmas. Here we describe a notch filter design with 105 GHz center frequency and 200 MHz rejection bandwidth. The design is based on a fundamental rectangular waveguide with cylindrical cavities coupled by narrow iris gaps, i.e. small elongated holes of negligible thickness. We use numerical simulations to study the sensitivity of the notch filter performance to changes in geometry and in material properties within a total bandwidth of ±10 GHz. The typical insertion loss in the passband is below 1.5 dB, and the attenuation in the stopband is approximately 40 dB.

1. Introduction

In magnetic confinement fusion devices mm-wave notch filters are used to protect diagnostic instruments from gyrotron stray radiation. Typical mm-wave diagnostics are electron cyclotron emission (ECE) spectroscopy, reflectometry, and collective Thomson scattering (CTS) [1]. Gyrotrons produce high power electromagnetic waves in the mm-wavelength range and are used for electron cyclotron resonance heating (ECRH) and current drive (ECCD), for mitigation of tearing modes, and as a probing radiation source in CTS experiments. Notch filters can make plasma diagnostics compatible to plasma experiments with gyrotron operation. The tokamak ASDEX Upgrade (AUG) is equipped with dual frequency 1 MW gyrotrons. Operation at 140 GHz is used for ECRH and ECCD [2]; operation at 105 GHz is used for fast ion measurements by CTS [3–5] which is the application of the notch filter presented here.

The required rejection bandwidth depends on whether stray radiation from a single gyrotron or from several gyrotrons is to be blocked. For a

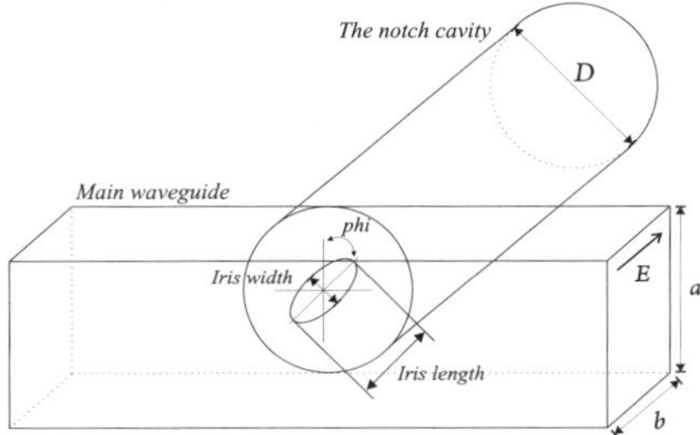

Fig. 1. T-junction aperture coupling using iris and horizontal cavity scheme

Fig. 2. 3-D sketch and constructed 105 GHz notch filter with a main waveguide and 6 resonator cavities

single gyrotron, a rejection bandwidth 130 to 200 MHz around the nominal gyrotron frequency is sufficient to accommodate gyrotron frequency chirp [3]. Depending on scattering geometry and bulk ion temperature the width of the measured CTS spectra can be as narrow as ± few hundred MHz. The rejection band must therefore be as narrow as possible since otherwise valuable information about the velocity distribution of bulk ions would be lost. If several gyrotrons are used, typical for ECRH or ECCD, the rejection band (stopband) must be wider due to slightly different center frequencies among the gyrotrons (up to 1 GHz [3,6]). This is compensated by broadening the rejection bandwidth of the notch filter. This type of filters will be presented elsewhere.

The CTS notch filters should have maximum 3 dB rejection bandwidth of 200 MHz, and they should have a low, frequency independent insertion loss, and the portion of reflected power should be low. F-band notch filters with notch frequencies up to 140 GHz with up to 100 dB rejection and typical bandwidths of few GHz have been designed [7,8]. The passband for notch filters for CTS is ± 5 GHz on AUG, so as to measure fast deuterons with energies up to 500 keV [3]. Several gyrotrons will be installed on ITER with frequencies of 170 GHz for ECRH and ECCD [9,10], of 120 GHz for machine start-up [11,12] and of 60 GHz for the CTS diagnostic [13–17]. ITER will also be equipped with reflectometry diagnostic [18] and ECE spectrometry [19] and therefore also with notch filters as presented here [20].

This paper will present the design of a compact and sensitive F-band notch filter with a possibility of tuning the center frequency ±2%. We assess the sensitivity to changes in physical parameters and material properties by numerical simulation finalizing the paper with S-parameter simulations.

2. Iris coupled T-junction in Circular Waveguide

The narrow 3 dB rejection bandwidth of notch filters for fusion diagnostics can be achieved using a symmetrical right-angle T-type junction of a rectangular and a circular guide coupled by a small elliptical aperture in a metallic wall of negligible thickness (iris). A great benefit of an iris design is the simple cavity adjustment with only a single tuning screw per cavity. The major axis of the elliptical aperture forms the angle ϕ with respect to vertical direction (fig. 1). The angle ϕ is restricted to 0 or $\pi/2$ to ensure that only the fundamental TE_{11} mode can propagate in the circular guide. This mode has optimum passband coverage and relatively low loss. Accordingly, the TE_{10} mode propagates in the rectangular waveguide which is also the fundamental mode for this transmission line. We choose an aperture in a metallic wall with 0.2 mm minimum thickness for the simulation as lower dimensions cannot be realized. The finite thickness introduces only a slight translation in frequency characteristics. The effect of the finite thickness is small because 0.2 mm is much smaller than the vacuum wavelength $\lambda_0 = c/140 \cdot 10^9 = 2.86$ mm. Another important parameter is the cut-off wavelength λ_c [21, p.460] that determines the lowest frequency in a waveguide. For the main guide $\lambda_c = 2a = 4.06$ mm, where $a = 2.032$ mm is the broad side of the guide. A waveguide with circular cross-section has a cut-off wavelength $\lambda_{c,nm} = \pi D/p_{nm}$ where $D = 2$ mm is the cavity diameter and p_{nm} are zeros in the Bessel functions of the first kind. For the TE_{11} mode we have $\lambda_{c,11} = 2\pi/1.841 = 3.41$ mm. One should always verify that $\lambda_c > \lambda_0$ for all types of waveguides. The distance between the resonators is $\Delta l = \lambda_g/4 + n\lambda_g/2$ where $\lambda_g = \lambda_0/(1 - (\lambda_0/\lambda_c)^2)^{1/2} = 5.2$ mm and $n = \{0, 1, 2\ldots\}$. In order to minimize insertion loss (IL) we set $n = 0$, so $\Delta l = 1.3$ mm.

3. Numerical Methods and Boundary Conditions

The electromagnetic field distribution inside a waveguide and resonator cavities can be described by Maxwell's equations with appropriate boundary conditions [21, pp.24-28]. If we know the fields existing in the structure, we can uniquely determine the scattering parameters. The simulator used for the filter characterizations is Computer Simulation Technology (CST) with a real time domain simulator based on the finite element method. The scattering parameters depend sensitively on the geometrical setup. For the simulation we assume air as medium; the surroundings are chosen to be silver. The test ports are placed at the beginning and at the end of the

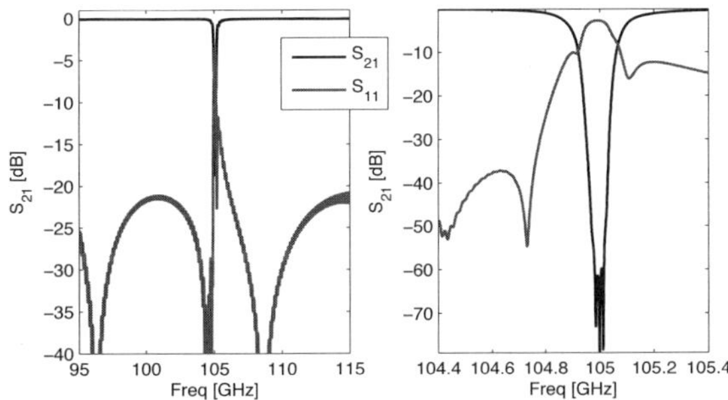

Fig. 3. Simulated insertion loss (S_{21}) and return loss (S_{11}) of a wideband notch filter centered around 105 GHz. Figure to the right has 1 GHz simulation bandwidth.

main rectangular waveguide.

4. Results

Figure 3 shows the computed S-parameters as function of frequency where the narrow, deep notch and its width at the center frequency become apparent. Figure 4 shows the sensitivity of the center frequency f_0, the 3 dB bandwidth BW_{3dB}, and the shape factor to changes in filter geometry and material conductivity. The new constellation for the W-band notch filter is found to differ from the design in [6] in one issue: the cavity length tuning shown in fig. 4(c) is relatively sensitive, i.e. -5.5 GHz/mm. However, this is not considered to be a problem since frequency shift is a linear function of the cavity length. The 3 dB rejection bandwidth is even less sensitive compared to the previous design [6]. The shape factor is at an acceptable level around 0.5. If required, shifting the center frequency by few GHz is not expected to degrade the passband coverage, 3 dB bandwidth or shape factor. Figure 4(b) shows that the iris width has a large impact on the 3 dB rejection bandwidth and thereby the shape factor. If we choose the iris width to be 0.2 mm, the 3 dB rejection bandwidth will become 370 MHz broad which is considerably more that the required 200 MHz. The simulations indicate that the iris width should be as small as possible, namely 0.1 mm, resulting in a 3 dB bandwidth of 190 MHz. The iris length in fig. 4(c) is less sensitive compared to the iris width and is particulary interesting since it shows the non-linear nature in center frequency drift. From fig. 4(c) we see that the iris length must be comparable to the cavity diameter ($\sim$ 1.9 mm) to meet the requirement of narrow rejection bandwidth. Fortunately, this is also the point with low sensitivity as seen from the curve gradient. The

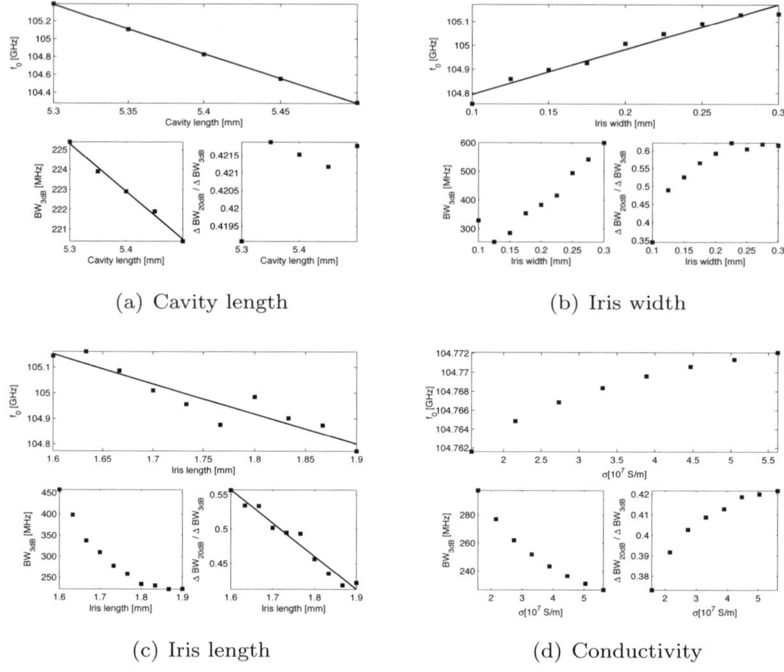

(a) Cavity length (b) Iris width

(c) Iris length (d) Conductivity

Fig. 4. Sensitivity of the center frequency f_0, the 3 dB bandwidth, and the shape factor to changes in the specified parameter. Symbols: simulations; line: least squares fit. Not all data sets have feasible point distribution for least squares fits.

material choice is an important factor when trying to minimize the overall passband transmission loss, especially at the points around the rejection band. A clear difference in shape factor from good to very good conductor appears in fig. 4(d). Commonly either copper ($\sigma = 5.96 \cdot 10^7$ S/m) or silver-coating ($\sigma = 6.3 \cdot 10^7$ S/m) are used as a construction material for frequencies above 110 GHz. For frequencies between 30 and 110 GHz it is recommended to use aluminium ($\sigma = 3.77 \cdot 10^7$ S/m) or copper leaving silver-coating as an option.

5. Discussion and Conclusions

A notch filter with broad passband coverage has been presented. The T-junction iris coupling in fig. 1 is suitable for narrow stopband design with dynamic notch frequency adjustment. The key design parameters such as center frequency f_0, rejection bandwidth BW_{3dB} and shape factor have been investigated by small changes in coupling geometry and resonator cavity lengths. From fig. 4 we see that several design parameters are sensitive to geometry changes. We also notice that center frequency, f_0, tuning is relatively sensitive to cavity length, i.e. -5.5 GHz/mm. We have shown that this notch filter design has rejection depth of $\sim$ 65 dB and rejection bandwidth of $\sim$ 200 MHz both satisfying the requirements. Return loss is better than 20 dB everywhere in the passband region.

Acknowledgments

This work, supported by the European Communities under the contract of Association between EURATOM / Risø DTU, was partly carried out within the framework of the European Fusion Development Agreement. The views and opinions expressed herein do not necessarily reflect those of the European Commission.

References

1. N. C. Luhmann *et al.*, *Fusion Sci. Tech.* **53**, 335 (2008).
2. D. H. Wagner *et al.*, *IEEE Trans. Plasma Sci.* **36**, 324 (2008).
3. F. Meo *et al*, *Rev. Sci. Instrum.* **79**, p. 10E501 (2008).
4. M. Salewski *et al.*, *Nucl. Fusion* **50**, p. 035012 (2010).
5. F. Meo *et al.*, *J. Phys.: Conf. Series, Proceedings of LAPD-14* (2010).
6. V. Furtula *et al.*, *Rev. Sci. Instrum, Proceedings of HTPD-10* (2010).
7. Y. Dryagin *et al.*, *Int. J. Infrared mm Waves* **17**, 1199 (1996).
8. T. Geist and M. Bergbauer, *Int. J. Infrared mm Waves* **15**, 2043 (1994).
9. B. Piosczyk *et al.*, *Fusion Engineering and Design* **66-68**, 481 (2003).
10. A. Kasugai *et al.*, *Nuclear Fusion* **48**, p. 054009 (2008).
11. E. M. Choi *et al.*, *J. Phys.: Conf. Series* **25**, p. 1 (2005).
12. K. Felch *et al.*, *J. Phys.: Conf. Series* **25**, p. 13 (2005).
13. M. Salewski *et al.*, *Rev. Sci. Instrum.* **79**, p. 10E729 (2008).
14. S. B. Korsholm *et al.*, *Burning Plasma Diagnostics* **988**, p. 118 (2008).
15. F. Leipold *et al.*, *Rev. Sci. Instrum.* **80**, p. 093501 (2009).
16. M. Salewski *et al.*, *Plasma Phys. Control. Fusion* **51**, p. 035006 (2009).
17. M. Salewski *et al.*, *Nucl. Fusion* **49**, p. 025006 (2009).
18. G. Vayakis *et al.*, *Nuclear Fusion* **46**, p. S836 (2006).
19. G. Vayakis *et al.*, *Fusion Engineering and Design* **53**, 221 (2001).
20. P. Woskov, *Proceedings of LAPD-13* (2007).
21. P. A. Rizzi, *Microwave Engineering - Passive Circuits* (Prentice-Hall, 1988).

Recent progress of the 20-channel grating polychromator on EAST

Liu Yong[1] and Li Erzhong[1] and Ling Bili[1] and Ti Ang[1] and Han Xiang[1] and Gary Taylor[2]

[1] *Institute of plasma physics, Chinese Academy of Sciences, Hefei 230031, China*
[2] *Plasma Physics Laboratory, Princeton University, Princeton, NJ 08543, USA*

A 20-channel grating polychromator transferred from PPPL, has been re-built for electron cyclotron emission measurements on EAST. This instrument measures the second harmonic electron cyclotron emission from plasma with frequency range from 90 GHz to 250 GHz, which corresponds to a central magnetic field (R_0=1.7 m) of 2-3.5 T. The radial resolution is around 2.5 cm. New pre-amplifiers are made and tested, based on the electronics of GPC-II on TFTR. These amplifiers have a gain of around 520, with a 400 kHz 3 dB roll off frequency.

1. Introduction

Electron cyclotron emission (ECE) measurement has been widely used on tokamak plasmas since 1960s [1], to provide the temporal evolution of local electron temperature [2,3] and the information on the electron velocity distribution [4–6]. Several kinds of frequency-resolving systems have been developed for ECE measurements, based on spectroscopy and microwave techniques, such as the Fourier-transform spectrometer, the Fabry-Perot interferometer, the grating polychromator and the heterodyne radiometer.

Experimental Advanced Superconducting Tokamak (EAST) is a full superconducting tokamak with a non-circular corss-section vacuum vessel [7]. The major and minor radius of the EAST device are separately 1.7 m and 0.4 m. Two ECE measurement systems have been developed on EAST: a 16-channel heterodyne radiometer [8,9] and a 20-channel grating polychromator [10], which is transferred from PPPL. This paper presents an introduction to the present state of the 20-channel polychromator.

2. Instrument

The 20-channel grating polychromator is based on the Ebert-Fastie grating monochromator and some other grating instruments used for plasma diagnostics. Figure 1 illustrates the schematic of the instrument. Electron cyclotron emission from plasma is diffracted at the entrance aperture, which is located at the focal point of the spherical collimating mirror M1. The grating disperses the plane wave reflected by M1 according to the relation

$$n\lambda_m = d(sin\theta_i + sin\theta_r) \tag{1}$$

where d is the grating constant, θ_i and θ_r are respectively the angles of incidence and reflection with respect to the grating normal. n is the order of interference. Note that $\theta_i - \theta_r = 2S_m$ is constant for each exit apertures. Hence, we obtain

$$n\lambda_m = 2dsin(\theta_i - S_m)cos(S_m) \tag{2}$$

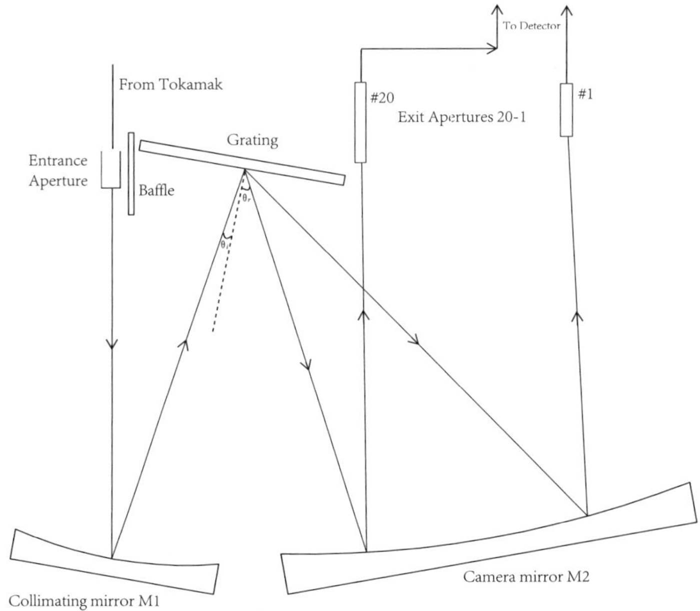

Fig. 1. Schematic of the 20-channel grating polychromator

2.1. *The transmission line*

The second harmonic electron cyclotron emission from the plasma is directly collected and introduced to the polychromator by 8-meters long corrugated waveguide. Three miter bends are employed in the transmission line. To improve the poloidal and toroidal resolution, a quasi-optic antenna composed of an elliptical mirror and a flat mirror, is going to be employed, which is based on experiences on TFTR [10], DIII-D [11], Alcator C-Mod [12] and some other tokamaks.

2.2. *Adjustment of the mirrors M1 and M2*

To adjust the mirrors M1 and M2, the grating is replaced with a flat mirror of the same size with the grating. An alignment laser passes through the center of the entrance aperture and then to the center of M1. M1 was adjusted to place the alignment laser spot on the middle of the flat mirror. The flat mirror was then rotated until the alignment laser spot was located at the middle of mirror M2. M2 was then aligned so the alignment laser spot was between exit apertures 10 and 11.

2.3. *Determination of the grating angles and the exit aperture positions*

To determine the grating angles, we first determine its zero. After the mirrors M1 and M2 are fixed, the flat mirror is rotated until the beam is reflected back on itself, where we call this position the zero position. Then, the exit aperture positions S_m is determined by rotating the flat mirror until the laser spot is on the corresponding exit aperture. The S_m values are shown in Table 1, and the frequencies associated with some typical grating angles are illustrated in Figure 2.

Table 1. The exit aperture positions S_m

Channel	1	2	3	4	5
S_m (degree)	30.39	29.62	28.89	28.15	27.41
Channel	6	7	8	9	10
S_m (degree)	26.70	25.97	25.25	24.48	23.79
Channel	11	12	13	14	15
S_m (degree)	23.13	22.51	21.90	21.35	20.81
Channel	16	17	18	19	20
S_m (degree)	20.20	19.64	19.06	18.49	17.91

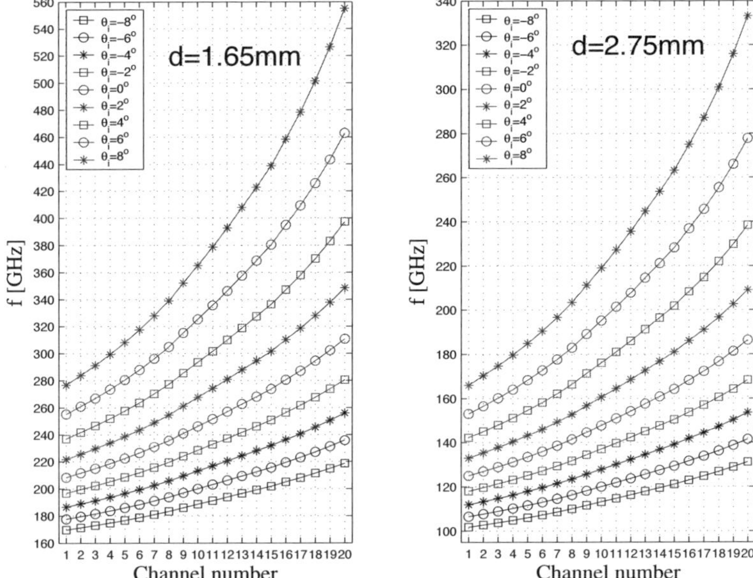

Fig. 2. The frequencies associated with some typical grating angles, for the exit apertures

2.4. *Detector and pre-amplifier*

Hot electron indium antimonide bolometers cooled to 4.2 K are used as detectors, which are mounted in a cryostat with a 6.5 L liquid-helium capacity. New pre-amplifier is designed based on the one used on GPC-II [13], to replace the original one [10], of which some are broken. The test results show that, the amplifier has a gain of around 520, with a 400 kHz 3 dB roll off frequency. Figure 3 illustrates the noise level of the pre-amplifier.

3. Test results and future plan

This polychromator was tested during the 2010 EAST experimental campaign. Figure 4 shows the waveform for the first 3 channels of GPC, together with the plasma current, loop voltage and the linear integrated electron density. The toroidal magnetic field for this shot is 1.97 T (@1.7 m), and the third channel of GPC observes emission from plasma core. The central electron temperature is around 850 eV, calculated from soft x-ray spectrum, and the signal to noise ratio for Channel 3 of GPC is around 20.

In near future, a quasi-optic antenna composed of an elliptical mirror and a flat mirror will be built to improve the poloidal resolution and

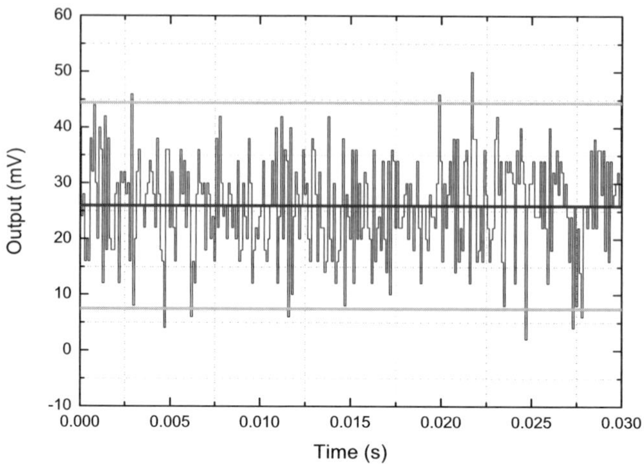

Fig. 3. Noise level of the new pre-amplifier

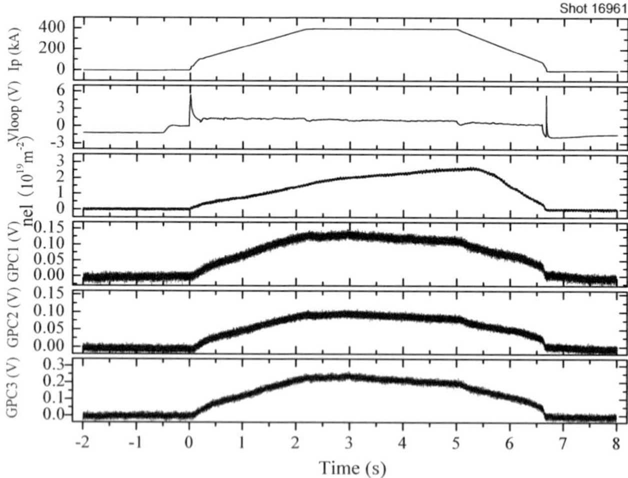

Fig. 4. Waveform for the first 3 channels of GPC, together with the plasma current, loop voltage, linear integrated electron density.

the signal to noise ratio. An in-situ absolute calibration system is also planned. This calibration system will employ a liquid nitrogen-cooled Eccosorb source. Due to the low throughput of the system, a lock-in amplifier and a chopper will be used to detect the ultra weak signal.

References

1. A. J. Lichtenberg, S. Sesnic and A. W. Trivelpiece, *Phys. Rev. Lett.* **13**, 387 (1964).
2. *TFR group, in Controlled Fusion and Plasma Physics (Proc. 7th Europ. Conf. Lausanne, 1975)* **1**, p. 14b (1975).
3. I. H. Hutchinson and D. S. Komm, *Nucl. Fusion* **17**, 1077 (1977).
4. K. Kato and I. H. Hutchinson, *Phys. Rev. Lett.* **56**, 340 (1986).
5. K. Kato and I. H. Hutchinson, *Phys. Fluids* **30**, 3809 (1987).
6. C. Sozzi, E. D. L. Luna, D. Farina, J. Fessey, L. Figini, S. Garavaglia, G. Grossetti, S. Nowak, P. Platania, A. Simonetto, M. Zerbini and J. E. contributors, *Buring plasma diagnostics* , 73 (2008).
7. W. Songtao and EAST Team, *Fusion Engineering and Design* **82**, 463 (2007).
8. A. Ti, B. L. Ling, Q. S. Fei and X. Gao, *Int. J. Infrared Milli. Waves* **28**, 243 (2007).
9. S. Sajjad, X. Gao, B. Ling, A. Ti and Q. Du, *Meas. Sci. Technol.* **19**, p. 075701 (2008).
10. A. Cavallo, R. C. Cutler and M. P. McCarthy, *Rev. Sci. Instrum.* **59**, 889 (1988).
11. R. F. Ellis, M. E. Austin and D. Taussig, *Proc. of the 14th ECE/ECRH Workshop, Santorini, Greece* (2006).
12. R. Chatterjee, P. E. Phillips, J. Heard, C. Watts, R. Gandy and A. Hubbard, *Fusion Engineering and Design* **53**, 113 (2001).
13. M. McCarthy, A. Janos, J. Faunce and Z. Wang, *15th IEEE/NPSS Symposium on Fusion Engineering* , 109 (1992).

A NEW ECE SYSTEM AND ELECTRON TEMPERATURE MEASUREMENTS ON KSTAR TOKAMAK*

SEUNG HO JEONG†

Korea Atomic Energy Research Institute, 1045 Daedeokdaero Daejeon, 305-353, Korea

KYU-DONG LEE[1], YUICHIRO KOGI[2], ATSUSHI MASE[3], YOSHIO NAGAYAMA[4], KAZUO KAWAHATA[4]

[1] *National Fusion Research Institute, 113 Gwahangno Daejeon, 305-333, Korea*
[2] *Fukuoka Institute of Technology, Higashiku Fukuoka, 811-095*
[3] *KASTEC, Kyushu University, Kasuga Fukuoka, 816-8580, Japan*
[4] *National Institute for Fusion Science, Toki Gifu, 509-5292, Japan*

The ECE diagnostic system on KSTAR tokamak has been upgraded in order to follow an increase of the operational magnetic field of KSTAR after the 1st plasma. A new 48CHs ECE radiometer with frequency range of 110 GHz to 162 GHz has been installed and the waveguide system was also replaced by corrugated circular over-mode waveguides. But the ECE collecting optics, which consists of two metallic mirrors, was not changed. With this radiometer, some measurements of the electron temperature at the high field side were carried out during the 2nd KSTAR campaign, when the tokamak was operated with 2T of toroidal magnetic field. The electron temperature as well as its radial profile was measured and sawtooth phenomena were also observed.

1. KSTAR ECE diagnostic system

1.1. *ECE collecting optics*

The KSTAR ECE collecting optics is designed by using Gaussian beam approximation. The optical system is composed of two OFHC mirrors, one is a concave ellipsoidal mirror and the other is plane mirror. The ellipsoidal mirror can be rotated, so it views the blackbody source in the opposite direction of the plasma center while in-situ calibration is being done. The beam size on the ECR layer is about 5 cm in diameter over the frequency range of 110 GHz to 156 GHz. The ECE cassette shown in Fig.1 is a support structure of 2 metal mirrors and 2 vacuum windows. Fig. 1. shows Gaussian beam ray trace for 112GHz.

---

* This work is supported by KSTAR operation project.

† shjeong2@kaeri.re.kr.

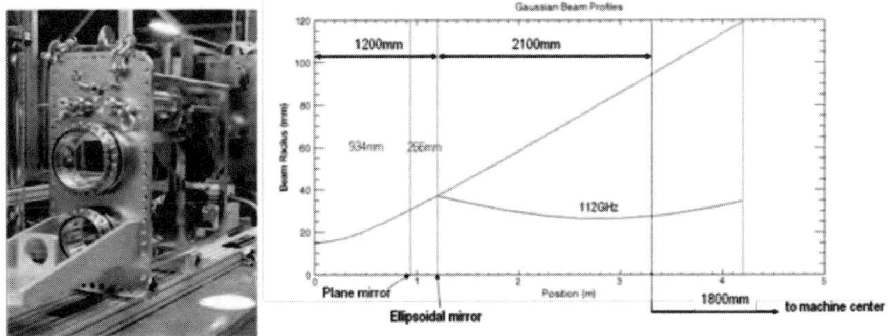

Figure 1. KSTAR ECE cassette and its Gaussian beam ray trace.

1.2. *ECE waveguide system*

In order to enhance the S/N ratio of the ECE system, the smooth circular over-mode waveguide was replaced by the corrugated circular over-mode waveguide which has a frequency range of 50-220 GHz. The waveguide system originally included an 84 GHz notch filter and 2 miter-bends. Then, a dichroic HPF and a WR5-WR6 combined HPF were added to prevent the 110 GHz ECH power. The total transmission loss is less than 3 dB for a length of 30 m.

1.3. *ECE heterodyne radiometer*

The new radiometer system consists of two down conversion stages, amplifiers, band-pass filter banks, and video detectors. The system detects ECE power in the frequency range of 110 to 162 GHz. In the pre-down converter, the ECE signal (RF) is converted to an IF signal in the frequency range up to 26.5GHz. The pre-down converter produces two output branches which are further fed to a second-down converter module. Through a power divider and BPFs, except the frequency range of 1-9.5 GHz, the signals with frequencies of 8-14 GHz, 14-19.5 GHz, and 18-26.5 GHz are fed to mixers and are converted again to signals with frequency range below 9.5 GHz by using other LOs and mixers, and fed to four output ports. The next detector module resolves the signal with the BPF banks and detects the amplitude of the resolved signal by video detectors. The center frequencies of the filter banks cover a range of 2-9 GHz with 1 GHz steps. Microwave integrated circuit (MIC) technology was employed to fabricate the detector module on a planar substrate. The detector signal is digitized after the video amplifier stage.

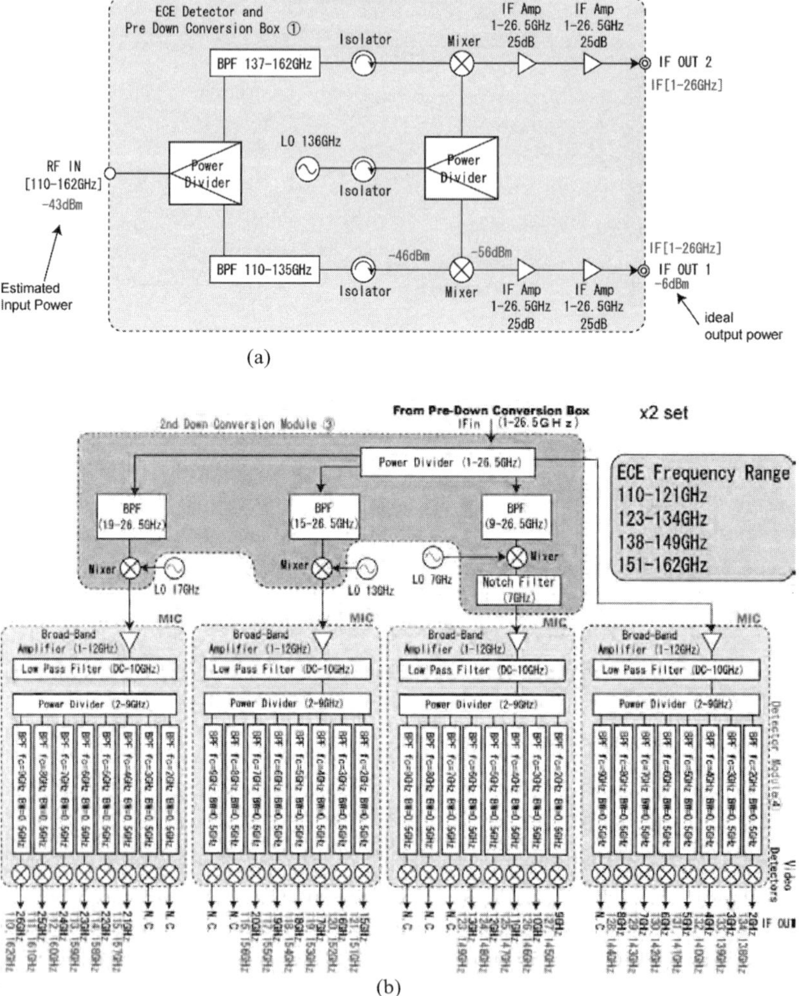

Figure 2. (a) shows a pre-down converter, (b) shows a second down converter.

2. KSTAR ECE system calibrations

Calibration of the ECE system is carried out by using a blackbody emission source, a slowly rotating chopper, digitizer, and the digital integration. The chopper periodically obstructs the line of sight to a high temperature calibration source. Then, the data corresponding to hot or cold temperature are acquired with 50 kHz sampling rate during 200 ms and averaged. The differences of two averaged data are again averaged. The total number of averaged data, k, is 1024.

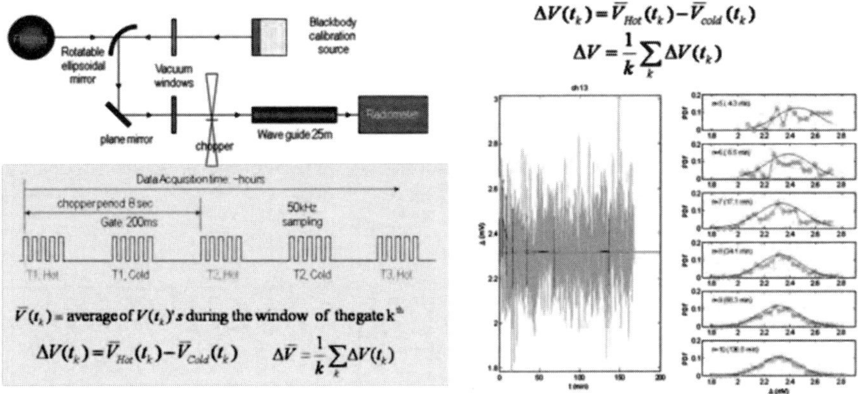

Figure 3. Calibration set up, time sequence of the calibration data acquisitions, and an example of a probability distribution function of a $\Delta V(t_k)$ data set

3. Experimental results

The plasma discharge started up with 250 kW ECH pre-ionization and the ECH power was turned off just before 2.5 s. The toroidal field was 2.0 T. The plasma current reached a flattop of 320 kA at 1.2 s and the flattop was maintained for 1.4 s. In this KSTAR campaign, ECE measurements were carried out for the high field side of plasma. Then, harmonic overlap appeared in the region of the minor radius, r >0.3 m.

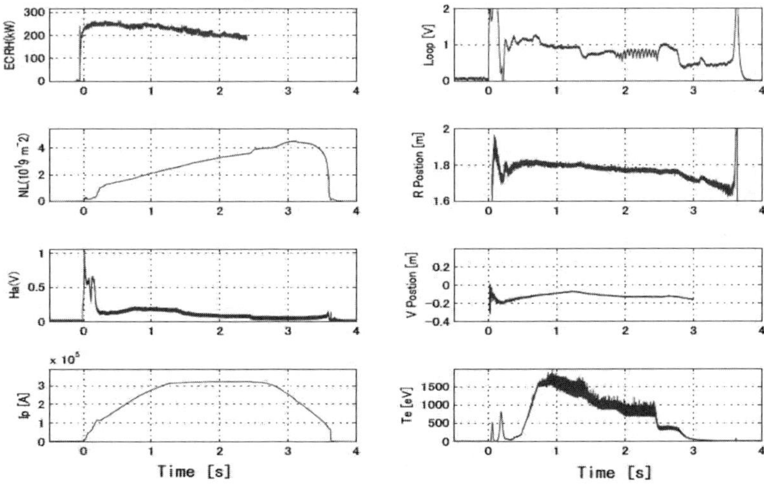

Figure 4. Plasma parameters of #2048 KSTAR tokamak discharge. Some parameters of ECH power, line density, H_a intensity, plasma current, loop voltage, and plasma positions are shown here. The electron temperature of one ECE channel in the core region is also shown

The electron temperature in the core region increased up to 1.5keV while ECH power was turned on. But there were a couple of peculiar behaviors of T_e profiles. First, T_e shown in Fig.5(a) shows the tendency to increase gradually as the radial position varies from r=0.1 m to r=0.5 m while ECH power ON but this is not observed in OH plasma shown in Fig.5(b). Second, the outermost 5 channels show that the edge plasma seems to be affected by ECH more than others which are in the region of r=0.1-0.45 m.

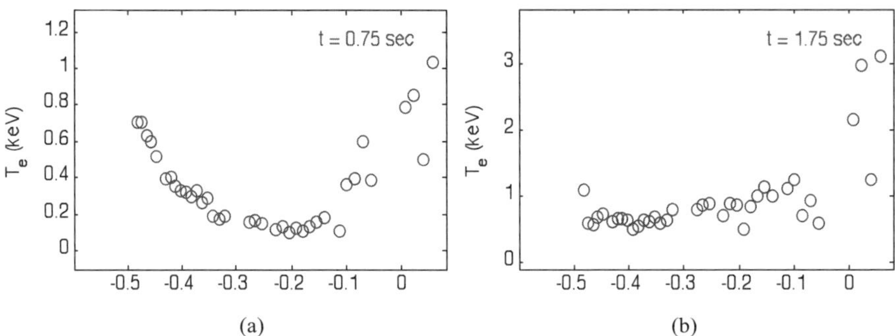

Figure 5. The electron temperature profiles shot#2048 at t=1.5s and shot#2202 at t=1.75s

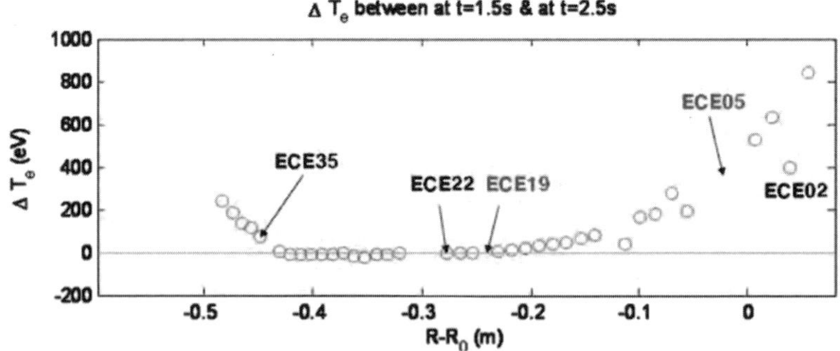

Figure 6. The difference of two electron temperature profiles, one is the profile at t=1.5s(ECH power ON), the other is the profile at t=2.5s(ECH power OFF).

Sawtooth-like energy relaxation phenomena were observed. The sawtooth started during ECH power ON. After ECH power OFF it lasted but the period of sawtooth became about twice longer than early one. The Fig. 7 show that the sawtooth behavior varies with time. The intensity of sawtooth is normalized and the blue color indicates a valley and the red color indicates a ridge. The X-axis is time axis, the Y-axis represents ECE channels. The frequency of top channel(ch#1) is 110Ghz from center region and bottom channel(ch#40) frequency is 156GHz from HFS edge region. The inversion layer(Bold line) moves to the plasma center and then it finally disappears, whereas another inversion layer appears from the edge and moves inward. We guess that the peculiar behavior of edge channels aroused from the 3rd harmonics which have same frequencies but their locations are in the core region.

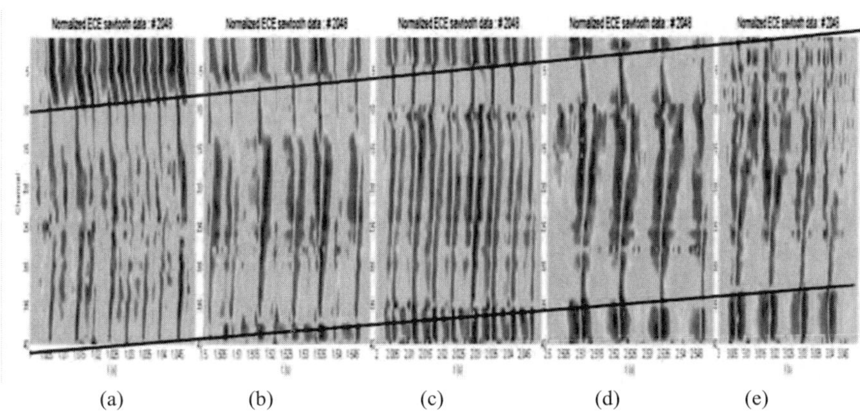

(a) (b) (c) (d) (e)

Figure 7. 2-dimentional(radial x time) behavior of sawtooth at t=1s-1.05s(a), t=1.5s-1.55s(b), t=2s-2.05s(c), t=2.5s-2.55s(d), t=3s-3.05s(e).

Acknowledgments

The new radiometer has been developed by NIFS and Kyushu Univ. as a part of the Korea/Japan KSTAR collaboration program.

References

1. 1.S. H. Jeong, I. Y. Kim, C. K. Hwang, *Rev. Sci. Instrum.*, **74**(3), 1, (2003).
2. 2.Y. Kogi, T. Sakoda, A. Mase, N. Ito, Y. Yokota, S. Yamaguchi, Y. Nagayama, S. H. Jeong, M. Kwon, K. Kawahata, *Rev. Sci. Instrum.*, **79** 10F115(2008).

III. Electron Bernstein Waves

2D EBW EMISSION STUDIES IN MAST

VLADIMIR F SHEVCHENKO, MAARTEN DE BOCK
EURATOM/CCFE Fusion Association, Culham Science Centre, Abingdon, Oxon, OX14 3DB, United Kingdom

SIMON FREETHY, RODDY VANN
University of York, York, YO10 5DD, United Kingdom

ALEXANDER N SAVELIEV
Ioffe Institute, Politekhnicheskaya 26, 194021 St. Petersburg, Russia

Angular scanning of EBW emission (EBE) has been conducted in MAST. From EBE measurements over a range of viewing angles the angular position and orientation of the B-X-O mode conversion (MC) window can be estimated giving the pitch angle of the magnetic field in the MC layer. The radial position of the corresponding MC layer is found from Thomson scattering measurements. Measurements at several frequencies can provide a pitch angle profile. Results of pitch angle profile reconstruction from EBE measurements are presented in comparison with motional Stark effect measurements. Microwave imaging of the B-X-O MC window is proposed as an alternative to the angular scanning. The proposed scheme is based on an imaging phased array of antennas allowing the required angular resolution. Image acquisition time is much shorter than MHD time scales so the EBE imaging can be used for pitch angle measurements even in the presence of MHD activity.

1. Angular Scanning of EBW Emission in MAST

Feasibility studies of thermal Electron Bernstein wave emission (EBE) measurements as a potential edge current diagnostic have been conducted on the MAST tokamak. A broadband radiometer covering the range from 7 GHz to 40 GHz was used as a receiver. In front of the radiometer antenna a fast rotating mirror (FRM) was installed providing a continuous scanning of the viewing direction. Fig. 1a illustrates the relative position of the Bernstein – eXtraordinary – Ordinary (B-X-O) mode conversion (MC) window in the MAST plasma and the range of viewing angles covered by the FRM. From EBE measurements for a range of viewing angles the optimum viewing for B-X-O MC can be found giving the pitch angle of the magnetic field in the MC layer, at the position where $\omega = \omega_{pe}$. EBE measurements at several frequencies allow a pitch angle profile to be reconstructed.

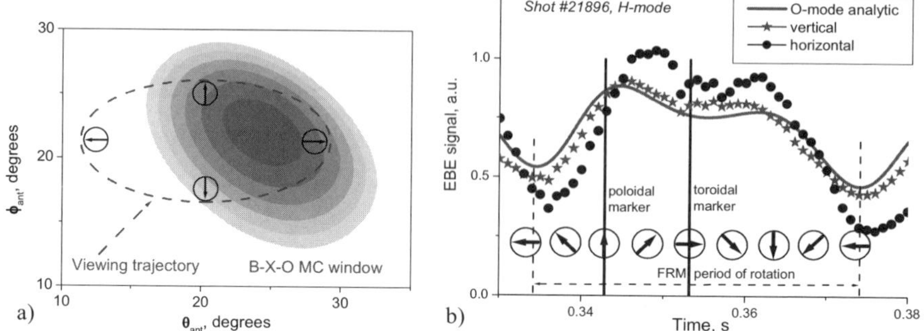

Figure 1. a) Angular range covered by the fast rotating mirror (FRM) and the B-X-O mode conversion window calculated for 16 GHz and pitch angle of 40°. b) EBE 16 GHz signals measured at perpendicular polarisations during one period of FRM rotation and the B-X-O mode coupling efficiency estimated analytically.

Preliminary experiments with the FRM have shown that thermal EC emission from the plasma is highly anisotropic [1, 2]. The resultant signal modulation due to FRM rotation is a combination of a number of factors: geometric modulation, the presence of stray radiation in the vessel and finally anisotropic EC and EBW emission. The range of viewing angles was limited in these first experiments because of the large horn antenna employed. The antenna opening of 100 mm was comparable with the FRM and vacuum window aperture in these preliminary experiments. As a result geometrical effects were responsible for a significant fraction of signal modulation.

In order to increase the range of viewing angles the large horn antenna has been replaced with a very compact antenna DP241-AB manufactured by Flann Microwave. The new antenna receives two linear polarisations simultaneously in the frequency range from 6 to 40 GHz. It has an aperture of 50 mm which is half the vacuum window size. Due to this fact the effect of geometrical modulation has been suppressed in the whole frequency range of the radiometer.

Fig. 1b illustrates radiometer signals measured during ELM-free H-mode in MAST. These signals were recorded simultaneously at two perpendicular polarisations at 16 GHz during one period of FRM rotation. The B-X-O MC efficiency is shown for comparison. MC was estimated using a WKB formula with magnetic shear effects taken into account [3]. The shape of both signals is very close to that predicted by WKB MC theory. One can see that both signals have well pronounced dual maxima which mean the optimum viewing angles for B-X-O MC are inside the ellipse scanned by the FRM. The amplitudes of the signals are similar near the maxima indicating that the polarisation of the registered emission is close to circular which corresponds to the oblique O mode. Near the minima the signal with vertical polarisation is stronger than horizontal indicating that the B-X tunnelling may become comparable with B-X-O MC for the viewing angles close to perpendicular.

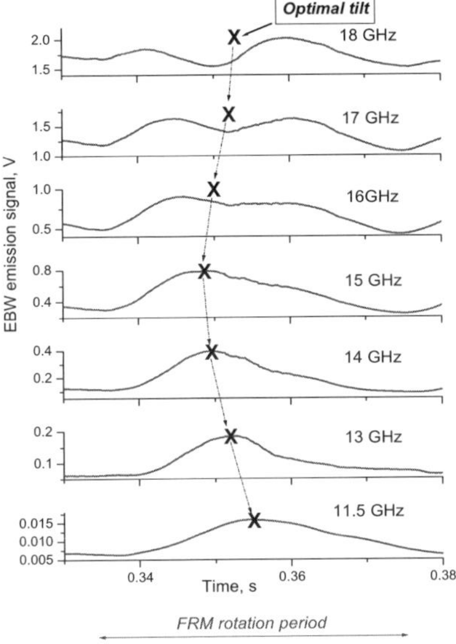

Figure 2. EBE signals in a range of frequencies measured during one FRM rotation period.

These observations prove that the predominant fraction of these signals can be attributed to the mode converted EBE. Indeed, the amplitudes of the radiometer signals correspond to radiative temperatures of several hundred eV indicating that emission comes from the plasma core. EC harmonics up to the third are blocked by cut-offs prohibiting ECE completely. At the same time the signals are dramatically suppressed during ELMs which disturb only the pedestal region of the plasma where MC occurs. All these observations support the EBE nature of the signals measured. However the magnitude of the contribution from stray radiation is not clear.

To estimate the stray radiation fraction we assume that stray emission is isotropic, is proportional to EBE and its fraction is independent of frequency. The maximum signal modulation due to FRM rotation, up to a factor of 4, is usually observed for frequencies from 13 to 15 GHz (see Fig. 2). At these frequencies the optimum direction is just outside the scanning range that makes modulation stronger. This requires the stray radiation fraction to be not higher than 10% otherwise the shape of the signals can not be reproduced by WKB MC simulations or full wave modelling. Two important conclusions can be drawn out from experimental signals presented in Fig. 1 and Fig. 2. The radiometer signals indeed correspond to the B-X-O mode converted EBE. The B-X-O MC window is wider than estimated using density gradients from Thomson scattering (TS) measurements directly. The edge gradient is usually smoothed due to the finite size of the scattering volumes. In order to account for the difference the edge TS profile have been fitted using a modified hyperbolic tangent (mtanh) function convoluted with the instrument function representing the spatial resolution of the diagnostic. The modified TS profiles agree well with EBE measurements.

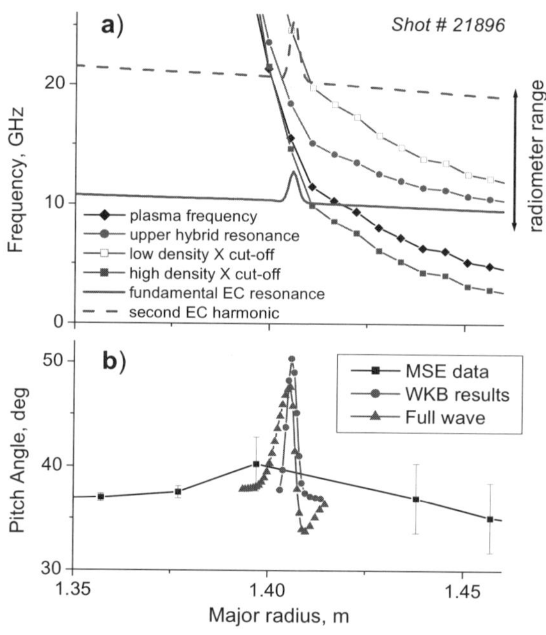

Figure 3. a) Midplane topology of resonances and cut-offs near the plasma boundary. b) Magnetic pitch angle measured with MSE and reconstructed from EBE in the framework of WKB approximation and full wave analysis.

Fig. 2 represents EBE signals obtained on MAST during ELM-free period of H-mode in a range of frequencies covering the fundamental EC resonance. The "X"s indicate viewing orientations when the viewing angle tilt was close to the pitch angle. The radial position of the corresponding MC layer is found from high resolution TS measurements (see Fig. 3a). WKB modelling of the MC process was used for reconstruction of the pitch angle profile from EBW emission data. The WKB model reproduces well the shape and general behaviour of the EBW signals. The magnetic pitch angles were reconstructed using the best fit method between the data and the model.

A full wave analysis has been employed later for verification of WKB results. Pitch angle reconstruction from EBW measurements in comparison with the motional Stark effect (MSE) diagnostics are shown in Fig. 3b. One can see all the EBE data lie in a very narrow ~1 cm layer and indicate dramatic changes in the magnetic pitch angle within this layer. Explanation of this phenomenon is outside the scope of the present paper and will be given elsewhere. The estimated total field variations in the pedestal region can be seen in Fig. 3a from the behaviour of the EC resonance curves. Similar but less accurate results have been obtained earlier from EBE spectroscopic measurements on MAST [4]. Measurements at higher frequency (15–37 GHz), corresponding to the layers located deeper into the plasma show good agreement with MSE data. The WKB results have been verified with full wave modelling. In general WKB reconstruction and full wave results are in good agreement. Full wave analysis gives smaller deviation of the pitch angle from MSE data in the outer region of the plasma. However it must be noted that not all the minor features of experimental signals have been reproduced by the full wave modelling.

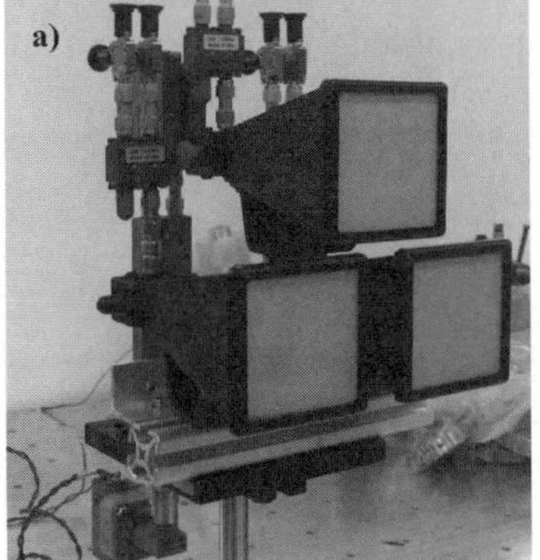

Figure 4. a) Triple antenna assembly prepared for EBE imaging lab tests with the noise source. b) Pilot experiment set-up with triple antenna imaging system on MAST.

Temporal resolution of the technique based on mechanical scanning of the radiometer viewing angle is limited by the maximum speed of FRM rotation. At present it is about 10 ms. Relative accuracy of the pitch angle reconstruction on a channel-to-channel basis is about 1° for the ELM-free H-mode plasma. Reconstruction becomes very difficult or even impossible in the presence of MHD activity. This is the main limitation of this technique.

2. EBW Emission Imaging Developments

Angular scanning EBE measurements have demonstrated the possibility of magnetic pitch angle reconstruction from the EBE angular distribution. Microwave imaging of the B-X-O MC window is proposed as an alternative to the mechanical angular scanning with FRM. The proposed scheme is based on the phased array technique allowing the required angular resolution to be achieved in MAST experiments. Image acquisition time can be much shorter than MHD time scales so the EBE imaging technique can potentially be used for pitch angle measurements in any kind of plasma.

To exploit this possibility on MAST a microwave imaging system (MIS) is being developed in collaboration with York University. MIS consists of an array of antennas and relies on two methods of microwave image acquisition widely used in radio astronomy, namely aperture synthesis and phased array techniques. A triple antenna prototype has been assembled to conduct initial tests with a microwave noise source (see Fig. 4a) in the lab. Each antenna receives two polarizations in the frequency range from 6 to 40 GHz. Microwave signals are down-converted using heterodyne techniques and the intermediate frequency (IF) signals are digitized in a full vector form at a sampling rate up to 500 Msps. The triple antenna prototype was tested with the noise source and then calibrated with a monochromatic source against phase and amplitude balance across all channels.

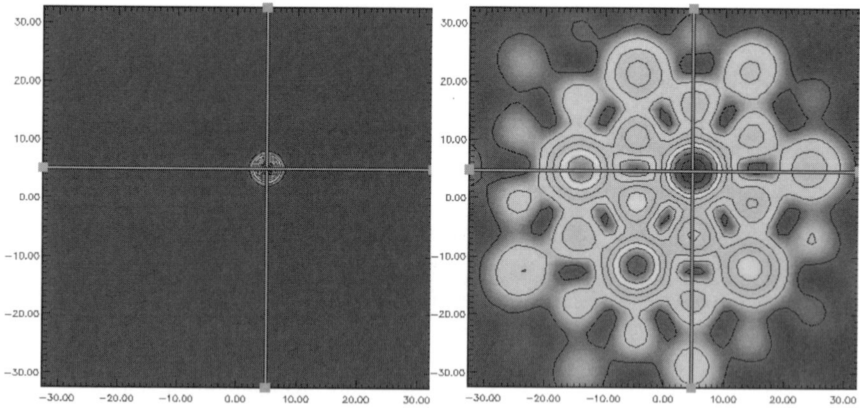

Figure 5. The microwave source (left) used in modelling and the response of a 4 antenna Y shaped array as reconstructed by SVD (right). Similar results can be obtained by the cross correlation method, taking into account the antenna gain diagram.

Finally the triple antenna system has been installed on MAST and pilot plasma experiments have been conducted at a fixed frequency of 17.4 GHz. The system showed extremely high resilience to electromagnetic pick-up despite the close proximity of the microwave components to the vacuum vessel (see Fig. 4b). Experimentally achieved broadband signal-to-noise ratios were about 8 for L-mode plasmas. This figure was typically a factor of two higher for plasmas in H-mode. Preliminary analysis showed a high degree of coherence of signals received by different antennas. Typical cross-correlation coefficients between channels are in the range of a few percent. Further analysis of the data is in progress.

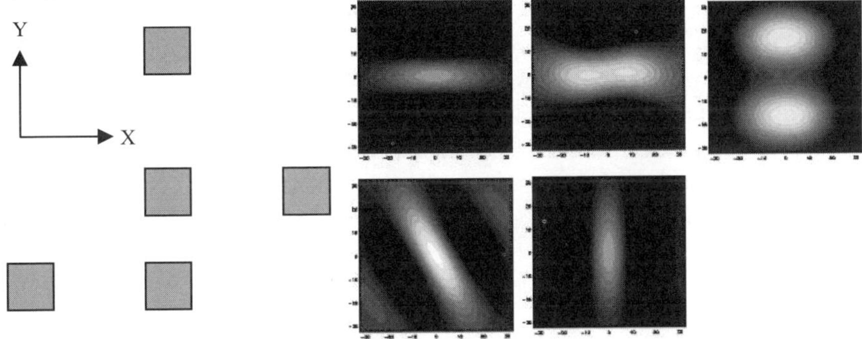

Figure 6. The real part of the 'dual beams' (right) for a 5 antenna array with various spacing (left). This array provides information on the source width and is most sensitive in the vertical direction.

Several methods for inverting the antenna signals into an image have been identified. Broadly they are:

- Singular Value Decomposition (SVD) of the linear system [5];
- Gaussian linear inversion from Bayesian statistics [6];
- Cross-correlation of antenna pairs as a function of phase difference.

Several codes have been developed allowing the modelling of arbitrary antenna configurations, subject to both near field and far field sources. These codes have been used to compare the inversion methods mentioned above in the far field for low antenna number arrays, in the presence of noise, incoherence and a variety of image shapes. The SVD and the mean of the Gaussian linear inversion produce indistinguishable results as far as we can tell. The cross correlation method has no inbuilt side lobe suppression like the SVD, but is far more resilient to noise. For many antenna arrangements it was found that the array was able to resolve the position of the centre of the source very well, better than a factor of 10 over the diffraction limit of the array as illustrated in Fig. 5. However, for some array configurations information about the width of the source could not be recovered using these techniques.

For the SVD, the image is a linear combination of what are known as 'dual beams' formed from the eigenfields of the system and these dual beams are a property of the array only (see Fig. 6). Analysis of these beams allows us to predict the information to which the array will be sensitive. This provides an effective way to design the array configuration. This method of design makes it clear that in order to have sensitivity to the source width, antenna pairs must have various spacing. With 3 antennas only the position of the source maximum can be reconstructed.

Additional information about the source shape can be gained by performing a deconvolution on measurements of power from each antenna. Since each antenna may be pointed at a slightly different angle and the emission may be modelled by a bivariate Gaussian, using the position measurements from the phased array allows resolution of the necessary 2 covariances and rotation angle to define this bivariate image Gaussian using only 4 antennas.

First experiments with MIS using a multiple antenna array are planned on MAST in 2011.

Acknowledgments

This work was funded partly by the United Kingdom Engineering and Physical Sciences Research Council under grants EP/G003955 & EP/H016732 and the European Communities under the contract of Association between EURATOM and CCFE. The views and opinions expressed herein do not necessarily reflect those of the European Commission.

References

1. V. Shevchenko et al., *Proc. RF-17 Topical Conf.*, Clearwater, Florida, USA, AIP Vol. 978, p. 323 (2007).
2. F. Volpe et al., *Proc. EC-15 Joint Workshop*, Yosemite, California, USA, p. 184 (2009).
3. R. A. Cairns and C. N. Lashmore-Davies, *Phys. Plasmas*, **7**, No 10, p. 4126 (2000).
4. J. Urban et al., *Proc. EPS-32 Conf. on Plasma Phys.* Tarragona, Vol. **29C**, P-1.121 (2005).
5. G. Saklatvala et al, *Journal of the Optical Soc. of America*, **25**, 4 (2008).
6. O. Ford et al, *Rev. Sci. Instrum*, **79**, 10 (2008).

PROSPECTS FOR EBW HEATING AND CURRENT DRIVE ON SPHERICAL TORI

J. Urban* and J. Preinhaelter
Institute of Plasma Physics AS CR, EURATOM/IPP.CR Ass., Prague, Czech Rep.
**E-mail: urban@ipp.cas.cz*

J. Decker and Y. Peysson
EURATOM-CEA, Cadarache, France

G. Taylor
Princeton Plasma Physics Laboratory, Princeton, NJ 08543, USA

L. Vahala
Old Dominion University, Norfolk, VA 23529, USA

G. Vahala
College of William & Mary, Williamsburg, VA 23185, USA

The electrostatic electron Bernstein wave (EBW) can provide localised on- and off-axis heating and current drive in typically overdense (high-β) spherical tori (ST) where the usual electromagnetic EC modes are cut-off. Hence, the EBW is a candidate for plasma control and stabilisation in such devices. We present here a modelling of EBW heating and current drive in realistic ST conditions, particularly in typical NSTX equilibria and in model equilibria for NHTX [1] and MAST Upgrade [2,3]. The EBW injection parameters are varied in order to find optimized scenarios and possible ways to control the deposition location and the driven current. It is shown that EBWs can be deposited and efficiently drive current at any radial location.

Keywords: Tokamak; Spherical torus; Current drive; Heating; EBW

1. Introduction and motivation

This paper pursues the prospects for electron Bernstein wave (EBW) heating and current drive on spherical tori by means of coupled ray-tracing and Fokker-Planck simulations. The electrostatic EBW is the only electron cyclotron wave that can be propagated and effectively absorbed in typically overdense (high-β) spherical tokamak plasmas. Its potential uses are similar to the standard EC O- and X-mode—besides fundamental heating and current drive, localised and controllable on- or off-axis heating and current drive can be used for plasma control and stabilisation. The key question is how to control and optimise the EBWs, which must be mode-converted at the upper hybrid resonance (UHR) layer from appropriately launched O- or X-modes and whose propagation in a plasma is intensively influenced by the plasma parameters (particularly the magnetic configuration and the electron temperature and density). To tackle this problem, we present and discuss here an extensive set of simulations involving four different target spherical tokamak scenarios, listed in Table 1, with various EBW injection parameters. As can be seen, the chosen scenarios differ in various parameters. Two of them are typical NSTX L- and H-mode experimental discharges, the other two are TRANSP model scenarios of the planned MAST Upgrade and of NHTX (a potential plasma facing component test facility).

Table 1. Target spherical tokamak scenarios.

Name	B_0 [T]	n_{e0} [10^{19}m^{-3}]	T_{e0} [keV]	I_p [MA]	Source
NSTX L-mode	0.5	2.6	2.9	0.6	shot 123435
NSTX H-mode	0.5	3.9	1.4	1.0	shot 130607
MAST Upgrade	0.78	3.5	2.4	1.2	TRANSP
NHTX	2.0	20.0	5.7	3.5	TRANSP

2. Simulation methods

We employ here two coupled simulation codes. The AMR (Antenna, Mode-conversion, ray-tracing) code [4,5] calculates the EBW ray trajectories using a kinetic non-relativistic electrostatic dispersion equation. The rays are launched with the optimum $N_\parallel$ for the particular case.

After an AMR run, the ray-trajectories and wave vector evolution are passed to the LUKE code [6], which calculates the quasilinear damping and driven current using identical plasma equilibrium and profiles. LUKE is a fully relativistic 3-D Fokker-Planck solver. It calculates the evolution of the electron distribution function for axisymmetric plasmas in the low-collisionality regime. LUKE can account for collisions, RF quasilinear diffusion due to RF waves, inductive toroidal electric field, and fast electron radial transport. The code uses a fully implicit 3-D time evolution scheme for a fast convergence to the time-asymptotic solution. It has been verified that the damping profile calculated by LUKE in the low power limit agrees with linear theory. In our simulations, the launched power is 1 MW unless specified otherwise.

The so-called O-X-B scenario, which involves a conversion from O-mode to slow X-mode and subsequently to the EBW, is considered here because of its practical feasibility. The optimum angle for a full O-X-B conversion is given by [7] $N^2_{\|\mathrm{opt}} = (1+\omega/\omega_{\mathrm{ce}})^{-1}$, $\mathbf{N}\cdot(\mathbf{B}\times\nabla n_{\mathrm{e}}) = 0$. This yields two optimum wave vectors for a given frequency and plasma parameters. Consequently, the only injection parameters that can be arbitrarily varied are the antenna position and the frequency with subsequent choice of two angles with $N_{\|} = \pm\sqrt{N^2_{\|\mathrm{opt}}}$. This is in contrast to the usual EC waves, for which the injection angle (and more or less also the $N_{\|}$ at the absorption layer) can be chosen at will. We do not discuss the O-X-B coupling in detail (e.g., we neglect 2D, incident beam k-spectra or fluctuation effects); instead, the optimum angle (for the central ray) is chosen for each case and the EBW behaviour is simulated, assuming 100% conversion. AMR full-wave calculations, performed for several cases, show >90% conversion efficiency, which would not have a large effect on the results in this paper. The incident Gaussian beam waist is positioned into to vicinity of the outboard plasma edge and the size is chosen such that the Rayleigh range (the distance in which the beam doubles its cross-section) is 0.5 m.

3. Results

In this section, we present the major results of our EBW simulations for the four spherical plasma scenarios, listed in Table 1. The simulated frequency ranges are marked in Fig. 1. First two harmonics have been selected for NSTX and MAST as higher harmonics will likely be overlapping because of the Doppler broadening. The same applies to NHTX, where, however, only the first harmonic is simulated since the second is only marginally overdense and the O-X-B conversion region occurs in the core plasma.

We now look at the results for the NSTX L-mode case in detail. In Fig. 2 is the total EBW driven current plotted for all the cases we have run, i.e., for various frequencies, antenna vertical positions and initial $N_{\|}$ sign. The envelope shows us the maximum current we can drive with a particular frequency. The peak value is ~ 0.2 A/W, which is rather high; the reason is the high $T_{\mathrm{e}}/n_{\mathrm{e}}$. The dependence on the launch parameters for a fixed frequency can be deduced from the various symbols that represent

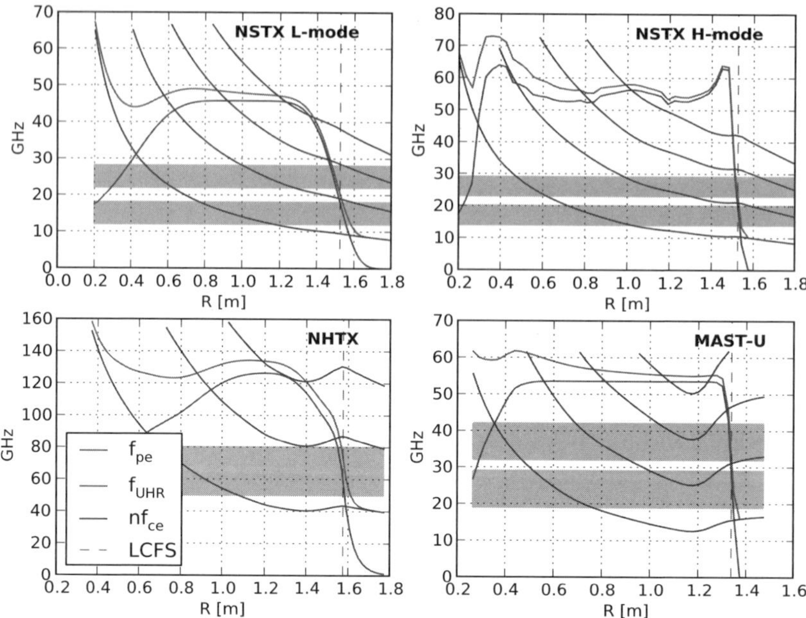

Fig. 1. Characteristic frequencies for the target scenarios. $f_{\rm pe}$ – electron plasma frequency, $f_{\rm UHR}$ – upper hybrid frequency, $nf_{\rm ce}$ – $n^{\rm th}$ EC harmonic. Filled areas represent the simulated frequency ranges.

different antenna positions and $N_{\|}$ signs. It is clear that we can achieve any absolute current drive efficiency (up to the ~0.2 A/W) with a variety of frequency and antenna position combinations. Figure 3 shows the driven current density radial location for the mid-plane and above mid-plane cases. The first harmonic fully covers the whole plasma, while the second harmonic is more localised (which is also an effect of smaller beam size) and leaves a minor white space at $\rho_{\rm pol} = 0.8 \ldots 1$ and $0 \ldots 0.1$. Figure 4 shows the current that can be driven at each radial position. The radial dependence of $|j|$ indicates that any decrease of the net current is caused by two current peaks with opposite directions. The decrease of I with ρ is caused by increasing collisionality. The results are similar for the other scenarios except that the absolute current differs depending on $n_{\rm e}$ and $T_{\rm e}$. NHTX tends to be less flexible in terms of accessibility.

The main result is presented in Fig. 5, where the normalised current drive efficiency ζ [8], which scales out collisional effects:

$$\zeta = \frac{e^3}{\epsilon_0^2} \frac{R n_{\rm e}}{k T_{\rm e}} \frac{I}{P}, \qquad (1)$$

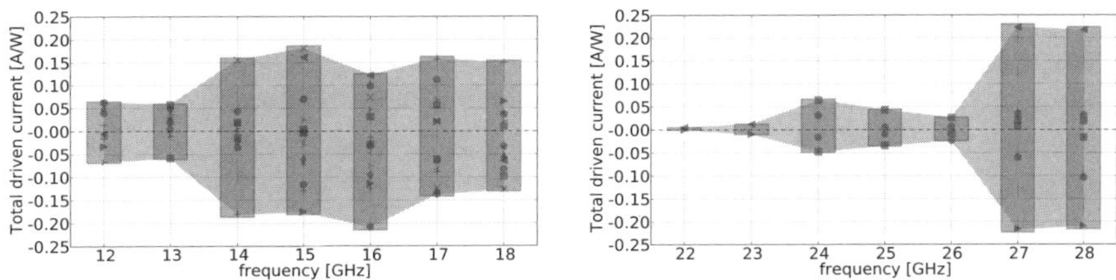

Fig. 2. Total EBW driven current for the NSTX L-mode case, first and second harmonic. The individual symbols represent various injection parameters.

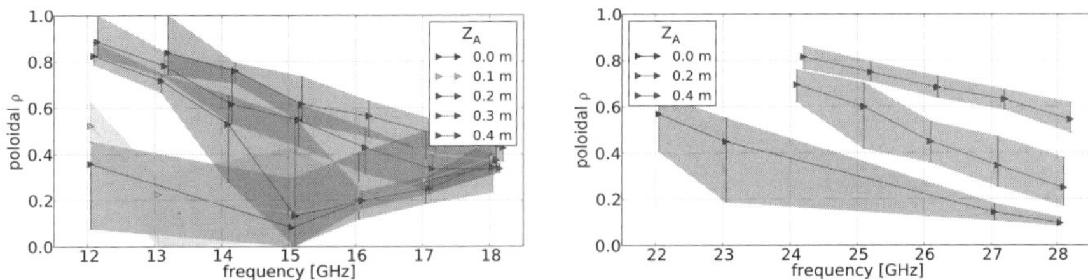

Fig. 3. Driven current density peak location (symbols) and full width at half maximum (error bars) of cases from Fig. 2 with vertical launch position $Z_{\mathrm{A}} \geq 0$ and initial $N_{\|} < 0$.

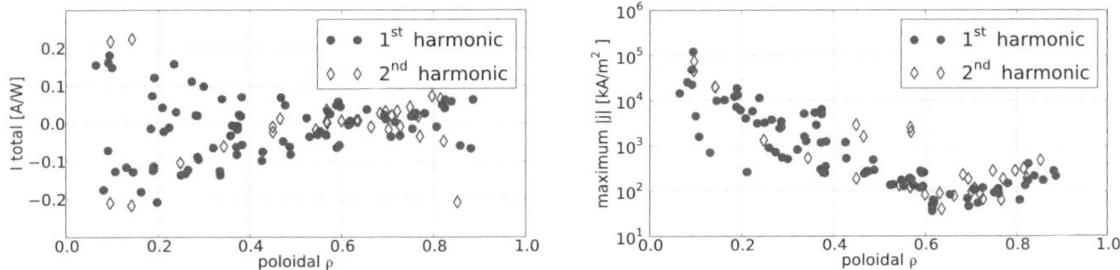

Fig. 4. Radial dependence of the total EBW driven current (left) and the maximum driven current density $|j|$ (right).

is plotted versus ρ_{pol}. From this figure we immediately see that EBWs can efficiently—with $\zeta \gtrsim 0.5$ for NSTX and $\zeta \gtrsim 0.4$ for MAST-U and NHTX—drive current at any radial position. By changing the injection parameters—frequency, antenna position and toroidal angle—we can select the deposition location and the current drive efficiency.

4. Conclusions

An extensive set of EBW heating and current simulations has been carried out for four spherical tokamak scenarios, which differ in the vacuum magnetic field, the plasma current, the electron density and temperature. It has

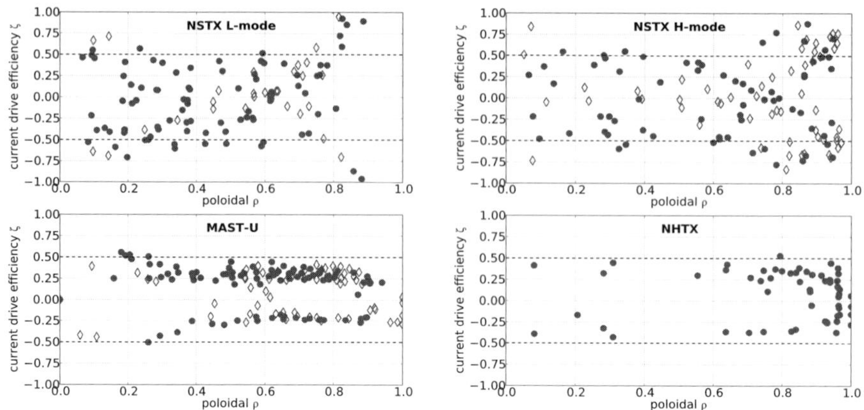

Fig. 5. Normalised current drive efficiency $\zeta\left(\rho_{\mathrm{pol}}\right)$ for all simulated scenarios, first (circles) and second (diamonds) harmonics.

been shown that current drive efficiency $\zeta \sim 0.4-0.5$ can be reached at any radial location by a proper choice of frequency and launch position. Therefore, a control mechanism for EBWs can be based either on frequency or vertical launch position adjustment. Such system would require a complex design that would profit from multiple-frequency sources (e.g., step tunable gyrotrons). However, EBW is the only candidate to provide features comparable to EC systems on conventional aspect ratio devices and hence EBW systems are highly desirable for spherical tori.

Acknowledgement

The work was partly supported by EFDA, EURATOM, GACR #202/08/0419, AS CR #AV0Z20430508, MSMT #7G09042, and U.S. DoE.

References

1. D. A. Gates et al., *Nucl. Fusion* **49**, 104016 (2009).
2. H. Meyer et al., *Nucl. Fusion* **49**, 104017 (2009).
3. CCFE (2010), `http://www.ccfe.ac.uk/MAST_upgrade.aspx`.
4. J. Urban et al., *J. of Plasma and Fusion Research SERIES* **8**, 1153 (2009).
5. J. Preinhaelter et al., *Plasma Phys. Control. Fusion* **51**, 125008 (2009).
6. J. Decker, and Y. Peysson, *DKE: a fast numerical solver for the 3-D relativistic bounce-averaged electron drift kinetic equation* (EURATOM-CEA, Cadarache, 2004).
7. J. Preinhaelter, and V. Kopecký, *J. Plasma Phys.* **10**, 1 (1973).
8. C. C. Petty et al., *Nucl. Fusion* **42**, 1366 (2002).

28 GHZ EBW HEATING, CURRENT DRIVE AND EMISSION EXPERIMENTS AT THE WEGA STELLARATOR

H.P. LAQUA, E. CHLECHOWITZ, M. GLAUBITZ, S. MARSEN, T. STANGE, M. OTTE, D. ZHANG

Max-Planck-Institut für Plasmaphysik, EURATOM Ass., D-17491 Greifswald, Germany

J. PREINHAELTER, J. URBAN

Institute of Plasma Physics, EURATOM/IPP.CR Association, 182 00 Prague, Czech Republic

This paper reports on detailed investigation of a fully 28 GHz EBW (electron Bernstein wave) heated plasma in the WEGA stellarator. The plasma shows a fast transition into the "OXB-state" when the threshold density is reached. The profiles become peaked. The EBW emission diagnostic measures a radiation temperature of several keV, which origins from a supra-thermal electron population. The angular dependence of the mode conversion could be confirmed with a movable launching mirror. The toroidal current and the plasma conductivity were measured for different microwave launch positions.

1. Introduction

The application of high power microwave systems for efficient heating of magnetically confined fusion plasmas is an established and very successful method. Nevertheless, electron cyclotron resonant heating (ECRH) is limited in density and high beta operation due to reflections of the heating wave at the associated density cut-off that prohibits further propagation into the central plasma region. However, over-dense operation in magnetically confined fusion plasmas without a density limit can be achieved by an alternative heating concept based on the conversion of the incident electromagnetic waves into electrostatic Bernstein waves (EBW) [1, 2, 3]. At WEGA this is performed by a two-step conversion process at the plasma edge. Here, an obliquely launched ordinary wave (O-wave) is converted first into an extraordinary wave (X-wave) at the O-mode cut-off density and finally into an EBW at the upper hybrid resonance (UHR) layer. The EBW can propagate into the over-dense plasma core, where it is absorbed by resonant cyclotron interaction. This so called OXB conversion process requires a minimum density of $0.97 \cdot 10^{19}$ m^{-3} (O-cut-off density) for the heating frequency of 28 GHz and an optimal oblique launch angle with respect to the magnetic field vector of 55° and a field strength of 0.5T. The propagation of the EBW is strongly correlated with the magnetic configuration. Stellarators are best suited for detailed investigations of this kind of heating. Their magnetic configuration is mostly generated by the vacuum magnetic field, which is well known. Additionally, they can be operated stationary without a density limit, which bases on MHD-stability. In the current-less steady state operation even small rf-generated plasma currents can be measured.

2. Experimental Setup

WEGA is a classical $l = 2$ stellarator with five field periods ($m = 5$), a major radius of $R = 72$ cm and a maximum effective plasma radius of $a = 11$ cm, respectively [4]. The toroidal magnetic field coils can be operated at 0.5 T for about 20 s. With additional vertical field coils for varying the radial plasma position and the magnetic shear the machine has a very flexible magnetic configuration. For plasma generation two microwave heating systems operating at a frequency of 2.45 GHz (altogether 26 kW, cw) and 28GHz (10 kW, cw), respectively, and a transformer with a capacity of 440 mVs for Ohmic heating are available. The 28 GHz microwaves are launched quasi-optically by a movable mirror system, which is inside the vacuum vessel. Thus the toroidal launch angle and the launch position could be varied. In addition both polarisation parameters (orientation and ellipticity) could be arbitrarily chosen. A specially designed waveguide converts the TE02 gyrotron mode into the HE11 mode for the quasi-optical transmission.

3. Experimental results

At WEGA, stationary (10 s) over-dense plasmas, which are heated by 28 GHz EBW only, can be established routinely in argon and helium [5]. The plasmas were generated by resonant absorption of the 28 GHz EC-waves as shown in Fig.1. Although the ECRH beam was already obliquely launched with O-wave polarisation, the plasma was supposedly heated by the second harmonic X-waves, which have been generated by depolarisation, when the waves are reflected at the metallic plasma vessel. This multi-pass heating generated a broad density profile, which was typically limited by the X-wave cut-off density at approximately $0.5 \cdot 10^{19}$ m^{-3}. A further density increase could be achieved when the 2.45 GHz waves were launched additionally. When the density threshold necessary for the OXB-conversion was reached, the OXB-conversion started instantly. The EBWs were well absorbed even at temperatures below 50 eV. Thus, with central deposition the central density was increasedand as a consequence the conversion efficiency also increased and a peaked pressure profile built up. The final density of $1.4 \cdot 10^{19}$ m^{-3} is only limited by the power balance. The transition from the non-OXB into the OXB-heated state occured on a fast time scale of 0.2 ms with a large increase of density and radiation. Simultaneously the 28 GHz stray radiation signal, which measures the non-absorbed part of the ECRH power, dropped. This indicates a strongly improved single pass absorption. With a 16-channel Si-diode array camera measuring the absolute XUV-emission the temporal development of the transition could be resolved as shown in Fig. 2. The central power deposition could also be proved by a fast modulation of the heating power and the coherent detection of the heat waves by the AXUV camera. The density and the bulk electron temperature profile were measured by a fast reciprocating Langmuir probe as shown in Fig. 3. Since the probe perturbed the plasma the temperature and density profiles were probably little underestimated. Nevertheless the probe density was consistent within error bars with the line-integrated density measured by the 80 GHz interferometer. The bulk electron temperature was relatively low (<50 eV) but the resonant absorption of the EBWs generated a strongly supra-thermal electron population with energies of the order of 10 keV, which was measured by X-ray emission. Since the plasma density of $\sim 1.4 \cdot 10^{19}$ m^{-3} was higher than the cut-off density of the EC-resonance frequency (28 GHz), no EC-emission from the over-dense plasma region could be measured anymore. However, EBW emission (EBE), which uses the inverse mode conversion process (BXO) was measured by a 12-channel radiometer and a spectrum analyzer, which could generate several emission spectrums during the discharge. The system was absolutely calibrated by the "hot-cold" method. Thus, at frequencies associated with the central region of the plasma the EBE showed a very high radiation temperature above 10 keV, which is remarkable since the total heating power is only 9 kW. The origin of this supra-thermal emission is still not well understood. In a bi-Maxwellian electron energy distribution, such a high radiation temperature should not exist since this radiation is well re-absorbed by the cold plasma periphery before leaving the over-dense plasma. The existence of suprathermal electrons was confirmed by X-ray emission measurement.

The knowledge of the density gradient at the mode conversion layer allowed an estimate of the angular acceptance window for OX-conversion using the analytic formula from Mjølhus [6]. This width was consistent with the angular window measured by the variation of the launch angle as shown in Fig. 3. The EBW-heating was simulated by 3D ray-tracing [7] shown in Fig. 4. Two different toroidal launch positions were investigated. In the tokamak-like symmetric plasma cross-section the waves were launched in the equatorial plane. Here, the ray-tracing predicted a minimal increase of the parallel wave refractive index $N_{//}$. Thus, the Doppler downshift in the absorption should only be 0.02 T, which could be reproduced experimentally by a magnetic field scan. For this launch position only low current drive efficiency of 1-5 A/kW was expected. Stellarators have 3D magnetic configurations, thus at different toroidal positions different plasma shapes exist. Out of the tokamak like symmetry position, the poloidal plasma cross-section is tilted and the EBW propagation is altered. Here, a larger increase of $N_{//}$ was predicted by the ray-tracing calculations, which should generate a current drive efficiency of about 50 A/kW and a Doppler down shift of 0.03 T. An even larger Doppler shift could be found in the experiments. With the help of a magnetic field strength scan during the discharge the maximum of both the EBE and the AXUV-emission intensity was found at 0.445 T. However, the plasma current remained always at a value below 14 A and could not be increased in the tilted plasma position. In order to get more information on the plasma conductivity the transformer was used to drive current in the OXB-phase. The plasma conductivity was ten times lower than in the case of non-OXB discharges at densities around $2 \cdot 10^{18}$ m^{-3}. The main difference in the plasma parameters was the much lower neutral gas (He) pressure of $1 \cdot 10^{-5}$ mbar in comparison with at least $8 \cdot 10^{-5}$ mbar in the OXB-heated plasma.

4. Discussion

The high neutral gas pressure is necessary to achieve the threshold density for OXB-conversion. Since the cross sections for inelastic collisions with HeI and HeII do not decay much with the impact energy of the electrons, we suppose that the fast electron population could not contribute to the electrical conductivity much and the plasma conductivity is determined by the low temperature bulk electrons. For operation in hydrogen complete ionisation should be achieved more easily and the radiated power should also be reduced. However, any attempt to establish OXB-heating in hydrogen failed because the heating power was not high enough to reach the 28 GHz OXB threshold density.

5. Summary and conclusion

In WEGA quasi-stationary plasmas fully sustained by electron Bernstein wave heating could be realized in argon and helium at 0.5 T operation. The EBWs were generated by launching an O-mode polarized beam under the optimum angle for the OXB mode conversion process. To overcome the density threshold for the O-mode cut-off, necessary for the propagation of the Bernstein waves, an additional non-resonant 2.45 GHz heating was used. Densities up to $1.4 \cdot 10^{19}$ m^{-3}, hence well above the critical density, and electron bulk temperatures of the order of a few ten eV could be achieved. Furthermore, a central power deposition of the Bernstein waves was observed from bolometer and Langmuir-probe data. The transition into the "OXB-state" was resolved with high temporal and spatial resolution. The EBE-diagnostic could detect extremely high radiation temperatures in the keV range in the central region. This non-thermal emission is assumed to originate from a fast supra-thermal electron population generated by the resonant electron cyclotron absorption of the EBWs and was supported by the detection of soft X-ray bremsstrahlung in the range of up to 10 keV. The conversion efficiency was optimized by the variation of the launch angle, beam polarisation, magnetic field strength, plasma position and neutral gas density. Unfortunately the expected enhancement of the EBW-driven plasma current could not be proved and is still a topic of investigation.

6. References

1. I. Bernstein, Phys. Rev. Lett. **109**, 10 (1958).
2. J. Preinhaelter, and V. Kopecky, J. Plas. Phys. **10**, 1 (1973).
3. H.P. Laqua et al., Phys. Rev. Lett. **78**, 18 (1997) 3467.
4. M. Otte et al., AIP Conf. Proc. **993** 1 (2008) 3-10.
5. H.P. Laqua et al., Proceedings of the 36th EPS Conference on Plasma Phys. Sofia, June 29 - July 3, 2009 ECA Vol.**33E**, O-4.047 (2009)
6. E. Mjølhus, J. Plasma Phys. 31 (1984) 7.
7. J. Preinhaelter et al., Plasma Phys. Control. Fusion **51** 12 (2009) 5008.

7. Figures

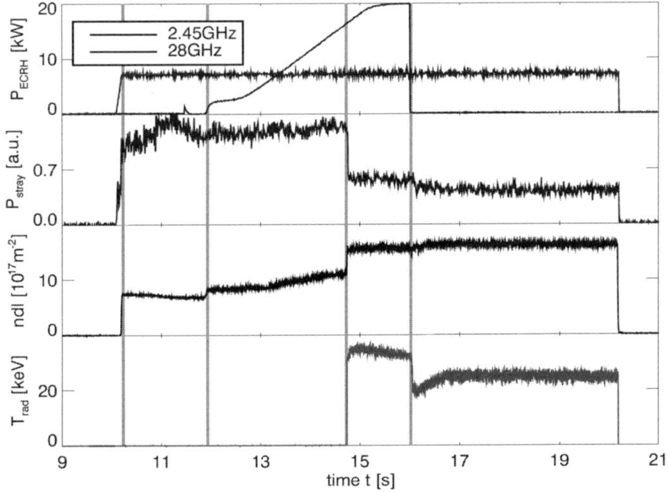

Figure 1: Time traces of an OXB-heated discharge at 0.48 T in He. Plasma ignition at 10 s with 28 GHz ECRH. Density increase by additional 2.45 GHz starting at 12 s. OXB-transition at 14.5 s associated with strong drop of stray radiation, increase of line density and appearance of supra-thermal EBW-emission (EBE). Switch-off of 2.45 GHz heating at 16 s. The discharge is purely sustained by 28 GHz OXB-heating for the remaining 4 s.

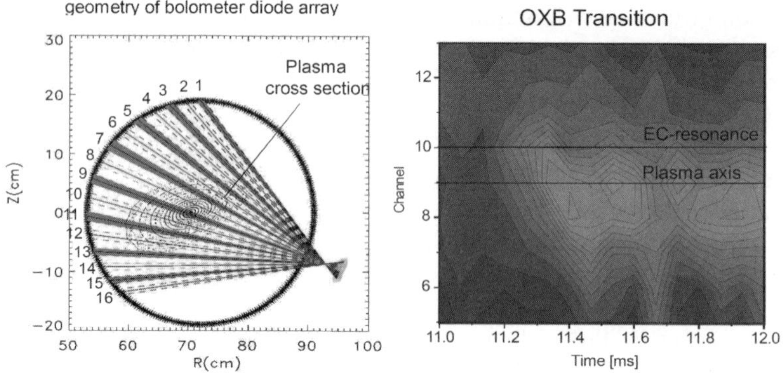

Figure 2: Left) Line of sight of Si-diode bolometer. Right) Temporal development of the transition into the OXB heated mode. The density rise started at the EC-resonance at channel 10. The profile developed within 0.2 ms into a peaked profile with the maximum at the magnetic axis between channel 9 and 8.

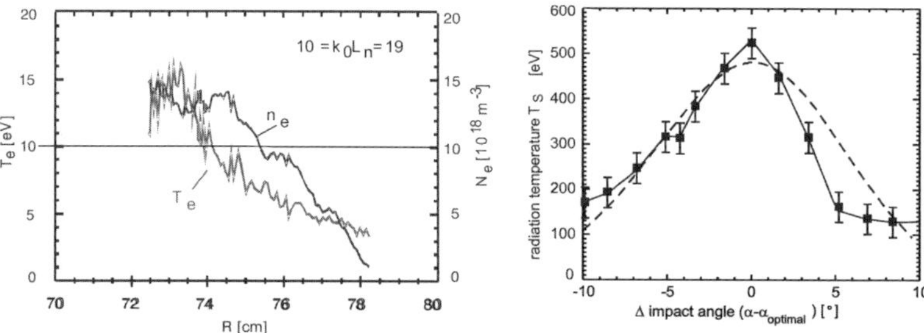

Figure 3: Left) Density and temperature profiles of the bulk electrons measured by a fast reciprocating Langmuir probe. Right) Central EBE radiation temperature as a function of the launch angle in comparison with the analytic OX-transmission function (dashed line) from Mjølhus [6] for a normalized density scale length $k_0L_n = 19$.

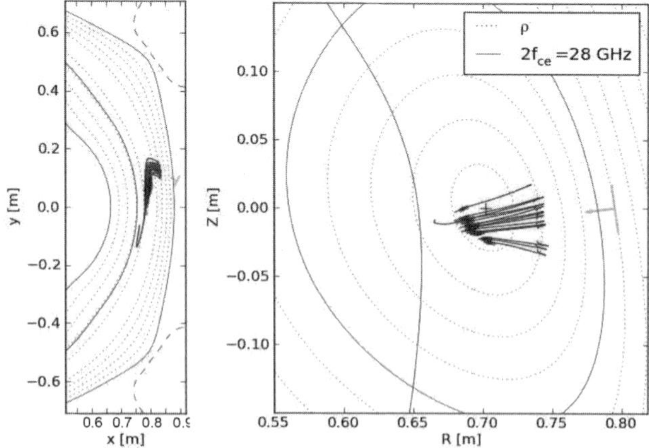

Figure 4: Ray–trajectories of EBWs as projections at the equatorial plane (left) and the poloidal cut plane at Φ=6.5°. Here, the rays were launched into the tilted plasma cross section.

Development of Phased-array Antenna System and Its Application to EBWH/CD Experiments in QUEST

H. Idei[1], M. Sakaguchi[2], E.I. Kalinnikova[2], K. Nagata[2], H. Zushi[1], K. Hanada[1], S. Tashima[2], M. Ishiguro[1], , H.Q. Liu[2], H. Igami[3], S. Kubo[3], K. Nakamura[1], A. Fujisawa[1], M. Sakamoto[1], M. Hassegawa[1], Y. Higashizono[1], R. Ogata[2], T. Ryokai[2], S. K. Sharma[2], S. Kawasaki[1], H. Nakashima[1], A. Higashijima[1], Y. Takase[4], T. Maekawa[5],O. Mitarai[6] and Y. Kishimoto[6]

[1] *Research Institute for Fusion Science, Kasuga, 816-8580, Japan*
[2] *Interdisciplinary Grad. School of Eng. Sci., Kyushu Univ., Kasuga, 816-8580, Japan*
[3] *National Institute for Fusion Science, Toki, 509-5292, Japan*
[4] *Department of Complexity Sci. and Eng., Univ. of Tokyo, Kashiwa, 277-8561, Japan*
[5] *Graduate School of Energy Science, Kyoto Univ., Kyoto, 606-8501, Japan*
[6] *Inst. of Ind. Sci. and Tech. Research, Tokai Univ., Kumamoto, 862-8652, Japan*

The prototype phased-array antenna system has been developed to control the incident angle and polarization to conduct the Electron Bernstein Wave Heating and Current Drive (EBWH/CD) experiments on the QUEST. The two orthogonal fields measured at the low power level were in excellent agreement with those evaluated by a developed Kirchhoff code. The elliptical polarization in two orthogonal fields can be controlled to excite pure O-mode in the oblique injection. The non-inductive plasma current of 10 kA was ramped up and sustained for 0.7 s. The phased-array antenna system for the reflectometry and the EBW radiometry has been also devleloped concerning the EBWH/CD experiments, and was confirmed to work well in the low power tests.

Keywords: Phased-array Antenna, EBW, Spherical Tokamak, QUEST

1. Introduction

The Q-shu University Experiment with Steady State Spherical Tokamak (QUEST) was proposed at Kyushu University to study plasma-wall interaction phenomena in the steady-state plasma. Electron Bernstein wave heating and current drive (EBWH and EBWCD) is one of attractive candidates of heating and current drive method to sustain the steady-state spherical tokamak (ST) configuration. In the EBWH/CD experiments, some mode conversions from the electron cyclotron (electromagnetic) wave to the electron Bernstein (electrostatic) wave are required. The planned target operation in the first stage of the QUEST is the steady state current drive at $\sim$ 20kA using the O-X-B mode conversion scenario in the rather low-density region. The O-X-B mode conversion process was analyzed using the TASK-WR ray tracing code [1] for an optimum incident angle [2]. In order to analyze the ray trajectory even in the non-optimum cases, the tunneling effect in the O-X mode conversion was taken into the TASK/WR code [3] using an analytical formula [4]. In order to conduct the EBWH/CD experiments with the mode conversion, the advanced antenna system with good directionality and polarization controllability has been required. A Phased-Array Antenna (PAA) system, which enables us to control the launching polarization and angle, was proposed, and the prototype antenna system was designed [5] and has been developed.

Detailed density profile measurements are essential to study the mode conversion phenomena. The density gradient where the ECW converts to the EBW is a key parameter to attain the high conversion efficiency. The phase time evolutions in the reflectometry were successfully measured in the QUEST, but the weak reflected-wave caused the small ratio of signal to noise in the measurements [6]. The PAA system was proposed also for the reflectometry to obtain the large reflected-wave. An EBW radiometry has been developed to measure time evolutions of electron temperature profiles in over-dense plasmas [7,8]. The EBW radiometry is not only a powerful tool to measure the electron temperature time evolutions, but also to investigate the mode conversion phenomena [9]. The oblique viewing EBW emission may be measured by the precise phase-array measurements between the PAA waveguide elements.

The EBWH/CD expriments have been conducted using the protype 8 [4x2] element PAA system in the QUEST. A new diagnostics 9 element [3x3] PAA system has been developed for the reflectometry and the EBW radiometry. In this paper, the prototype antenna performance is described in Sec.2, and the non-inductive plasma current experiments using the prototype PAA system are shown in Sec.3. In Sec.4, the new diagnostics PAA system are described, and the summary is finally given in Sec.5.

2. Prototype EBWH/CD Antenna Performance

The 8.2 GHz system was prepared for the lower hybrid current drive experiments in the previous TRIAM-1M tokamak, and has been used to the EBWH/CD experiments in the QUEST. In the system, the power of 200 kW was available using 16 rectangular-waveguide (WR-137) transmission lines. In order to control the launching polarization, an orthomode trans-

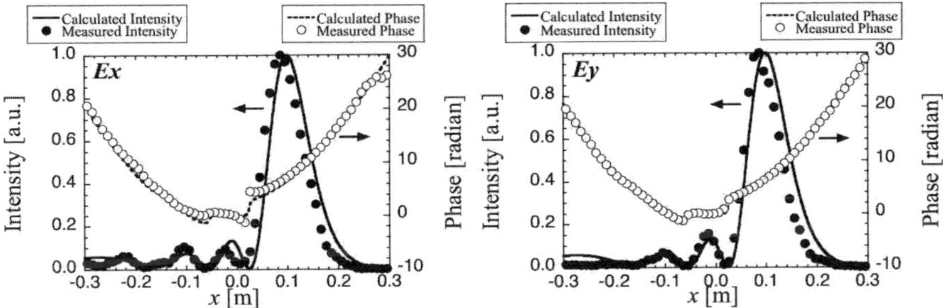

Fig. 1. Calculated and measured field $E_x(x)$ and $E_y(x)$ intensity and phase profiles at z=0.2 m, obliquely radiated from the prototype PAA System.

ducer has been developed to mix two orthogonal electric-field components. The arbitrary polarized field was expressed using the two orthogonal components with a phase difference. The 8 square-waveguide outputs from the orthomode transducers were led to the PAA with a radiation aperture of the [4 × 2] waveguide array. The phase differences between waveguide elements were adjusted to obtain a launched beam steered in the x-direction. Figure 1 shows the field intensity and phase profiles, measured at the low power test facilities, at a radiated position of z=0.2 m. In the figure, the calculated field intensity and phase profiles were also plotted. The calculated field profiles $E_{x,y}(x, y, z = 0.2\text{m})$ from the PAA were evaluated with a Kirchhoff integral as follows,

$$E_{x,y}(x,y,z) = \frac{ik}{2\pi z} \int_{-a/2}^{a/2} \int_{-a/2}^{a/2} \mathrm{d}x'\mathrm{d}y' E_{x,y}(x',y')[\exp(-ikr)/r], \qquad (1)$$

where the coordinates of (x', y') and (x, y) were at the antenna aperture position(z=0) and at the radiated z position. Here, r was $\sqrt{(x-x')^2 + (y-y')^2 + z^2}$, and a was a side of the square-waveguide, respectively. The two orthogonal fields measured at the low power level were in excellent agreement with those evaluated by a developed Kirchhoff code. The pure O-mode beam with the elliptical polarization in two orthogonal fields can be exited in the oblique injection.

3. Non-Inductive Plasma Current Experiments with Prototype Antenna System

The non-inductive plasma current experiments have been begun using the developed prototype PAA system in the last experimental campaine. The

O-mode beam was injected in the oblique injection with the refractive index in parallel to the magnetic field, $N_{//} \sim 0.4$. Figure 2 shows time evolutions, $P_{\mathrm{RF}}(t)$, $B_{\mathrm{v}}(t)$ and $I_{\mathrm{p}}(t)$, of the incident power, the central vertical field and the plasma current in the oblique O-mode injection, respectively. The each vertical magnetic field of two pairs of the poloidal magnetic coils (PFC17 and PFC26) was ramped up to about 2 mT. The toroidal magnetic field was 0.133T. The plasma current jumped up to 5 kA was observed at P_{RF}=18 kW, and its jump timing became early in the higher power injection. The plasma current was ramped up to about 10kA with taking the plasma equilbrium by the vertical fileds at the power level of P_{RF}=30 kW. The obtained 10 kA discharge was analyzed to be a low-aspect ellitipical ST configuration from the flux loop measurements. The aspect ratio and the ellipicity were 1.6 and 1.7 respectively. The 10 kA plasma current was sustained for 0.7s with the constatnt B_{v}, as shown in Fig.3. The plasma current or discharge was terminated due to the recycling enhancement. The more wall-conditioning process was required to obtain a longer discharge. The hard X-ray emission was detected even in a high energetic range more than 40 keV, indicating generation of high energetic electrons accelerated by the RF injection. The density was lower than the O-mode cutoff density. The power absorption in the oblique O-mode injection due to the bulk (<100eV) electrons was small at the 2nd and the peripheral fundamental resonances, and that due to the high energetic electrons was less than 10%. In the lower density plasma, the 2nd harmonic X-mode beam is rather effective to obtain the bulk electron absorption. The 2nd harmonic X-mode and fundamental O-mode experiments will be conducted in the lower and higher density plasmas.

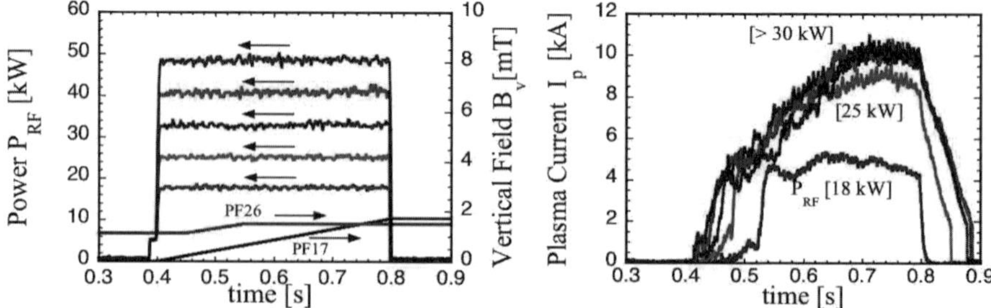

Fig. 2. Time evolutions, $P_{\mathrm{RF}}(t)$, $B_{\mathrm{V}}(t)$ and $I_{\mathrm{p}}(t)$, of the incident power, the central vertical field and the plasma current in the non-inductive plasma current experiments.

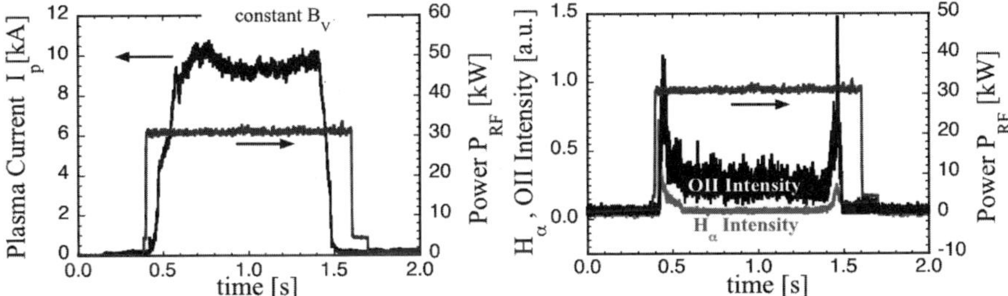

Fig. 3. Time evolutions, $P_{\rm RF}(t)$, $I_{\rm P}(t)$, $H_\alpha(t)$ and $OII(t)$, of the incident power, the plasma current and the $\rm H_\alpha$ and OII intensities in the non-inductive plasma current sustainment.

4. New Plasma Diagnostics Antenna System

The far field radiation is important for the reflectometry in the larger propagating length. In the low operating frequency around 10GHz, the reflected-wave signal became small due to the bad antenna directivity. It was expected the PAA system might focus the radiated field with good directionality even at the larger propagating length for the reflectometry. The oblique viewing EBW emission can be measured by the precise phase-array measurements between the PAA waveguide elements. A new 9 element [3x3] PAA system has been developed for plasma diagnostics of the reflectometry and the EBW radiometry. Figure 4 shows the field $E_x(x)$ intensity and phase profiles, measured at the low power test facilities, at radiated positions of z=0.2 and 0.6 m. The phase differences between the 9 elements were adjusted to obtain a straight beam of 10 GHz for the reflectometry

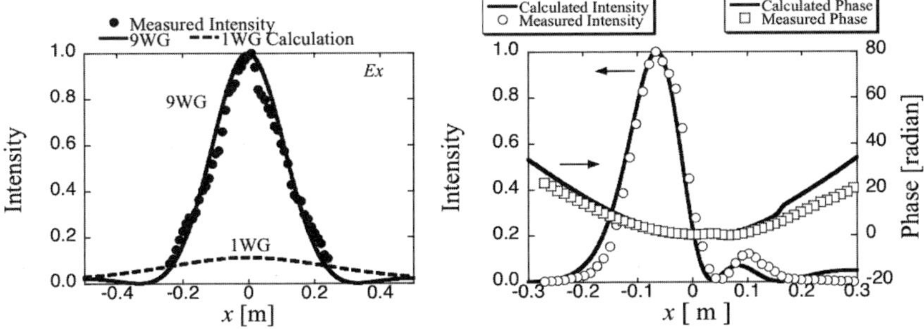

Fig. 4. (a) Measured and calculated field $E_x(x)$ intensity and phase profiles at a radiation position of z= 0.6 m for the reflectometry. In the calculation. the radiation field was evaluated with 1 and 9 elements in the PAA. (b) Measured and calculated field $E_x(x)$ intensity and phase profile at a radiation position of z=0.2 m for the EBW radiometry.

at z=0.6 m and were done to obtain a oblique viewing beam of 8 GHz for the EBW radiometry z= 0.2 m. In Fig.4 (a), the calculated field intensity and phase profiles, radiated from the PAA with 1 waveguide element and 9 elements, were also plotted. The well-focused beam was attained at the large propagation length with the 9 element phase-array for the reflectometry, as calculated by the Kirchhoff integral code. The measured fields in the oblique propagation were also in good agreement with the calculated fields from the Kirchhoff integral, as shown in Fig.4 (b). The developed PAA will be co-operated to measure the EBW emission to investigate the mode conversion phenomena, together with the reflectometry measurements.

5. Summary

The 8 element prototype PAA system has been developed for the EBWH/CD experiments in the QUEST. The measured field intensity and phase profile in two orthogonal field components were in excellent agreement with the Kirchhoff integral code calculation, indicating the pure O-mode oblique beam can be launched from the antenna. The non-inductive plasma current of 10 kA was ramped-up and sustained for 0.7 s at 0.133 T with the RF injection of 30 kW. The 9 element PAA system for the reflectometry and the EBW radiometry has been also developed concerning the EBWH/CD experiments. The well-focused straight and oblique viewing beams were obtained for the reflectometry and the EBW radiometry along the code calculation.

Acknowledgments

This research was supported by the Ministry of Education, Science, Sports and Culture, Grant-in-Aid for Scientific Research (B) [21360452], the NIFS LHD/bi-directional collaborations (NIFS09KOAH024/NIFS09KUTR046).

References

1. A. Fukuyama, *Fusion Eng. and Design* **53**, 71 (2001).
2. H. Idei, *et al.*, *J. of Plasma Fusion Res. Sers.* **8**, 1104 (2009).
3. E.I. Kalinnikova*et al.*, in *this Proceedings* (World Scientific, Singapore, 2010).
4. E. Mjølhus, *J. Plasma Phys.* **31**, 7 (1984).
5. H. Idei, *et al.*, in *Procs. of IRMMW-THz 2007 Conference*, WedP5-66 (2007).
6. H. Idei, *et al.*, *J. of Plasma Fusion Res. Sers.* **9**, *in print* (2010).
7. H. Laqua *et al.*, Phy. Rev. Lett.,**81** 2060 (1998).
8. S.J. Diem *et al.*, Phy. Rev. Lett.,**103** 015002 (2009).
9. V. Shevchenko *et al.*, in *this Proceedings* (World Scientific, Singapore, 2010).

Multiple ray-tracing analysis for EBWH/CD experiments in QUEST

E.I. Kalinnikova*[1], H. Idei[2], H. Zushi[2], K. Hanada[2], H. Igami[3], S. Kubo[3], A. Fukuyama[4] and H. Nuga[4]

[1]*Interdisciplinary Graduate School of Engineering Sciences, Kyushu University,Kasuga 816-8580, Japan*
[2]*Advanced Fusion Research Institute of Applied Mechanics,Kasuga 816-8580, Japan*
[3]*National Institute for Fusion Science, Toki 509-5292, Japan*
[4]*Department of Nuclear Engineering , Kyoto University,Kyoto 606-8501, Japan*
**E-mail: kalinnikova@triam.kyushu-u.ac.jp*

The power deposition profiles were analyzed with a multiple ray tracing code for the Electron Bernstein Wave Heating and Current Drive (EBWH/CD) experiments in the QUEST. In the EBWH/CD experiments in the QUEST, the O-X-B mode conversion scenario was selected for the plasma current sustainment in the rather low-density case. The algorithm for the wave penetration through evanescent layer beyond a O-mode cutoff position was developed for the multiple-ray analysis. The launching antenna positions were considered to obtain the significant wave absorption in the specific propagating direction using the developed ray-tracing code.

Keywords: EBW, O-X mode conversion, ray-tracing

1. Introduction

The Electron Bernstein Wave Heating and Current Drive (EBWH/CD) experiments have been conducted in the Q-shu University Experiments with Steady-State Spherical Tokamak (QUEST). In the spherical tokamaks like the QUEST, the plasma frequency(ω_{p}) may become larger than the electron cyclotron frequency(ω_{c}) in the operating density range due to the rather low magnetic field, and the electron cyclotron wave can not penetrate into the plasma due to the cutoff layers. In the O-X-B mode conversion, an obliquely incident electromagnetic O-mode wave, that reaches to the cutoff, is converted into an X-mode wave, and propagates as X-mode wave. The X-mode wave reaches the UHR and converts to the electrostatic Bernstein (B-mode) wave. In order to study the wave propagation and absorption, the wave trajectory has been calculated including the mode conversion processes using

TASK/WR ray-tracing code [1]. The real antenna beam can be described like superposition of plane wave with different phases and amplitudes. In order to describe experimental antenna beam [2], the multiple ray propagation and absorption were calculated.

This paper is organized as follows. Section 2 described the ray-tracing equation and parameters of calculations. The details of O-X conversion are shown and the results of calculations are presented. The summary is finally given in Section 3.

2. Ray-tracing calculation

The wave propagation and absorption were calculated using the TASK/WR code. The following ray-tracing equations were used:

$$\frac{\mathrm{d}\mathbf{r}}{\mathrm{d}s} = \frac{\partial D}{\partial \mathbf{k}}, \quad \frac{\mathrm{d}\mathbf{k}}{\mathrm{d}s} = -\frac{\partial D}{\partial \mathbf{r}}, \quad \frac{\mathrm{d}t}{\mathrm{d}s} = -\frac{\partial D}{\partial \omega}, \tag{1}$$

where $\mathbf{r}(s)$, $\mathbf{k}(s)$, $t(s)$ are the position, the wave vector and the time of the wave trajectory, respectively. The wave propagation satisfied the local dispersion equation $D(\mathbf{r}, \mathbf{k}, \omega) = 0$. The dispersion relation of $D(\mathbf{r}, \mathbf{k}, \omega)$ was described with non-relativistic hot plasma expression without cold and electrostatic approximations.

The geometrical coordinates were taken as a simple tokamak configurations with a circular poloidal cross-section. The major and minor radii were 0.64 m and 0.36 m, respectively. The profiles of the electron density and the plasma temperature and current were assumed to be parabolic. The central electron density and the temperature were $n_{\mathrm{e0}} = 0.2 \cdot 10^{19}$ m^{-3} and $T_{\mathrm{e0}} = 100$ eV, respectively. The total plasma current was 20 kA. The launching antenna frequency was 8.2 GHz. The plasma current and toroidal magnetic field were in the clockwise direction as viewed from the top of the device.

2.1. *O-X conversion*

The O-mode launching from the low-field side antenna propagated in the plasma, and reached to the O-mode cut-off layer. The wave can propagate inside plasma with no reflection, if the parallel refractive index with respect to the magnetic field line, $n_{||}$, satisfies an optimum condition:

$$n_{||}^2 = n_{||,\mathrm{opt}}^2 = \frac{\omega_{\mathrm{c}}}{\omega_{\mathrm{c}} + \omega} \tag{2}$$

If this condition is not valid, then a wave meets an evanescent region ($n_{x'}^2 < 0$). (Here and below we used the Cartesian coordinate system $x' - y' - z'$,

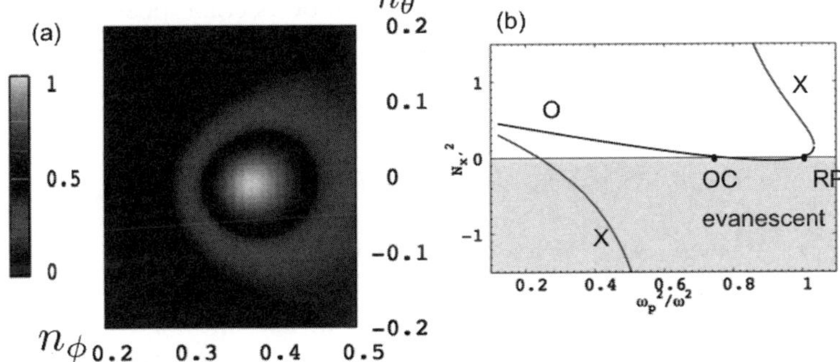

Fig. 1. (a)Counter plot of the O-X conversion efficiency as a function on the initial refractive index components for 1200 rays. (b) $N_{x'}^2$ of the wave for non-optimal case. *OC* is a O-mode cutoff position, *RP* is a restarting point.

where $\mathbf{e}_{x'}$ is in the ∇n_e direction, and $\mathbf{e}_{z'}$ is in the $\mathbf{B}$ direction.) If this region is comparable with the wavelength, the wave partially reflects back and partially penetrates inside plasma due to the tunneling effect. In order to evaluate the wave power transmitted through this region, the analytical expression for O-X conversion efficiency was used [3]:

$$T(n_{||}, n_{y'}) = \exp\left[-\pi k_0 L_n \sqrt{\frac{\omega_\mathrm{c}}{2\omega}}\left(2\frac{\omega_\mathrm{c}+\omega}{\omega}(n_{||,\mathrm{opt}} - n_{||})^2 + n_{y'}^2\right)\right] \quad (3)$$

Here $L_n \equiv n_\mathrm{p}/(\mathrm{d}n_\mathrm{p}/\mathrm{d}x)$ is the scale length of the density gradient at the O-mode cutoff point, $k_0 = 2\pi/\lambda_0$ and finally

$$P_{\text{after O-X}} = TP_{\text{O-mode}}. \quad (4)$$

The expression (3) is reasonable estimation if $k_0 L_n \geq 10$ [4]. Our calculations satisfied this condition.

In order to describe the antenna launching spectrum, the multiple ray trajectories were calculated for different launching refractive index components until O-mode cutoff positions and conversion efficiencies were found by Eq.(3). Fig. 1(a) shows the counter plot of the O-X conversion efficiency T as a function on the initial refractive index components. The suitable conversion efficiency was obtained in the window where the toroidal (n_ϕ) and poloidal (n_θ) refractive indexes were within the range ± 0.1 with respect to the optimal case ($T = 1$).

Figure 1(b) shows a dependence of the squared refractive indexes on $\omega_\mathrm{p}^2/\omega^2$ in a non-optimal case with the evanescent layer where $n_{x'}^2 < 0$.

The squared refractive index $n_{x'}^2$ for the O-mode became positive after the tunneling again, at the restarting position of the ray trajectory with the larger $\omega_{\rm p}^2/\omega^2$. A new algorithm was included to the TASK/WR code to find the restarting position, using the cold plasma dispersion:

$$Sn_\perp^4 - \left((S - n_{||}^2)(S + P) - D^2\right) n_\perp^2 + P\left((S - n_{||}^2)^2 - D^2\right) = 0, \qquad (5)$$

where $n_\perp^2 = n_{x'}^2 + n_{y'}^2$, $n_{||} = n_{z'}$ and $S = S(\omega, \omega_{\rm p}, \omega_{\rm c})$, $P = P(\omega, \omega_{\rm p})$, $D = D(\omega, \omega_{\rm p}, \omega_{\rm c})$ are the Stix parameters [5]. Let us find the position where Eq.(5) was satisfied along the Ox' direction at the fixed y' and z' coordinates. The refractive indexes $n_{y'}$ and $n_{z'}$ were kept as those at the cutoff position. The wave power transmitted through evanescent layer was calculated by Eq.(4). The wave trajectory was calculated with Eq.(1) after the tunneling again.

2.2. *Bernstein's wave propagation and absorption*

After penetrating the evanescent layer, the launching O-mode reached the turning point and converted to the X-mode [6]. The wave reached to the UHR layer where the X-mode converted to the B-mode. The ray trajectory was calculated until the wave was completely absorbed by plasma. The B-mode was absorbed before it reached a cold resonance layer (at $\omega = \omega_{\rm c}$) due to the Doppler-shifted resonance effect:

$$\omega - N\omega_{\rm c} - k_{||}v_{||} = 0. \qquad (6)$$

Here $N = \pm1, \pm2, ...$; $v_{||}$ is a parallel component of the velocity of plasma electrons absorbing the wave. It was shown that B-mode is totally absorbed by plasma electrons corresponding the tail of the Maxwellian distribution function.

The some examples of wave trajectories are shown in Fig.2(a). The shifting of the resonance position from the cold resonance position increased with the magnitude $n_{||}$ of EBW. Figure 2(b) shows the power deposition profile using the multiple ray tracing. The broader power profile was evaluated in multiple ray-tracing compared to one ray-tracing with the optimal case [7]. The sharp peaks in the multi-ray tracing corresponded to the contributions of the narrow power profiles for rays with small $n_{||}$, closely propagating along $R = const.$ surface (see Fig. 2(a)). The more smoothly radial profile can be calculated by increasing the total rays number.

In the Fig.2(a), the two types of wave trajectories are presented with the different signs of $n_{||}$. The sign of $n_{||}$ in the absorption region is important to consider the direction of driven current. In order to estimate

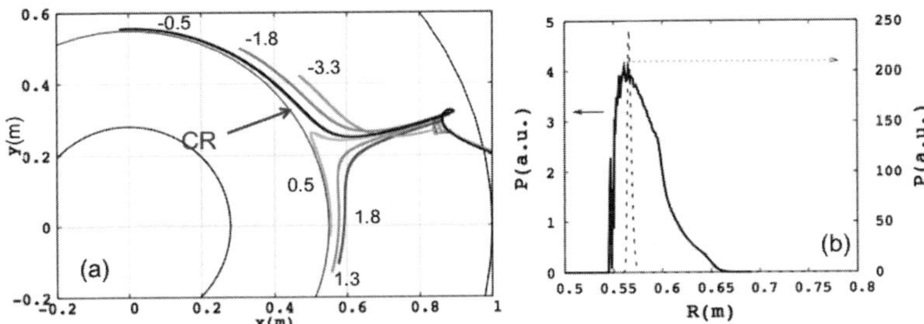

Fig. 2. (a)Toroidal cross section of the wave trajectories with initial $n_\phi = 0.37$ and different n_θ. The $n_{||}$ of EBW are shown for each trajectory.(b) The power deposition profiles for 1200 rays with $n_\phi = 0.2$ to 0.5 and $n_\theta = -0.2$ to 0.2 Dashed line corresponds to the normalized power deposition profile for the optimal case of O-X conversion. The cold resonance (CR) position was located at $R = 0.55$ m.

the contribution of opposite direction of absorbed rays, let us consider the absorbed power fraction $P^s_{\rm abs}$, taking the $n_{||}$ sign into account. Figure 3(a) shows the power fraction major radius profile of $P^s_{\rm abs}$ in comparison with the power deposition profile in Fig.2(b). The contour plot of $P^s_{\rm abs}$ as functions of n_ϕ and n_θ is also shown in the figure. The comparable fraction power was found in the opposite directions, indicating non-effective current drive experiments. The launching antenna shifting along z-axis was considered to increase the contribution of power deposition in the specific $n_{||}$ direction [8]. The antenna shifting is limited by a experimental installation on the ± 0.08 m. The power fraction profile and contour plot of $P^s_{\rm abs}$ for the antenna positions of $z = \pm 0.08$ m are shown in Figs.3(b) and (c). The power fractions $P^s_{\rm abs}$ with positive sign was evaluated for the antenna position of $z = -0.08$ m only. The more significant $P^s_{\rm abs}$ with negative sign was evaluated for the antenna position of $z = +0.08$ m. The antenna should be moved above or below middle plane in the experimental installation to generate the plasma current.

3. Summary

The O-X-B conversion was calculated by ray-tracing code for multiple waves. The O-X conversion process for non-optimum conditions was included in ray-tracing code. The broader absorption power profiles were calculated compared the optimum case. The significant power absorption in chosen direction was found by shifting launching antenna positions above torus middle plane.

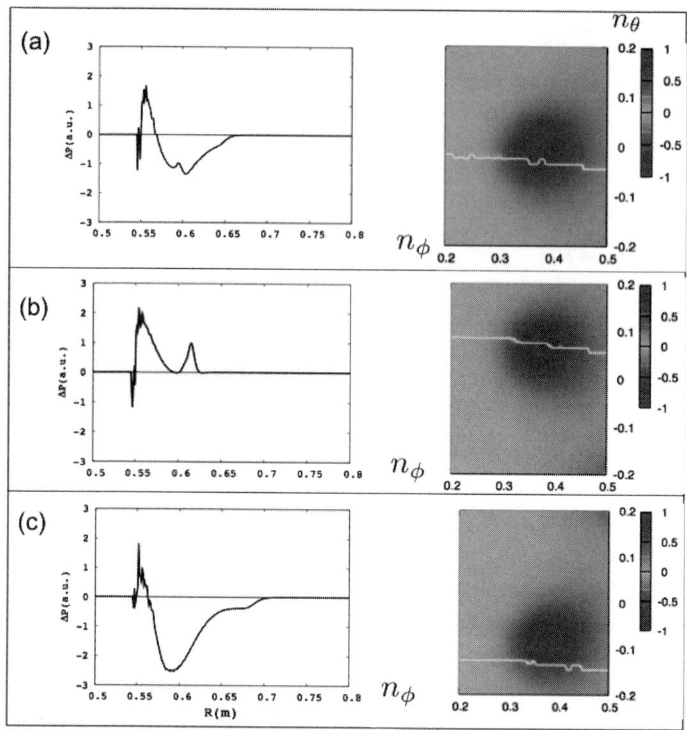

Fig. 3. The power fraction R-profiles $\Delta P = P^{+}_{\rm abs} - P^{-}_{\rm abs}$ (left) and contour plots of $P^{s}_{\rm abs}$ (right) for vertical antenna shifting$z = 0$ (a), -0.08 (b), $+0.08$ (c) m.

4. Acknowledgments

The authors thank Vladimir F. Shevchenko and Heinrich P. Laqua for helpful discussion. This research was supported by Ministry of Education, Science, Sports and Culture, Grand-in-Aid for Sc. Res. (B) [21360452].

References

1. A. Fukuyama, *Fusion Eng. And Design* **53** , 71 (2001).
2. H. Idei *et al*, in *this Proceedings* (World Scientific, Singapore, 2010).
3. E. Mjølhus, *J. Plasma Phys.* **31** 1 (1984),
4. H.P. Laqua, *Plasma Phys. Control Fusion* **49**, R1-R24 (2007).
5. Stix T.H., Waves in plasma (New York, Springer, 1992).
6. T. Maekawa*et al*, *Phys. Review Letters*, Vol 40, N21 (1978).
7. H. Idei *et al*, *J.Plasma Fusion Res. SERIES*, Vol 8 (2009).
8. C.B. Forest *et al*, *Physics of plasmas*, Vol 7, N4 (2000).

IV. Electron Cyclotron Theory

SUMMARY OF THEORY PRESENTATIONS

G. GIRUZZI

CEA, IRFM, 13108 Saint-Paul-lez-Durance, France

The presentations specifically devoted to the theory of Electron Cyclotron waves are summarized here, although many other presentations included theoretical aspects. The main topics covered are: wave propagation and absorption, cyclotron emission, current drive effects, control of MHD activity.

At the EC-16 workshop, 10 presentations were devoted to theoretical subjects: an invited talk, two oral talks and seven posters. The subjects can be classified as follows: ECCD, Cyclotron power loss, MHD control, wave propagation theory and applications. The use of sophisticated theories and computational tools in many experimental studies should be remarked as a general feature of this research field.

Electron Cyclotron Current Drive. *N.B. Marushchenko (IPP, Greifswald)* has given an invited talk on advanced ECCD calculation techniques, including momentum conservation and finite collisionality. Use of collisional operators imposing electron momentum conservation was shown to be essential in high temperature plasmas, generally leading to enhancement of the current drive efficiency, in particular for small values of the parallel refractive index. For high density stellarator applications, finite collisionality effects can also be important, and techniques to properly take them into account were discussed. The advanced computation packages TRAVIS and NEO-2 were presented. A semi-relativistic form of the Rosenbluth potentials for the collision operator, which could be quite useful in Fokker-Planck codes, was discussed by *Y.J. Hu (ASIPP, Hefei)*. Comparison of the ECCD efficiencies computed using various forms of the collision operator were presented by *Y.M. Hu (ASIPP, Hefei)*. Extensive calculations of ECCD efficiencies in the framework of a feasibility study for an ECRH system for JET were orally presented by *D. Farina (IFP, Milano)*. These results were obtained by means of the beam-tracing code GRAY, coupled with a linear CD efficiency calculation by the adjoint technique. A compact representation of the results in the form of operation diagrams was of particular interest and could be adopted in other contexts.

Electron Cyclotron Power Loss. *F. Albajar (F4E, Barcelona)* has presented a study on electron cyclotron emission as a power loss for tokamak reactors, performed using the RAYTEC code. The sensitivity of the emitted power in the presence of superthermal tails was discussed, as a function of the distribution function characteristics.

MHD control. This is the primary application of ECCD on ITER. *S. Nowak (IFP, Milano)* presented a set of detailed calculations for JET, showing that the 12 MW ECRH system being considered now should be able to suppress both 3,2 and 2,1 Neoclassical Tearing Modes, for the most relevant magnetic field range. The calculations have been performed using a generalized Rutherford's Equation. A control oriented study was performed by *G. Witvoet (University of Technology, Eindhoven)*, applied to sawtooth instabilities. A numerical model of sawteeth has been developed in order to design an appropriate feedback controller for this function. This type of studies is going to be more and more important for ITER.

Wave propagation theory and applications. This subject, although of a general interest, finds its most peculiar and attractive applications to Electron Bernstein Wave studies. *Z. Yu (ASIPP, Hefei)* developed a particular form of the gyrokinetic theory (called gyrocenter-gauge kinetic theory) allowing a kinetic description uniformly valid for arbitrary wave frequency, and presented applications to the EBW mode conversion problem. *E. Kalinnikova (Kyushu University)* presented a set of combined multiple ray-tracing and Fokker-Planck calculations for EBW heating and CD on the spherical tokamak QUEST, using the O-X-B mode conversion scheme. Sophisticated prediction tools of this type are necessary, in view of the complicated mechanism of EBW generation, with a very narrow window of operation. Even more sophisticated and computationally demanding (in this frequency range) is the full-wave approach, presented by *V. Vdovin (Kurchatov Institute, Moscow)* in an oral presentation. In some of these simulations, the real wavelength is used. In most of them, the plasma parameters are rescaled and an accordingly rescaled wavelength is used in order to reduce the computation time. A general feature found in these simulations is a coupling of the ordinary and extraordinary modes, for interaction at the fundamental EC harmonic, as well as a broadened power deposition profile, with respect to that obtained by the standard ray-tracing approach. Applications to the O-X-B scenario were also discussed.

ECCD MODELS WITH PARALLEL MOMENTUM CONSERVATION, FINITE COLLISIONALITY, AND HIGH T_e

N.B. MARUSHCHENKO[1)†], C.D. BEIDLER[1)], S.V. KASILOV[2,3)],
W. KERNBICHLER[3)], H. MAASSBERG[1)], Y. TURKIN[1)], R. PRATER[4)]

[1)] *Max-Planck-Institut für Plasmaphysik, IPP EURATOM, Greifswald, Germany*
[2)] *Institute of Plasma Physics, NSC KIPT, Kharkov, Ukraine*
[3)] *Inst. for Theoretical and Comput. Physics, TU Graz, Austria, EURATOM-ÖAW*
[4)] *General Atomics, PO Box 85608, San Diego, CA 92186-5608, USA*
[†] *E-mail: nikolai.marushchenko@ipp.mpg.de*

The recent progress in electron cyclotron current drive calculations with the adjoint technique is reviewed. The main attention is focused on such points as parallel momentum conservation in the like-particle collisions and finite collisionality effects. For high-temperature plasmas with significant relativistic effects, the effectiveness and accuracy of the developed numerical models are demonstrated by ray-tracing calculations.

Keywords: current drive, adjoint approach, Spitzer problem, ray-tracing

1. Introduction

The adjoint technique is a most advanced and convenient method for calculation of the current drive (CD) in plasmas [1,2]. The central idea is exploiting the self-adjoint properties of the linearized collision operator to express the current through the adjoint Green's function, which is proportional to the linear plasma response in presence of parallel electric field that is formally identical to the solution of the Spitzer-Härm problem [3–5]. The effectiveness of this method was demonstrated first for calculation of the current driven by a neutral beam in a homogeneous magnetic field [6], and then with the toroidicity included [7]. This technique was subsequently applied to determine the current generated by RF sources [8–10] and for the electron cyclotron current drive (ECCD) generated by asymmetric reflecting wall (passive ECCD) in toroidal plasmas [11]. At present, the adjoint technique is commonly used for calculations of the ECCD in different ray- and beam-tracing codes (see for example Ref. 12 and the references therein).

The key point of the adjoint technique is the choice of model for the corresponding adjoint Green's function. Formally, in toroidal plasmas (not necessarily axisymmetric) adjoint 4D drift kinetic equation must be solved. Different approaches which take into account such factors as small and finite collisionality, relativity, conservation of parallel momentum, and reduce the problem to the solvable level are considered by a number of authors [13–20].

In the present work, recent progress in electron cyclotron current drive calculations is reviewed. Considering the ITER scenarios, the role of parallel momentum conservation in like-particle collisions in high-temperature plasmas is illustrated. The applied models have been benchmarked against other ray-tracing codes [12] including the Fokker-Planck code CQL3D [21]. Also finite collisionality effects have been checked. Both momentum correction technique [19] and the field-line-tracing code NEO-2 [22] have been applied for calculations of ECCD. Importance of finite collisionality effects is illustrated by calculations of ECCD efficiency for the W7-X stellarator where the transition between the collisional and collisionless regimes is demonstrated.

2. Adjoint technique

The current driven by the RF source, $j_\parallel = -e \int \mathrm{d}^3u \, v_\parallel \delta f_e$, can be formally calculated (in linear approach) by solving the drift kinetic equation (DKE), which describes the linear response of electrons to the RF source,

$$v_\parallel \nabla_\parallel \delta f_e - C^{\mathrm{lin}}(\delta f_e) = -\frac{\partial}{\partial \mathbf{u}} \cdot \mathbf{\Gamma}_{\mathrm{RF}}(f_{eM}), \tag{1}$$

where $\nabla_\parallel \equiv \partial/\partial s$ is the derivative along the field-line, $\delta f_e = f_e - f_{eM}$ is the distortion of the electron distribution function from the Maxwellian, $f_{eM} = \frac{\mu n_e}{4\pi c^3 K_2(\mu)} \mathrm{e}^{-\mu\gamma}$ with $\gamma = \sqrt{1 + u^2/c^2}$ and $\mu = mc^2/T_e$, C^{lin} the linearized collision operator, $\mathbf{\Gamma}_{\mathrm{RF}}$ the quasi-linear diffusion flux in $\mathbf{u}$-space, and $u = v\gamma$ the momentum per unit rest mass.

Exploiting self-adjoint properties of C^{lin}, it's possible to express the current drive through the convolution of the RF source with the adjoint Green's function χ, which is formally identical to the solution of the (generalized) Spitzer-Härm problem [3–5]. The final expression for a practical use is [2,8–10,14]

$$\langle j_\parallel \rangle = \frac{e n_e v_{\mathrm{th}}}{\nu_{e0}} \cdot \frac{\langle b \rangle}{\langle b^2 \rangle} \cdot \left\langle \int \mathrm{d}^3u \, \mathbf{\Gamma}_{\mathrm{RF}} \cdot \frac{\partial \chi}{\partial \mathbf{u}} \right\rangle, \tag{2}$$

where $\langle ... \rangle$ denotes the averaging over the magnetic surface, $\nu_{e0} = 4\pi n_e e^4 \ln\Lambda/(m_{e0}^2 v_{\mathrm{th}}^3)$ is the collision frequency, $b = B/B_{\max}$ with $B_{\max}$

the maximum of the magnetic field at the given flux-surface, and χ satisfies the 4D "adjoint" drift kinetic equation (aDKE):

$$v_{\|}\nabla_{\|}(\chi f_{eM}) + C^{\rm lin}(\chi f_{eM}) = \nu_{e0}\,\frac{u_{\|}}{\gamma v_{\rm th}}\,b\,f_{eM}\,. \tag{3}$$

Since no additional assumption was made, the expression for the driven current Eq. (2) is general and applicable for any configuration and arbitrary collisionality. The limitation is only a validity of the linear approach and of the neoclassical ordering (it's assumed that all particle trajectories remain within the magnetic surface).

Here, we do not specify the structure of the quasi-linear diffusion term in Eq. (2) (it's assumed only that this term is known). It's sufficient to say only that due to of high localization of $\mathbf{\Gamma}_{\rm RF}$ in phase space, a knowledge of the complete $\chi(s,\mathbf{u})$ is necessary (contrary to the conductivity and bootstrap current calculations, where only the 1st moment of the response function is required). To date, there are no tools for precise solving the generalized Spitzer problem which can be routinely applied for ECCD. In order to solve Eq. (3), one needs to reduce it to the solvable level by additional simplifications. The main attention must be focused on the interplay of geometrical effects (magnetic configuration) and the collisional response, which are described by the Vlasov and collision operators, respectively. Note also that in this paper the relativistic formulation is applied while in some cited papers the non-relativistic model is considered.

3. Collisional and collisionless limits

In Eq. (3), different time-scales exist: while the first term (Vlasov operator) is characterized by the transit time τ_t, i.e. $v_{\|}\nabla_{\|} \propto \tau_t^{-1}$, the collision operator is characterized by the collision time τ_c, i.e. $C^{\rm lin} \propto \tau_c^{-1}$. For ordering Eq. (3), we take into account that the ratio of $\tau_t/\tau_c \equiv \nu_e^* = \nu_e(u)\gamma R/\iota u$ can vary significantly (here, $\nu_e(u) = \nu_{ee}(u) + \nu_{ei}(u)$, R major radius and ι rotational transform).

When collisionality is very high, $\nu_e^* \gg 1$, i.e. $\tau_t \gg \tau_c$, Eq. (3) reduces to the local problem with the straight magnetic field (formally, the 1st term in Eq. (3) becomes negligible). In this case, only the 1st Legendre harmonic of χ is necessary, i.e. $\chi = \xi\chi_1(u)$ with $\xi = u_{\|}/u$ and $\chi_1 = \frac{3}{2}\int_{-1}^{1}\chi d\xi$. Instead of Eq. (3), it's sufficient to solve the 1D integro-differential equation for χ_1,

$$\widehat{C}_1^{\rm lin}(\chi_1) = \nu_{e0}\,\frac{u}{\gamma v_{\rm th}}\,. \tag{4}$$

Here, $\widehat{C}_1^{\rm lin}(\chi_1) \equiv C_1^{\rm lin}(\chi_1 f_{eM})/f_{eM}$ with $C_1^{\rm lin}$ the 1st Legendre harmonic of the linearized collision operator. This is the classical Spitzer problem for calculation of the plasma conductivity which has been thoroughly studied for both non-relativistic [3,4] and relativistic [5] approaches (apart from the normalization, χ_1 coincides with the classical Spitzer function). Find that this approach gives the upper limit for CD efficiency. But practically, this limit is not of high interest for the hot plasmas in toroidal devices.

In the opposite (collisionless) limit, $\nu_e^* \ll 1$, i.e. $\tau_t \ll \tau_c$, the impact of the trapped particles is important. In this case, dimensionality of the problem can also be reduced to 2D due to the coupling between the pitch, ξ, and the local magnetic field, $b(s)$, through the (normalized) magnetic moment, $\lambda = (1-\xi^2)/b$. By averaging Eq. (3) over the magnetic surface, the Vlasov operator is annihilated and the problem is reduced to a 2D equation [2,14],

$$\langle \frac{b}{|\xi|}\, \widehat{C}^{\rm lin}(\chi)\rangle = \mathrm{sgn}(\xi)\, \nu_{e0}\, \frac{u}{\gamma v_{\rm th}}\, \langle b^2\rangle. \tag{5}$$

This is the basic model for calculation of CD in the different ray- and beam-tracing codes. The description chosen for the collision operator is very important for the solution.

The general solution of Eq. (5) can be represented as a series of the eigenfunctions of the bounce-averaged Lorentz-operator [11,15] and the problem is reduced to a system of coupled ordinary differential equations. Since the series converges rapidly, this method is sufficiently precise and can be easily applied for ECCD calculations. Actually, this technique was successfully applied for simulation of another scenarios (not ECCD).

There is an alternative (and somewhat easier) way to consider the problem with almost the same accuracy. Since for the Spitzer problem only the 1st Legendre harmonic is necessary, it's sufficient to approximate the collision operator as follows [2,13,14]:

$$\widehat{C}^{\rm lin}(\chi) \simeq \xi \widehat{C}_1^{\rm lin}(\chi_1) + \xi \nu_e(u)\chi_1 + \nu_e(u) L(\chi), \tag{6}$$

Here, $L = \frac{1}{2}\frac{\partial}{\partial\xi}(1-\xi^2)\frac{\partial}{\partial\xi}$ is the Lorentz operator. In this approximation, Eq. (5) can be solved analytically,

$$\begin{aligned} \chi(u,\lambda) &= -\mathrm{sign}(\xi)\, \mathcal{H}(\lambda)\, \mathcal{K}_1(u), \\ \mathcal{H}(\lambda) &= \frac{\langle b^2\rangle}{2f_c}\, h(1-\lambda) \int_\lambda^1 \frac{d\lambda}{\langle\sqrt{1-\lambda b}\rangle} \text{ with } f_c = 1 - \frac{3}{4}\langle b^2\rangle \int_0^1 \frac{\lambda d\lambda}{\langle\sqrt{1-\lambda b}\rangle}, \end{aligned} \tag{7}$$

where f_c and $f_{\rm tr} = 1 - f_c$ are the fractions of circulating and trapped particles, respectively, and $h(x)$ is the Heaviside function. The function $\mathcal{K}_1(u) = \frac{3}{2}\int_0^1 d\lambda\, \chi(u,\lambda) = \chi_1(s,u)/b(s)$, which is proportional to the

Spitzer function, must be found as the solution of a 1D integro-differential equation [2,14],

$$\widehat{C}_1^{\mathrm{lin}}(\mathcal{K}_1) - \frac{f_{\mathrm{tr}}}{f_c}\,\nu_e(u)\mathcal{K}_1(u) = \nu_{e0}\,\frac{u}{\gamma v_{\mathrm{th}}}. \tag{8}$$

Most important for a precise calculation of ECCD is the model chosen for the relativistic operator $\widehat{C}_1^{\mathrm{lin}}$. Generally, the model must include the diffusion and drag of the test-particle in the Maxwellian background, the pitch-angle scattering, and the reaction coming from the Maxwellian which guaranties parallel momentum conservation. In this case [5,14]

$$\widehat{C}_1^{\mathrm{lin}}(\mathcal{K}_1) = \frac{1}{u^2}\frac{d}{du}\left(u^2 D_{uu}^{e/e}\frac{d\mathcal{K}_1}{du}\right) + F_u^{e/e}\frac{d\mathcal{K}_1}{du} - \nu_e\mathcal{K}_1 + I^{e/e}(\mathcal{K}_1). \tag{9}$$

Here, $D_{uu}^{e/e}$ and $F_u^{e/e}$ are diffusion and friction Coulomb coefficients, respectively, and $I^{e/e}$ is the 1st Legendre harmonic of the integral part of the collision operator, which is responsible for parallel momentum conservation (*pmc*) in e/e-collisions (here and below, this approach is called the *pmc*-model). Historically, nevertheless, the high-speed-limit (*hsl*) approach [1,2,13] was commonly accepted as the standard approach for calculation of ECCD in the ray-tracing codes [12]. This approach is based on the assumption that only the supra-thermal electrons with $\mu(\gamma-1) \gg 1$ (or, simply, $u \gg v_{\mathrm{th}}$) are involved in the cyclotron interaction. For supra-thermal electrons, the e/e drag and the pitch-angle scattering are dominant in the collision operator while other processes such a diffusion and parallel momentum conservation are neglected due to of its minor importance:

$$\widehat{C}_1^{\mathrm{hsl}}(\mathcal{K}_1) \simeq F_u^{e/e}\frac{d\mathcal{K}_1}{du} - \nu_e\mathcal{K}_1. \tag{10}$$

In this case, Eq. (8) can be analytically solved [2]. Unfortunately, in high temperature plasmas, the main contribution in ECCD is produced by the bulk electrons and applicability of the *hsl* approach fails even for highly oblique launch.

Recently, a sufficiently fast and accurate tools for solving Eq. (8) in *pmc*-approach was developed [20]. This solver (green_func) is based on the variational principle [6,16] and the solution is defined as the minimum of the relativistic functional. (Optionally, also the non-relativistic and *hsl* approaches are included in green_func.) Apart from this, the solver synch from Ref. 11 for solving Eq. (5) was revised in order to make it applicable for arbitrary magnetic configurations. Both these solvers have been successfully implemented in the ray-tracing code TRAVIS [23] and are routinely used now for ECCD calculations [24].

In Fig. 1, the results of solving Eq. (8) for the different approaches is shown. The Spitzer function, $\mathcal{K}_1(u)$, was calculated with the *pmc*-model by solver green_func [20] is shown. The Spitzer function, $\mathcal{K}_1(u)$, was calculated with the *pmc*-model by solver green_func [20] in both the weakly relativistic and non-relativistic approaches. One can find that for 25 keV the weight of relativistic effects is significant. Apart from this, precise fully relativistic solution obtained by the solver synch [11] is shown. It's important to mention that (apart from the supra-thermal electrons) the *hsl*-solution [2] is significantly less than the *pmc*-solutions, since primarily the range $u < 4v_{\text{th}}$ contributes in ECCD. It is also found that the discrepancy between the weakly and fully relativistic *pmc*-solutions is rather small in this range (this discrepancy is the fault of the rather simple trial function chosen in Refs. 6,20 for fitting, but not the approach itself). From these results one can conclude that the *pmc*-model is preferable for CD calculations, while the *hsl*-model can lead (depending on scenario) to significant underestimation of the CD efficiency.

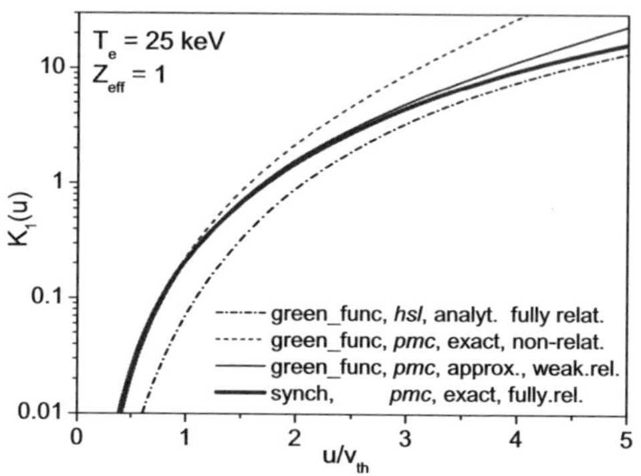

Fig. 1. Spitzer functions, calculated with different approaches for circular tokamak with $\epsilon = 0.25$ and $Z_{\text{eff}} = 1$.

In Fig.2, the results of ray-tracing calculations for ITER reference scenario-2 are presented. The angle-scan for the equatorial launcher (top-mirror) is depicted. Three different codes were applied, TORAY [25] (*hsl*-model), TRAVIS [24] (both *hsl*- and *pmc*-models) and Fokker-Planck code CQL3D [21]. One can see that the *hsl*-model sig-

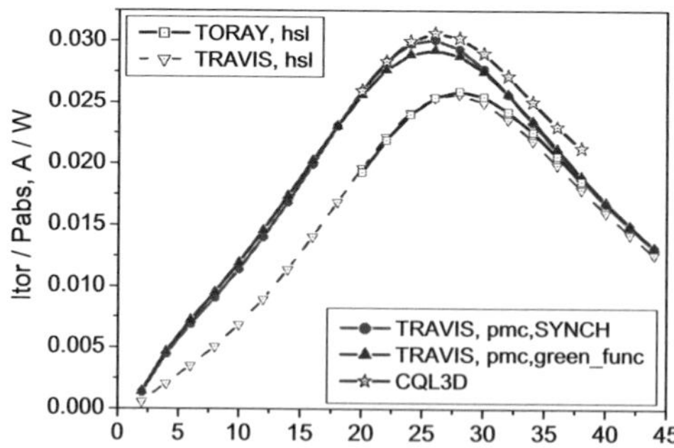

Fig. 2. ITER, equatorial launcher (top mirror): ECCD efficiency calculated by the different codes with the different approaches.

nificantly underestimate the ECCD efficiency and in the main range of interest (between 20° and 40°) the discrepancy varies from 10% to 30%. Note also that the results for *hsl*-model calculated by the different codes coincide well. Additionally, accuracy of the *pmc*-model is well confirmed by Fokker-Planck calculations.

4. Finite collisionality effects

When the plasma parameters and magnetic configuration are such that finite collisionality effects appear to be non-negligible the (generalized) Spitzer function cannot be represented in terms of only the global invariants of motion. Formally, this means that the solution of Eq. (3) becomes to be local, i.e. (similar to Eq. (7)) $\chi(s,u,\lambda) = -\mathrm{sign}(\xi)\,\mathcal{H}(s,\lambda)\,\mathcal{K}_1(u)$. As such, the problem is still a 4D one for stellarators and 3D for tokamaks.

Importance of the collisional effects for calculation of ECCD was recognized first in Ref. 17. Considering the axisymmetric configurations, it was shown that the contribution from the barely trapped electrons may be scaled as $\Delta j_{cd} \propto \sqrt{\nu_e^*}\, j_{cd}^c$ (here, j_{cd}^c is the current in the collisional limit). The semi-analytical model for calculation of the generalized Spitzer function, which can be applied for ECCD calculations, was proposed [18]. There was clearly demonstrated the diffusion of the current from the passing electrons into the barely trapped ones. From other side, this model was developed only for tokamaks with large aspect ratio and cannot be directly generalized to stellarators. Apart from this, the collision operator for the barely trapped electrons was approximated by the Lorentz term that limits the model by only low collisionality when the layer of barely trapped electrons is sufficiently thin.

The momentum correction technique [19] is based on use of the mono-energetic parallel conductivity coefficient $D_{33}(\nu_e^*)$ precalculated by the DKES code which solves the drift-kinetic equation Eq. (3) for an arbitrary collisionality in the mono-energetic approach (with the collision operator is approximated by only the Lorentz term). Formally, D_{33} being proportional to $\langle b\chi_1\rangle \propto \langle \mathcal{H}(s,\lambda)\rangle$ is actually the total "measure" for the electrons driving the current (including the barely trapped ones). This circumstance makes it possible to introduce the "effective" trapped particle fraction which depends (through the collisionality) on the energy. Indeed, the definition

$$f_{\mathrm{tr}}^{\mathrm{eff}}(u) = 1 - f_c^{\mathrm{eff}}(u) = 1 - \frac{3}{\langle b^2\rangle}\frac{B_0^2}{B_{\mathrm{max}}^2}\frac{\nu_e(u)\gamma^2}{u^2}D_{33}(\nu_e^*) \qquad (11)$$

guaranties that both limits, $f_{\mathrm{tr}}^{\mathrm{eff}}(u \to 0) = 0$ and $f_{\mathrm{tr}}^{\mathrm{eff}}(u \to \infty) = f_{\mathrm{tr}}$ hold (here, $f_{\rm tr}$ is the "geometrical" trapped particles fraction and B_0 is reference field for DKES. One finds that despite the originally non-relativistic formulation applied in DKES, the mono-energetic transport coefficient $D_{33}(\nu_e^*)$ can be interpreted as a function of the relativistic collisionality (in DKES, ν_e^* is only a parameter). Next, substituting $f_{\rm tr}^{\rm eff}(u)$ and $f_c^{\rm eff}(u)$ instead of $f_{\rm tr}$ and f_c in Eq. (8), we obtain a new equation for $\mathcal{K}_1^{\rm eff}(u)$ [19],

$$\widehat{C}_1^{\rm lin}(\mathcal{K}_1^{\rm eff}) - \frac{f_{\rm tr}^{\rm eff}}{f_c^{\rm eff}}\,\nu_e(u)\mathcal{K}_1^{\rm eff}(u) = \nu_{e0}\,\frac{u}{\gamma v_{\rm th}}, \tag{12}$$

which includes already the integral effect of the barely trapped particles and guaranties conservation of parallel momentum. The obtained then the "effective" adjoint Green's function, $\chi^{\rm eff}(u,\lambda) = -{\rm sign}(\xi)\,\mathcal{H}(\lambda)\,\mathcal{K}_1^{\rm eff}(u)$, can be applied for ECCD calculations.

This technique is well benchmarked against other techniques and analytical models [19] and is sufficiently accurate and inexpensive for calculation of the current drive. Contrary to the model of Lin-Liu [17,18], this approach is applicable for arbitrary magnetic configurations. Nevertheless, since the local dependencies of the solution are lost ($\chi^{\rm eff}$ contains only the finite collisional effects averaged within the magnetic surface), this model also has a limitations for ECCD calculations. In particular, an applicability to the scenarios with well pronounced Ohkawa effect is rather critical.

The numerical tool for approximate solving of Eq. (12) [20] being sufficiently accurate and inexpensive is already implemented in the TRAVIS code, making it possible to calculate the current drive for arbitrary collisionality. Only the neoclassical transport coefficients must be precalculated by DKES for a number of flux surfaces in the given magnetic configuration. Strictly speaking, $D_{33}(\nu_e^*, \rho, E_r\gamma/u)$ is defined in 3D parameter space, i.e. the collisionality, flux-surface label and the radial electric field (the last is not influential for the conductivity and $E_r = 0$ gives sufficient accuracy).

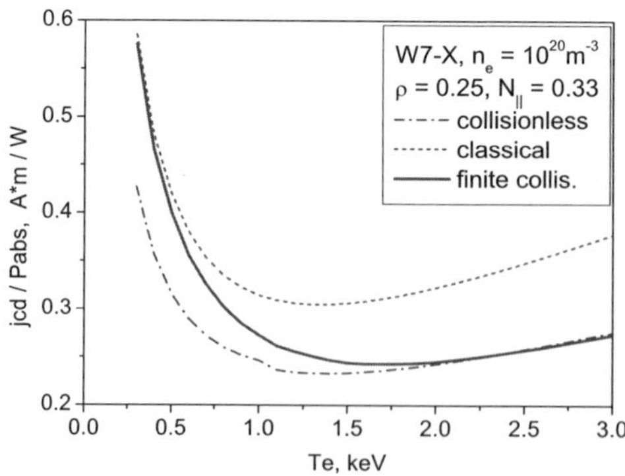

Fig. 3. Temperature dependence of ECCD efficiency (X2-mode) in W7-X for the fixed direction of the beam at the point $B = B_{\rm max}$.

In Fig. 3, the results of application of the heuristic model to the W7-X stellarator are shown. Since the launch ports for RF beams are located near the maximum of B, the Ohkawa effect is expected to be negligible even for small launch angle. The efficiency of ECCD for X2-mode for the given point and given direction was calculated. The scan over T_e reflects in fact the scan over the collisionality. Both collisional and collisionless limits are shown. As expected, the efficiency calculated with the generalized Spitzer function from Eq. (12), perfectly covers both limits for low and high temperatures. Transition from highly collisional case to the collisionless is observed in the range between 0.5 keV to 2.0 keV. This result is obtained for the fixed launch angle near the maximum of efficiency ($\approx 29°$). For smaller angles, transition to the collisionless limit appears for higher temperatures.

Recently, the code NEO-2 [22] has been developed. This code solves a non-relativistic drift kinetic equation (Eq. (3) with $\gamma = 1$) by field-line-integration technique, taking into account parallel momentum conservation with arbitrary collisionality. To date, applicability of the code for ECCD calculations has been tested for computing of the generalized Spitzer function only in circular tokamaks for non-relativistic plasmas. The preliminary results of these calculations which demonstrate specific features of the generalized Spitzer function with finite collisionality are presented in Ref. 26.

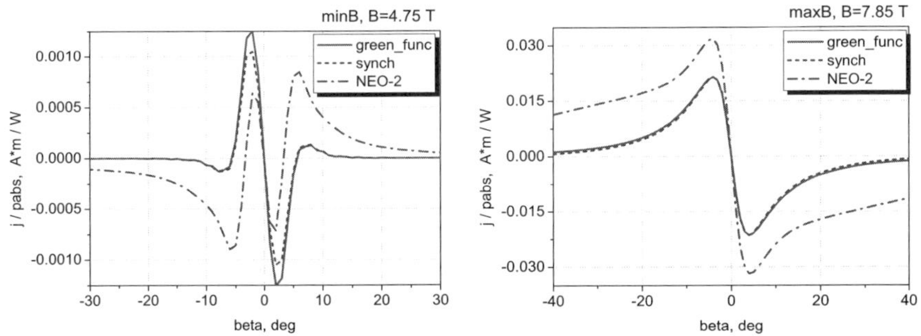

Fig. 4. Angle dependence of the ECCD efficiency (X2-mode) in the circular tokamak with $\epsilon = 0.25$, calculated with NEO-2 solver. Magnetic field chosen guaranties existence of both Fisch-Boozer and Ohkawa effects. Left: $B = B_{\text{min}}$ point. Right: $B = B_{\text{max}}$ point.

In Fig. 4, the results of calculations for a (compact) circular tokamak with $R/a = 4$, where the fraction of trapped is large, are shown. Plasma parameters are chosen to have a significant collisionality: for $n_e = 6.65 \times 10^{19}$ m^{-3} and $T_e = 1$ keV, $\nu_e^*(v_{\text{th}}) \simeq 0.26$ ($Z_{\text{eff}} = 1$). Spitzer function for the given point was precalculated by NEO-2. The ECCD efficiency for the X2 mode was calculated for two opposite spatial points in the equatorial plane which correspond to the minimum (left) and maximum (right) of B. The scan over the angle between **k** and the radial vector was performed for fixed values of B, chosen in such a way that the cyclotron interaction does not disappear. One can see that the finite collisionality effect is well pronounced in both cases. For the minimum of B point (left), the Ohkawa effect is visibly reduced while the Fisch-Boozed effect is increased in comparison with the collisionless model. This happens due to the collisional detrapping of the barely trapped electrons. In the maximum of B point (right), only the passing electrons drive the current. Nevertheless, due to reduction of friction over the trapped electrons and collisional response of the barely trapped electrons, the resulting ECCD efficiency is significantly higher than in the collisionless limit even for the large angles.

5. Summary

The first important conclusion from detailed consideration of the different approaches applied for ECCD is the following: parallel momentum conservation in the like-particle collisions is mandatory and the high-speed-limit is not sufficiently accurate (especially in hot plasmas). Simple and fast numerical solvers developed recently for solving the Spitzer problem in the collisionless limit are suitable for common use in the ray-tracing codes. These solvers are well benchmarked against the Fokker-Planck code for the ITER scenarios. It was shown that the model with parallel momentum conservation reproduces well the Fokker-Planck results, while the commonly used high-speed-limit in ray- and beam-tracing codes can significantly underestimate the current drive.

Concerning finite collisionality effects, the models developed to date are still of limited applicability. In particular, the semi-analytical model of Lin-Liu [17,18] is well applicable only for tokamaks with large aspect ratio and low collisionality, while the momentum correction technique being applicable for arbitrary configurations perfectly describes the scenario with only passing electrons involved in the cyclotron interaction. This problem can be handled by the code NEO-2, which calculates the generalized Spitzer function for arbitrary collisionality and includes the collisional detrapping affecting the current drive efficiency. Currently, the code NEO-2 is too time-consuming for routine calculations, but it can be applied for verification of other models' applicability for different scenarios and magnetic configurations. Since finite collisionality effects are of high interest and might be even more pronounced in stellarators, this problem must be given special attention.

This work, supported by the European Community under the contract of Associations between EURATOM, IPP Greifswald, and Austrian Academy of Sciences, was carried out within the framework of the European Fusion Development Agreement. The views and opinions expressed herein do not necessarily reflect those of the European Commission.

References

1. N. J. Fisch, *Rev. Mod. Physics* **59**, p. 175 (1987).
2. Y. R. Lin-Liu, V. S. Chan and R. Prater, *Phys. Plasmas* **10**, p. 4064 (2003).
3. L. Spitzer and R. Härm, *Phys. Rev.* **89**, p. 977 (1953).
4. F. L. Hinton and R. D. Hazeltine, *Rev. Mod. Physics* **48**, p. 239 (1976).
5. B. J. Braams and C. F. F. Karney, *Phys. Fluids B* **1**, p. 1355 (1989).
6. S. P. Hirshman, *Phys. Fluids* **23**, p. 1238 (1980).
7. M. Taguchi, *J. Phys. Soc. Jpn.* **51**, p. 1975 (1982).
8. T. M. Antonsen and K. R. Chu, *Phys. Fluids* **25**, p. 1295 (1982).
9. M. Taguchi, *J. Phys. Soc. Jpn.* **52**, p. 2035 (1983).
10. T. M. Antonsen and B. Hui, *IEEE Trans. Plasma Sci.* **PS-12**, p. 118 (1984).
11. S. V. Kasilov and W. Kernbichler, *Phys. Plasmas* **3**, p. 4115 (1996).
12. R. Prater *et. al.*, *Nuclear Fusion* **48**, p. 035006 (2008).
13. R. H. Cohen, *Phys. Fluids* **30**, p. 2442 (1987), and **31**, p.421 (1988).
14. M. Taguchi, *Plasma Phys. Control. Fusion* **31**, p. 241 (1989).
15. C. F. F. Karney, N. J. Fisch and A. H. Reiman, in *AIP Conf. Proc.*, (190)1989.
16. M. Romé, V. Erckmann, U. Gasparino and N. Karulin, *Plasma Phys. Control. Fusion* **40**, p. 511 (1998).
17. Y. R. Lin-Liu *et. al.*, in *Proc. 26th EPS Conf. on Contr. Fusion and Plasma Physics, Maastricht, 14 - 18 June 1999 ECA Vol.**23J** (1999) 1245 - 1248*, June 1999.
18. Y. R. Lin-Liu *et. al.*, in *Proc. 14th Topical Conf. on RF Power in Plasmas, Oxnard, California, 7 - 9 May, 2001, p. 438 - 441*, May 2001.
19. H. Maaßberg, C. D. Beidler and Y. Turkin, *Phys. Plasmas* **16**, p. 072504 (2009).
20. N. B. Marushchenko, C. D. Beidler and H. Maassberg, *Fusion Sci. Technol.* **55**, p. 180 (2009).
21. R. W. Harvey and M. G. McCoy, in *Proc. IAEA Tech. Comm. Meeting 1992*, (IAEA, Vienna, 1993).
22. W. Kernbichler *et. al.*, *Jpn. J. Plasma Fusion Res.* **3**, p. S1061 (2008).
23. N. B. Marushchenko *et. al.*, *Jpn. J. Plasma Fusion Res.* **2**, p. S1129 (2007).
24. N. B. Marushchenko, H. Maassberg and Yu. Turkin, *Nuclear Fusion* **48**, p. 054002 (2008), and **49** p.129801 (2009).
25. K. Matsuda, *IEEE Trans. Plasma Sci.* **17**, p. 6 (1989).
26. W. Kernbichler *et. al.*, *Contrib. Plasma Phys.* (2010), accept. for publication.

A RELATIVISTIC THEORY OF ELECTRON CYCLOTRON CURRENT DRIVE EFFICIENCY

Y. M. HU and Y. J. HU

Institute of Plasma Physics, Chinese Academy of Sciences, Hefei
and
Center for Magnetic Fusion Theory, Chinese Academy of Sciences, Hefei

Y. R. LIN-LIU

Department of Physics and Center for Mathematics and Theoretical Physics, National Central University, Taiwan

A fully relativistic theory of electron cyclotron current drive (ECCD) efficiency based on Green's-function techniques is considered. Numerical calculations of the current drive efficiency in a uniform magnetic field are performed. The numerical results with parameter regimes relevant to ITER operations are compared with those of two simplified models in which the electron-electron Coulomb collision operator is respectively approximated by its high-velocity limit and a semi-relativistic form. Our results indicate that the semi-relativistic approximation of the collision operator should be appropriate for modeling the ECCD efficiency under ITER conditions.

1. Introduction

Owing to its ability to drive local current on- and off-axis, electron cyclotron current drive (ECCD) has been considered the best tool for current profile control to achieve high beta and high confinement in tokamak operation. The EC heating and current drive system is planned for ITER. Although relativistic effects have been known to play important roles in the wave absorption and current drive, existing ECCD models rely on either a semi-relativistic Coulomb collision operator or the high-velocity model collision operator to describe the electron-electron interaction [1]–[3]. The predictions of the ECCD efficiency by these models were confirmed by the experiments in the present-day tokamks. Their validity for the ITER plasmas remains to be established. In this work, we present a fully relativistic theory of the ECCD efficiency. The relativistic collision operator of Baliaev and Budker is used to model the Coulomb collision among electrons. The relativistic generalization of the local Kennel-Engelmann wave-induced diffusion operator [4] in velocity space is used to describe the wave-particle interaction. The ECCD efficiency is calculated using the Green's-function techniques. In the parameter regime relevant to ITER operation, the current drive efficiencies without considering trapped electron effects will be numerically evaluated and compared with the predictions of the existing models. Our results can be used as a guide for upgrading Fokker-Planck codes and approximate current drive packages for ITER applications.

2. ECCD Efficiency

Following Cohen [2], or equivalently Lin-Liu, Chan, and Prater [3], we use Green's function techniques to calculate electron cyclotron current drive efficiency. The wave-particle interaction is described by the relativistic generalization of the local Kennel-Engelmann wave-induced diffusion operator [4] in velocity space, $S_w(f)$ whose explicit expression is given in Ref. [1]. The dimensionless ECCD efficiency defined in Ref. [3] can be expressed as

$$\zeta^* = \frac{e^3 n_e}{2\pi\varepsilon_0^2 T_e}\frac{\langle j_\parallel\rangle}{Q} = -\frac{1}{\Theta}\frac{2}{\log\Lambda^{e/e}}\left\langle\frac{B}{B_{max}}\right\rangle\frac{\int\gamma^2 d\gamma D_l f_M \tilde{L}\tilde{f}_r}{\int\gamma^2 d\gamma D_l f_M} \tag{1}$$

where

$$j_\parallel = -q_e B\left\langle\int d\Gamma f_r S_w(f_M)\right\rangle \tag{2}$$

$$Q = \left\langle\int d\Gamma w S_w(f_M)\right\rangle \tag{3}$$

$$\langle(\cdots)\rangle = \frac{\oint\frac{dl_p}{B_p}(\cdots)}{\oint\frac{dl_p}{B_p}}, \tag{4}$$

$$w = (\gamma-1)mc^2,\quad \Theta = \frac{T_e}{mc^2},\quad \tilde{L}\equiv\frac{\partial}{\partial\gamma}+n_\parallel\frac{\partial}{\partial u_\parallel},\quad \gamma=\sqrt{1+u^2},\quad u=\gamma v,$$

$$\tilde{f}_r = \frac{B_{max}\nu_c}{c}f_r,\quad \nu_c = \frac{\Gamma^{e/e}}{c^3},$$

$$\Gamma^{e/e} = \frac{n_e q_e^4\log\Lambda^{e/e}}{4\pi\epsilon_0^2 m^2},\quad f_M = \frac{n_e}{4\pi m^3 c^3\Theta K_2(\Theta^{-1})}\exp\left(-\frac{\gamma}{\Theta}\right),$$

$$D_l = |\psi_l|^2 \left(\frac{u_\perp}{\gamma}\right)^2 ,$$

$$\psi_l = J_{l+1}(x)\frac{(E_1 - iE_2)}{2} + J_{l-1}(x)\frac{(E_1 + iE_2)}{2} + \frac{u_\parallel}{u_\perp} J_l(x) E_\parallel$$

$$x = l\frac{n_\perp u_\perp}{y}, \; y = \frac{l\omega_c}{\omega}, \; \omega_c = \frac{q_e B}{m}, \quad u_\parallel = \frac{\gamma - y}{n_\parallel}, \quad n_\parallel = \frac{k_\parallel c}{\omega}, \; n_\perp = \frac{k_\perp c}{\omega} .$$

Here $j_\parallel$ is the wave driven current, Q is the absorbed wave power density, w is the particle energy, dl_p is the line element along the poloidal circumference, B_p is the poloidal magnetic field, $d\Gamma$ is the volume element in velocity space; $T_e, m, c, u, n_e, q_e, \log \Lambda^{e/e}, B, B_{max}, K_2, J_l, l, E, k, \omega_c, \omega$ are, respectively, electron temperature, electron mass, light speed, momentum per unit mass normalized to c, electron density, electron charge, the Coulomb logarithm, magnetic field, the maximum of B on the flux surface, the 2nd-order modified Bessel function of the second kind, the Bessel function of order l, cyclotron harmonic number, wave electric field, wave number, electron cyclotron frequency and wave frequency. The subscript $\parallel$ ($\perp$) denotes the parallel (perpendicular) component respect to magnetic field. E_1 and E_2 are the perpendicular components of wave electric field. f_M is electron's Maxwellian distribution.

The perturbed electuon distribution f_1 and the corresponding current drive response function f_r satisfy, respectively, the linearized Fokker-Planck equation and the adjoint equation given as follows:

$$v_\parallel \mathbf{b} \cdot \nabla f_1 - C_e^l f_1 = S_w(f_M) , \tag{5}$$

$$v_\parallel \mathbf{b} \cdot \nabla f_r - C_e^{l+} f_r = \frac{v_\parallel B}{\langle B^2 \rangle} , \tag{6}$$

where $\mathbf{b} = \mathbf{B}/B$. $v_\parallel$ is the velocity component along the magnetic field. C_e^l denotes the linearized Coulomb collision operator. C_e^{l+} is the adjoint collision operator and

$$\int d\Gamma f C_e^l g = \int d\Gamma g C_e^{l+} f . \tag{7}$$

For the case of a homogeneous magnetic field $B = B_0$, Eqs. (2) and (6) are reduced to, respectively,

$$j_\parallel = -B_0 q_e \int d\Gamma f_r S_w(f_M) , \tag{8}$$

$$C_e^{l+} f_r = -\frac{v_\parallel}{B_0} . \tag{9}$$

From Eq. (9), we find f_r has the form of $f_r^{(1)}(u)\cos\theta$, where θ is the angle between the velocity and the magnetic field. So the equation can be reduced to a one-dimensional integro-differential equation.

3. Collision Models

We consider three collision operators for modeling the electron-electron collisions. The first one is the fully relativistic Coulomb collision operator of Beliaev and Budker [5]. A differential form of this operator was constructed by Braams and Karney [6]. Their formulation will be extensively used here. The second operator to be considered is a semi-relativistic form of the collision operator first used by Karney and Fisch to model rf and EC current drive [7]. The third one is Fisch's relativistic high-velocity model [3, 8].

The linearized adjoint collision operator is given as $C_e^{l+}f_r = f_M^{-1}C_e^l(f_r f_M)$. By substituting $f_r = f_r^{(1)}(u)\cos\theta$ into Eq. (9) and dividing the both sides of the equation by $\cos\theta$, the right-hand side of the equation can be expressed as

$$\begin{aligned}\frac{1}{\cos\theta}C_e^{l+}(f_r^{(1)}(u)\cos\theta) = \frac{1}{f_M\cos\theta}&[C^{e/e}(f_M f_r^{(1)}(u)\cos\theta, f_M) \\ &+ C^{e/e}(f_M, f_M f_r^{(1)}(u)\cos\theta) \\ &+ C^{e/i}(f_M f_r^{(1)}(u)\cos\theta, f_{iM})]\,,\end{aligned} \tag{10}$$

where f_{iM} is the ion's Maxwellian distribution. Assume that the ions are infinitely massive relative to electrons, the electron-ion collision term becomes relatively simple,

$$\frac{C^{e/i}(f_M f_r^{(1)}(u)\cos\theta, f_{iM})}{f_M\cos\theta} \approx -\frac{Z\gamma\chi}{u^3}\,, \tag{11}$$

where Z is the effective ion charge and $\chi = \frac{B_0\nu_c}{c}f_r^{(1)}(u)$. The test-particle part of the electron-electron collision term can be written as

$$\begin{aligned}\frac{C^{e/e}(f_M f_r^{(1)}(u)\cos\theta, f_M)}{f_M\cos\theta} = D_{fuu}^{e/e}\frac{\partial^2\chi}{\partial u^2} &+ \left(\frac{2D_{fuu}^{e/e}}{u} + \frac{\partial D_{fuu}^{e/e}}{\partial u} + F_{fu}^{e/e}\right)\frac{\partial\chi}{\partial u} \\ &- \frac{2}{u^2}D_{f\theta\theta}^{e/e}\chi\,,\end{aligned} \tag{12}$$

and the field-particle contribution gives

$$\frac{C^{e/e}(f_M, f_M f_r^{(1)}(u)\cos\theta)}{f_M \cos\theta} \equiv I_f^{e/e} = \frac{1}{\gamma} f_m(u)\chi + \frac{1}{\Theta}\int_0^u D_1 \frac{u'^2}{\gamma\gamma'} f_m(u')\chi du' + \frac{1}{\Theta}\int_u^\infty D_2 \frac{u'^2}{\gamma\gamma'} f_m(u')\chi du' , \quad (13)$$

with

$$D_{fuu}^{e/e} = \frac{\gamma}{u}\left[\int_0^u A_1 \frac{u'^2}{\gamma'} f_m(u')du' + \int_u^\infty A_2 \frac{u'^2}{\gamma'} f_m(u')du'\right],$$

$$D_{f\theta\theta}^{e/e} = \frac{1}{u\gamma}\left[\int_0^u B_1 \frac{u'^2}{\gamma'} f_m(u')du' + \int_u^\infty B_2 \frac{u'^2}{\gamma'} f_m(u')du'\right],$$

$$F_{fu}^{e/e} = \frac{\gamma}{u}\left[\int_0^u C_1 \frac{u'^2}{\gamma'} f_m(u')du' + \int_u^\infty C_2 \frac{u'^2}{\gamma'} f_m(u')du'\right],$$

where

$$\gamma' = \sqrt{1+u'^2}, f_m = \frac{4\pi m^3 c^3}{n_e} f_M ,$$

$$A_1 = \frac{2\gamma^2}{u^2} j'_{0[2]02} - \frac{8}{u^2} j'_{0[3]022}, \quad A_2 = \frac{2\gamma'^2}{uu'} j_{0[2]02} - \frac{8}{uu'} j_{0[3]022} ,$$

$$B_1 = \frac{\gamma^2}{2} j'_{0[1]2} - \left[1 + \frac{\gamma^2}{u^2}\right] j'_{0[2]02} + \frac{4}{u^2} j'_{0[3]022} ,$$

$$B_2 = \gamma^2 \frac{u}{u'}\left[\frac{1}{2}\frac{\gamma'^2}{\gamma^2} j_{0[1]2} - \frac{u'^2}{u^2}\left(\frac{1}{u'^2} + \frac{1}{\gamma^2}\right) j_{0[2]02} + \frac{4}{\gamma^2}\frac{1}{u^2} j_{0[3]022}\right],$$

$$C_1 = -\frac{\gamma}{u} j'_{0[1]1} + \frac{2}{\gamma u} j'_{0[2]11}, \quad C_2 = -\frac{4\gamma'}{\gamma u'} j_{0[2]02} ,$$

$$D_1 = \frac{1}{\gamma}\left[\frac{1}{u^2}\left(2j'_{1[1]1} + \frac{j_{1[1]2}}{\Theta} - 10\frac{j'_{1[2]02}}{\Theta}\right) + \frac{\gamma}{u^2}\left(-2\frac{j'_{1[1]1}}{\Theta} + 4\frac{j'_{1[2]11}}{\Theta} + 6\frac{j'_{1[2]02}}{\Theta^2} - 24\frac{j'_{1[3]022}}{\Theta^2}\right) + \frac{j'_{1[1]0}}{\Theta} + 2\gamma\frac{j'_{1[2]02}}{\Theta^2}\right],$$

$$D_2 = \frac{1}{\gamma u'^2}\left(2j_{1[1]1} + \frac{j_{1[1]2}}{\Theta} - 10\frac{j_{1[2]02}}{\Theta}\right)$$

$$+ \frac{\gamma'}{\gamma u'^2}\left(-2\frac{j_{1[1]1}}{\Theta} + 4\frac{j_{1[2]11}}{\Theta} + 6\frac{j_{1[2]02}}{\Theta^2} - 24\frac{j_{1[3]022}}{\Theta^2}\right)$$

$$+ \frac{1}{\gamma}\frac{j_{1[1]0}}{\Theta} + \frac{1}{\gamma}2\gamma'\frac{j_{1[2]02}}{\Theta^2}\,.$$

The special functions j_* are related to generalized Legendre functions, the subscript $*$ represents a string of indices, $j_* = j_*(u)$ and $j'_* = j_*(u')$. The explicit expressions of j_* are given in the Appendix of Ref. [6]. So the normalized one-dimensional integro-differntial equation for the fully relativistic model is writen as

$$D^{e/e}_{fuu}\frac{\partial^2\chi}{\partial u^2} + \left(\frac{2D^{e/e}_{fuu}}{u} + \frac{\partial D^{e/e}_{fuu}}{\partial u} + F^{e/e}_{fu}\right)\frac{\partial\chi}{\partial u}$$

$$-\frac{2D^{e/e}_{f\theta\theta} + Z\gamma/u}{u^2}\chi + I^{e/e}_f + \frac{u}{\gamma} = 0\,. \tag{14}$$

The semi-relativistic Collision Operator used by Karney and Fisch [7] can be obtained from that of Baliaev and Budker [5] by assuming one of colliding particles is non-relativistic. The normalized adjoint equation is given as

$$\frac{1}{u^2}\frac{\partial}{\partial u}u^2 D^{e/e}_{suu}\frac{\partial\chi}{\partial u} - \frac{u}{\gamma\Theta}D^{e/e}_{suu}\frac{\partial\chi}{\partial u}$$

$$-\frac{2D^{e/e}_{s\theta\theta} + 3Z\gamma/4\pi u}{u^2}\chi + 3I^{e/e}_s(\chi) + \frac{3u}{4\pi\gamma} = 0\,, \tag{15}$$

where

$$D^{e/e}_{suu} = \int_0^u \frac{\gamma^3 u'^4}{\gamma'^2 u^3} f_m(u')du' + \int_u^\infty \gamma' u' f_m(u')du',$$

$$D^{e/e}_{s\theta\theta} = \int_0^u \frac{u'^2}{2u^3}\left(3\gamma\gamma'^2 u^2 - \gamma^3 u'^2\right) f_m(u')du' + \int_u^\infty \gamma' u' f_m(u')du',$$

$$I_s^{e/e} = \frac{\chi(u) f_m(u)}{\gamma} + \frac{1}{5}\int_0^u u'^2 f_m(u')\chi(u')\frac{1}{\Theta}\left[\frac{\gamma}{u^2}\frac{u'}{\gamma'^4}\left(\Theta\left(4\gamma'^2+6\right) - \frac{1}{3}\left(4\gamma'^3 - 9\gamma'\right)\right) + \frac{\gamma^2}{u^2}\frac{u'}{\gamma'^4}\left(\frac{u'^2}{\Theta}\gamma' - \frac{1}{3}\left(4\gamma'^2+6\right)\right)\right]du'$$
$$+\frac{1}{5}\int_u^\infty u'^2 f_m(u')\chi(u')\frac{1}{\Theta}\left[\frac{\gamma'}{u'^2}\frac{u}{\gamma^4}\left(\Theta\left(4\gamma^2+6\right) - \frac{1}{3}\left(4\gamma^3 - 9\gamma'\right)\right) + \frac{\gamma'^2}{u'^2}\frac{u}{\gamma^4}\left(\frac{u^2}{\Theta}\gamma - \frac{1}{3}\left(4\gamma^2+6\right)\right)\right]du'.$$

For the relativistic high-velocity model, the corresponding equation is given as [3]

$$\frac{\gamma^2}{u^2}\frac{\partial\chi}{\partial u} + (Z+1)\frac{\gamma}{u^3}\chi - \frac{u}{\gamma} = 0\,. \tag{16}$$

We note that Eq. (16) is a simple differential equation; there is no field-particle contributions and it does not conserve momentum.

We have numerically solved equations (14), (15) and (16) for χ of various collision models. Their predictions of the current drive efficiency are easily obtained by using Eq. (1).

4. Comparison of ECCD Efficiencies Between Collison Models

For the convenient application in the ray-tracing code TORAY-GA [9]–[12], we use Cohen's definition [2] of dimensionless ECCD efficiency

$$\zeta_c = \frac{\langle j_{||}\rangle}{en_e v_t}\left(\frac{Q}{\nu_t n_e m v_t^2}\right)^{-1} = -(2\Theta)^{-1}\frac{\int \gamma^2 d\gamma D_l f_M \tilde{L}\tilde{f}_r}{\int \gamma^2 d\gamma D_l f_M}\,, \tag{17}$$

where

$$\nu_t = \frac{\Gamma^{e/e}}{v_t^3}, \Gamma^{e/e} = \frac{n_e e^4 \log\Lambda^{e/e}}{4\pi\epsilon_0^2 m^2}, v_t = \sqrt{\frac{2T_e}{m}}\,. \tag{18}$$

Hence we have

$$\frac{\zeta_c}{\zeta^*} = \frac{\log\Lambda^{e/e}}{4}\,. \tag{19}$$

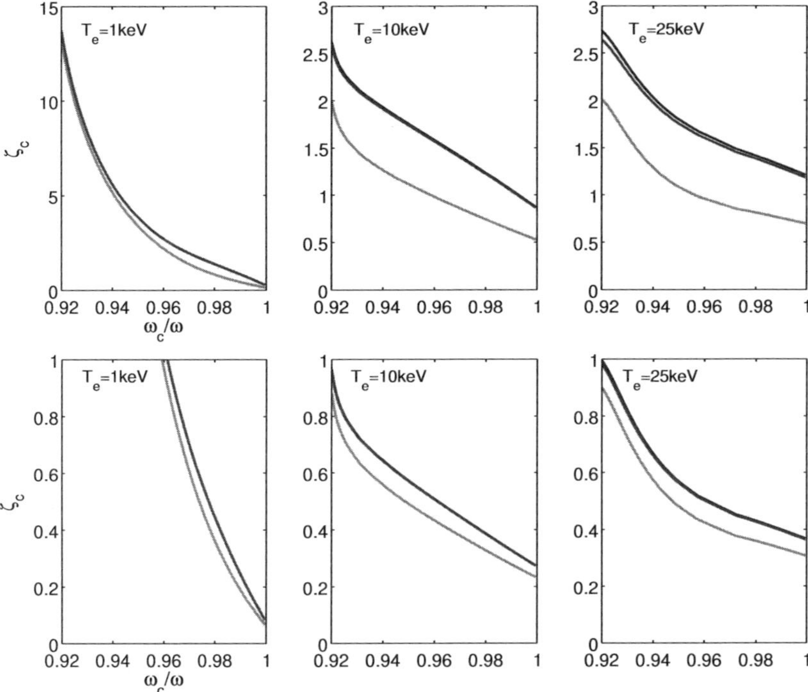

Fig. 1. Dimensionless efficiency ζ_c , as a function of $\frac{\omega_c}{\omega}$, evaluated for O-mode with $n_{\parallel} = 0.4$, $Z = 1.6$ (upper three figures) and $Z = 10$ (lower three figures) with different electron temperatures T_e. The green, blue and magenta line, represent the results from the high-velocity limit, semi-relativistic and fully relativistic collision models, respectively.

Figures 1 shows, under lower temperature ($T_e \leq 1$ keV) conditions, the three models agree well with each other. Under ITER conditions ($T_e \approx 25$ keV), the efficiency given by the semi-relativistic model still agree well with the one given by the full-relativistic model. However, in this case, the efficiency given by the high-velocity limit is much smaller than the one given by the other two collision models. These results indicate that the semi-relativistic collision model may be precise enough for the current drive applications under ITER conditions while the high-velocity limit model tends to underestimate the drive efficiency notably, especially for low value of Z, which can explain the benchmark results by R. Prater [13].

Figures 2 plots normalized response function χ as a function of u for different values of Z and T_e. It can be found that χ given by the full-relativistic model agrees with each other while the χ given by the high-

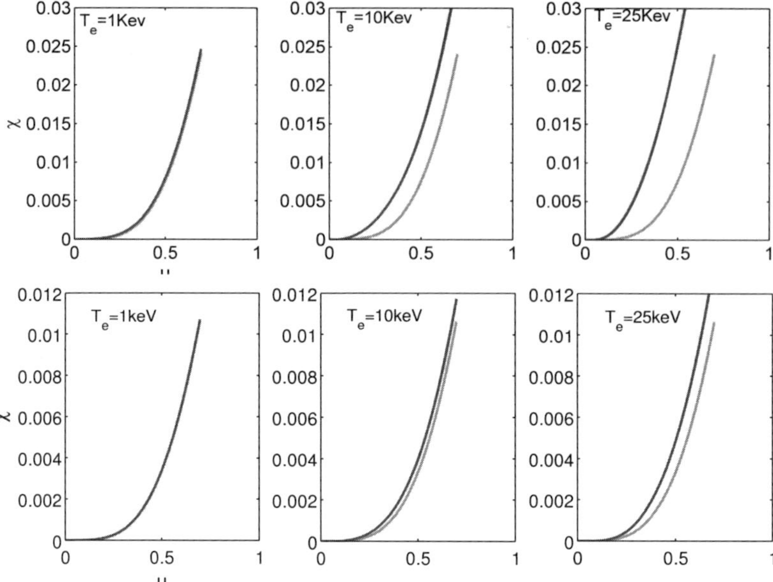

Fig. 2. Normalized response function χ , as a function of u, evaluated for $Z = 1.6$ (upper three figures) and $Z = 10$ (lower three figures) with different electron temperatures T_e. The green, blue and magenta line, represent the results from the high-velocity limit, semi-relativistic and fully relativistic collision models respectively.

velocity limit model is much smaller than the one given by the other two collision models.

Figures 3 compares the drive efficiencies given by the momentum-conservation collision model with the collision model which does not conserve momentum. The momentum conservation property of the collision operator is ensured when the field particle contribution term $C^{e/e}(f_M, f_M f_r^{(1)} \cos\theta)$ is retained. When this term is discarded, the momentum conservation property of the collision operater is lost. The results is Fig. 3 indicate that the efficiency given by the non-conservation collision model is smaller than that given by momentum conservation model, especially for the cases of high temperature and low value of Z . These results may be generally applicable to current drive problem and may be used to explain why the high-velocity limit collision model, which does not conserve momentum, tends to underestimate the drive efficiency.

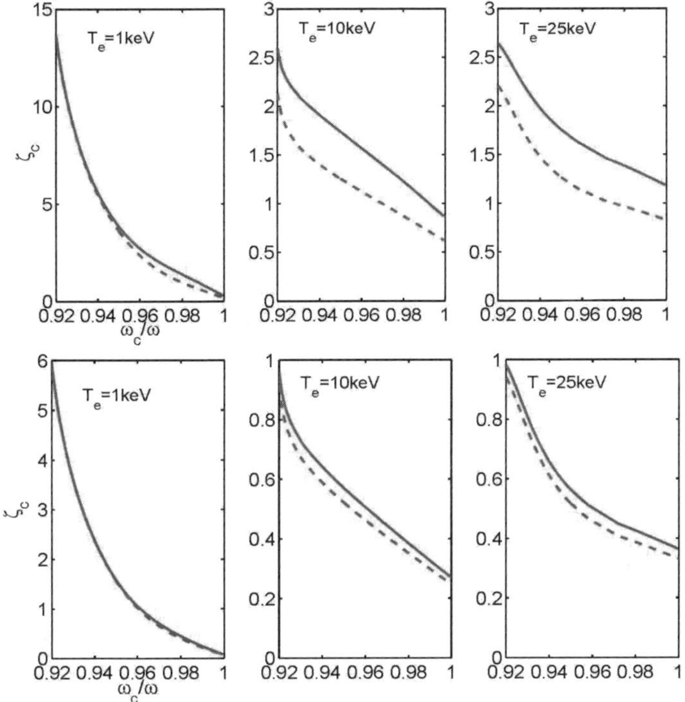

Fig. 3. Dimensionless efficiency ζ_c , as a function of $\frac{\omega_c}{\omega}$, evaluated for O-mode with $n_\parallel = 0.4$, $Z = 1.6$ (upper three figures) and $Z = 10$ (lower three figures) with different electron temperatures T_e by using the fully relativistic collision operator with momentum-conservation (solid line) and without momentum-conservation (dash line).

5. Conclusion

For the core plasma, the ECCD efficiencies were calculated by using the full-relativistic, semi-relativistic and high-velocity limit collision models. It is found that, for the lower temperature ($T_e \leq 1$ keV), the three models agree well with each other, but under ITER conditions ($T_e \approx 25$ keV), the efficiency given by the semi-relativistic model still agree well with that given by the full-relativistic models, while the efficiency given by the high-velocity limit tends to be underestimated notably. It can also be found that the efficiency tend to be underestimated when the momentum conservation of the collision model is switched off. So it can be conclued that the high-velocity limit collision model will be invalid but the semi-relativistic collision model should be still precise enough for ECCD applications under the ITER conditions.

Acknowledgments

This work is supported by the National Nature Science Foundation of China Under Grant No. 10975156.

References

[1] R.W. Harvey and M.G. McCoy, in Proceedings of the IAEA Technical Committee Meeting, Montreal, 1992 (IAEA, Vienna, 1993), p. 498.
[2] R.H. Cohen, Phys. Fluids 30, 2442 (1987), 31, 421 (1988).
[3] Y.R. Lin-Liu.,V.S. Chan and R. Prater, Phys. Plasma 10, 4064 (2003).
[4] C.F. Kennel and F. Engelmann, Phys. Fluids 9, 2377 (1966)
[5] S.T. Beliaev and G.I. Budker, Sov. Phys. Dokl. 1, 218 (1956).
[6] B.J. Braams and C. F.F. Karney, Phys. Fluids B 1, 1355 (1989).
[7] C.F.F. Karney and N. J. Fisch, Phys. Fluids 28, 116 (1985).
[8] N.J. Fisch, Phys. Rev. A 24, 108 (1981).
[9] A.H. Kriz, H. Husan, R.C. Goldfinger, and D.B. Batchelor, in Proceedings of the 3rd Joint Varena-Grenoble International Symposium on Heating in Toroidal Plasma (Commission of the European Communities, Brussels, 1982), Vol. II, p. 709.
[10] K. Matsuda, IEEE Trans. Plasma Sci. 17, 6 (1989).
[11] E. Mazzucato, I. Fidone, and G. Granata, Phys. Fluids 30, 3745 (1987).
[12] K. Matsuda ans J.Y. Hsu, Phys. Fluids B 3, 414 (1991).
[13] R. Prater, D. Farina, Yu. Gribov, et al., Nucl. Fusion 48, 1 (2008).

O-MODE AND X-MODE COUPLING EFFECTS IN TOROIDAL PLASMAS AT FUNDAMENTAL AND SECOND EC HARMONICS AND IMPLICATION FOR ITER

VDOVIN V.L.

RRC Kurchatov Institute, Institute of Tokamaks Physics Moscow, Russia

We present recent numerically well resolved ECRF 3D STELEC toroidal full wave code [1] modelling results for fundamental and second harmonics scenarios in several tokamaks and ITER. Improved numerical resolution for middle size tokamaks further solidly confirms previously discovered O- and X- modes strong coupling at fundamental harmonic leading to broadened power deposition profiles, in compare with ray tracing predictions, due to influence of *Upper Hybrid Resonance* (UHR). For the T-10/DIII-D tokamaks we consider O-mode outside launch cases with EC resonance in plasma with UHR usual "moon serp" surface and out off plasma EC resonance at High Field Side when UHR surface is in-plasma internally closed one with Electron Bernstein Waves (EBW) being excited inside of it due to mode conversion process. Combined self consistent dynamic O-mode, X-mode and EB waves structure is intriguing one and is shown. This out off plasma EC fundamental resonance scenario was discovered in WEGA stellarator [2].. In second harmonic scenarios we are concentrating on X-mode outside launch at sufficiently large plasma densities for JET, some times close (but lower) to this mode density cut off. Exact boundary problem EC wave solution again shows that simultaneously is exciting also the O-mode (with smaller amplitude), presumably due to reflection effects and wave depolarization at the wall.

1. Fundamental harmonic modeling

1.1 Out of plasma EC resonance scenario

ECH out of plasma fundamental resonance O-mode launch high resolution modelling at low frequency strongly validate that EBWs play crucial role in toroidal plasmas. Unusual out of plasma cold electron fundamental cyclotron resonance at HFS in WEGA stellarator has shown [2] efficient ECRF heating with two groups heated electrons: fast electrons group and warm electrons. Our idea was that main Heating role play Electron Bernstein Waves born at Upper Hybrid Resonance and well trapped in core plasma. Fig.1 shows O-X-B eigen modes coupling at EC resonance out off plasma fundamental harmonic in DIII-D/WEGA-like oblique $N_{//}$=0.32 O-mode outside launch with F=6 GHz, Bo=0.16 T, Ne(0)=2.3 10^{17} m^{-3}, Te(0)=9.2kV at Ne(0) < $N_{crit|O\text{-}mode}$, The MC EB Waves are shown by red colour in expanded scale in Fig.1.b. The X-mode, experiencing its density cut off, appears automatically, presumably due to plasma toroidicity, wave reflection and O- and X-mode coupling through boundary conditions at the conducting wall surrounding plasma.

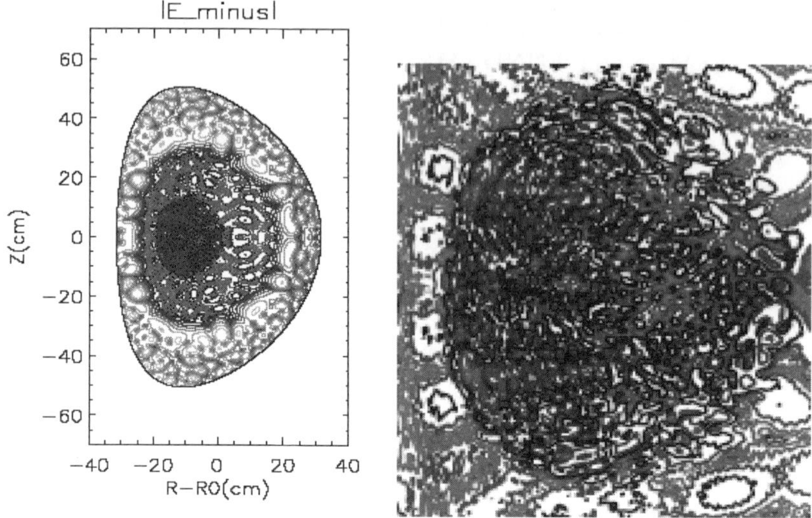

Figure 1a,b. Contour plots of |E_minus| e.m. (yellow-green) and EB Waves (red) at 6 GHz for DIII-D/Wega-like scenario. O-mode outside launch, $N_{//}$=0.32

Radial power deposition profile for this DIII-D/WEGA-like scenario is given in Fig.2 with power absorption at second and fundamental harmonics: $Pe(2\omega_{ce})$ = 61% (red), $Pe(\omega_{ce})$ = 39%.The O-X-B modes coupling at EC out off plasma fundamental harmonic in T-10-like ***O-mode*** *quasi perpendicular* $N_{//}$=0.016 outside launch at Ne < $N_{crit|O\text{-}mode}$ |E_minus| is displayed in Fig.3 and shows O-, X-mode and EBW interplay. Power is mainly absorbed at second harmonic due to relativistic effects.

1.2 *DIII-D/T-10 fundamental harmonic modelling at 60 GHz*

Similar EC heating efficiencies at fundamental harmonic in DIII-D for O-mode and X-mode outside launches at 60 GHz were reported [3]. Their nice

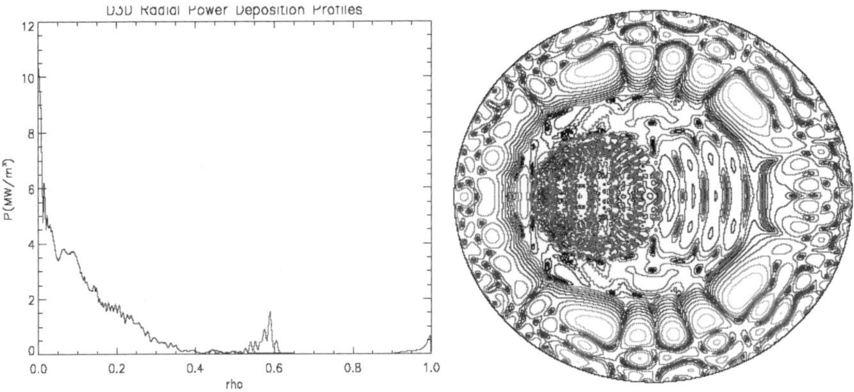

Fig. 2 Radial power deposition, $N_{//}$=0.3.

Fig.3. |E_| in T-10-like at $N_{//}$=0.016

explanation argued to X-mode reflection from X's cut off and conversion to O-mode during reflection from the walls. STELEC treats this effect automatically. Modelling was performed for O- and X-mode antenna polarizations at F=60 GHz, Bo=2.14 T, Ip=0.14-0.28 MA, Ne(0)=2.3 10^{19} m^{-3}, Ne(s)=2.3 10^{18} m^{-3}, α_n =1.0, Te(0)=36.8kV (to increase EBW resolution), α_T =2.0, outside launch spectrum $N_{//}(0)$=0.016 – 0.3, RF power 1 MW. Fig.4 displays |E_psi|| contour plots for O- launch (left) and X-mode launch (right) in T-10/DIII-D at $N_{//}$=0.16. One sees the X-mode appearance (with its cut off), even at O-mode antenna polarization, UH resonance broadly borning the EBW. Both scenarios look similar ones but waves amplitudes for X-antenna are remarkably higher ones: A(X-mode)~7A(O-mode) – as it is predicted by WKB (rays) treatment.

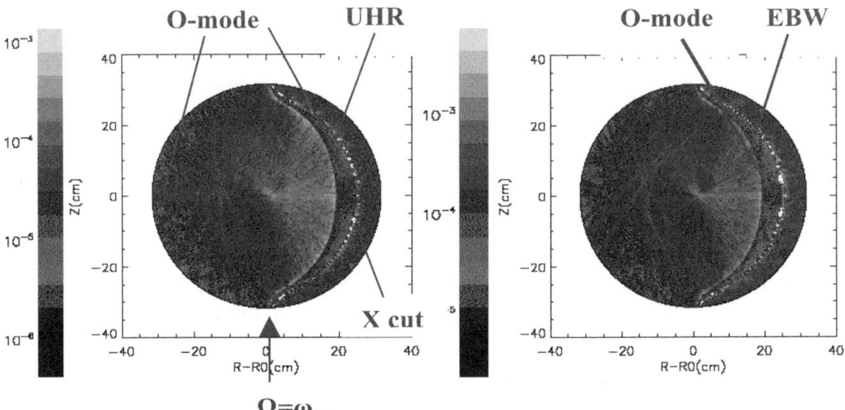

Fig.4a Radial field |E_psi| for O-mode antenna.

Fig.4b |E_psi| for X-mode antenna.

Radial power deposition profiles for both cases are shown in Figs.5a,b.

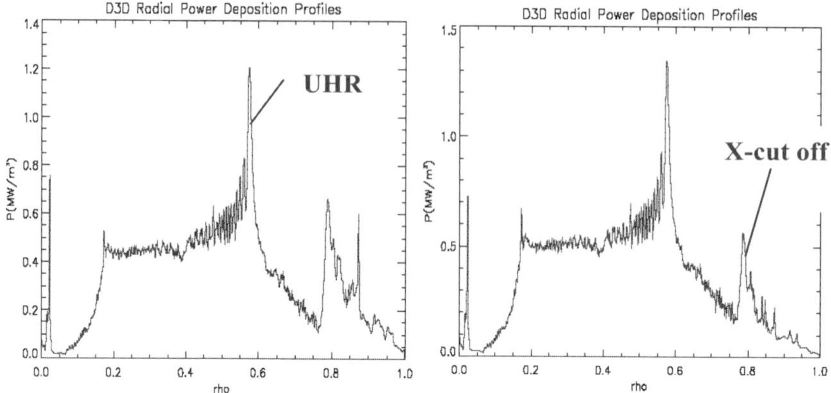

Fig.5a Power deposition for O-antenna, $N_{//}$=0.16 Fig.5b Power deposition for X-antenna

Power deposition for O- and X-mode in T10/DIII-D at 60 GHz at quasi perpendicular launch $N_{//}$=0.016 are again very similar while X-mode excited amplitudes are ~10 times higher of O-mode case (Figs.6a,b). This modelling

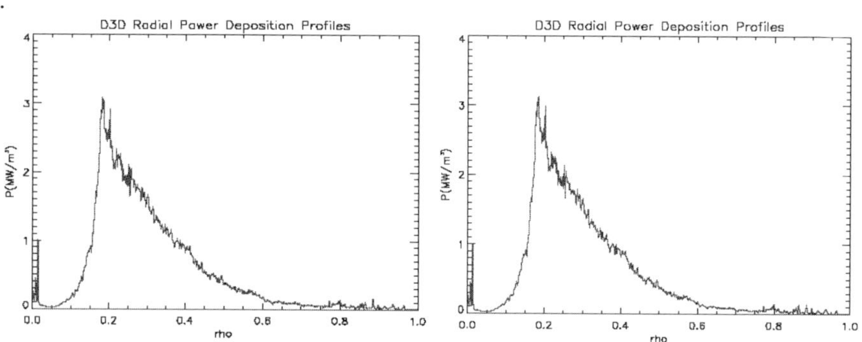

Fig.6a Power deposition for O-antenna, $N_{//}$=0.016 Fig.6b Power deposition for X-antenna

strongly supports very similar heating efficiencies and temperature profiles reported by DIII-D [2] for both antenna polarizations thus confirming O- and X-modes coupling in toroidal plasma.

ITER fundamental harmonic modelling was performed reduced at reduced frequency11.15 GHz due to lack of adequate world Super Computer to model so large plasma at specified 170 GHz . This approach makes use similarity laws formulated in [1]. Machine and plasma parameters were: Bo=0.33-0.39 T, Ip=0.634 MA, Ne(0)=3 10^{17} m^{-3}, Ne(s)=3 10^{16} m^{-3}, α_n =1.0, Te(0)=25.8kV, α_T=1.0, outside launch spectrum $N_{//}(0)$=0.016 – 0.5, RF power 1 MW. The O-mode fundamental harmonic quasi perpendicular equatorial launch at $N_{//}$=0.017, Bo=0.381 T, is demonstrated by |Re(E_minus)| contour plots in Fig.7 and power deposition in Fig.8. The EC and UHR resonances positions in equatorial plane are X(ωce)= -27 cm, X(UHR)=38 cm.

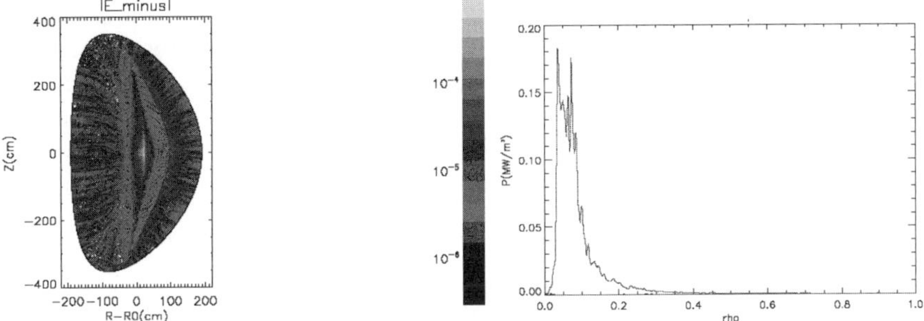

Fig.7 |Re(E_minus)| in ITER at $N_{//}$=0.017 Fig.8 Power deposition in ITER at $N_{//}$=0.017

|Im(E_z)| field contour plots and radial power deposition P(rho) at O-mode fundamental harmonic oblique equatorial launch at $N_{//}$=0.49, at increased Bo=0.391 T with X(ω_{ce})= -11 cm, X(UHR)=53.5 cm, are given by Figs.9,10.

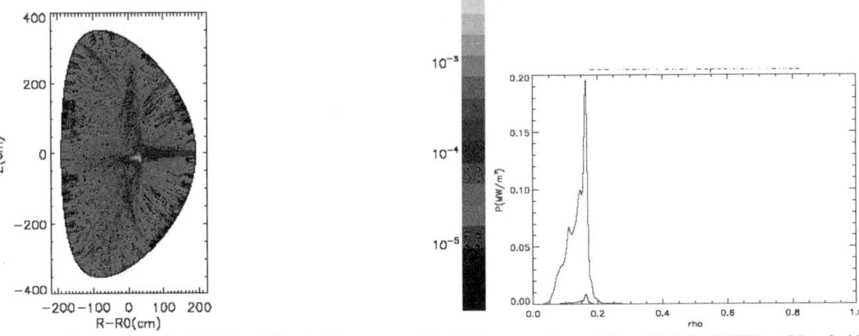

Fig. 9 |Im(E_z)| in ITER at $N_{//}$=0.49 Fig.10 power deposition P(r) in ITER at $N_{//}$=0.49

For both cases we see that power deposition is governing by UHR position, large amplitudes small scale EBW activity near of it. Important to stress that EBW are interacting with small group of very energetic electrons which can contribute to highly localised and efficient Current Drive, e.g. for NTM suppression.

2. Second harmonic ECRF modelling

There are no Upper Hybrid resonance at typical tokamaks plasma densities and O- and X-modes are more weakly coupled, as STELEC shows, and ray tracing method still can be used in rare plasma. However full wave code modelling is important at middle and high densities near of X- and O-modes cut offs, to describe interplay of refraction, reflection, interference and diffraction

effects. Due to limited paper space we will consider only example for JET ECH second harmonic X-mode outside F=27.5 GHz equatorial oblique launch to over dense L-mode plasma: N(0)=5 10^{18} m^{-3}, Te(0)=9.8 keV, $N_{//}(0)$=0.2, displayed in
Figs.11a,b for |E_psi| and |Im(E_z| e.m. field components. Fig.11a shows X-mode cut off with following wave bouncing in poloidal direction between X-cut off layer and the wall, while simultaneously excited O-mode well focalised beam pattern propagating to second harmonic layer near magnetic axis, Fig.11b, not experiencing O-mode cut off. Respective radial power deposition is in Fig.12.

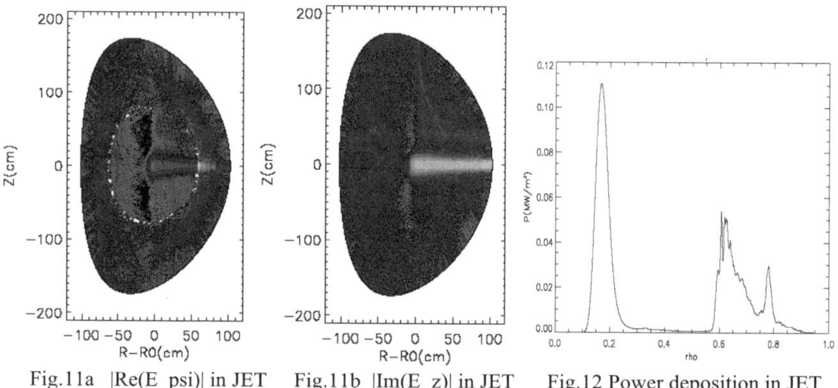

Fig.11a |Re(E_psi)| in JET Fig.11b |Im(E_z)| in JET Fig.12 Power deposition in JET

Thus we see a possibility of double peaks power deposition with a single gyrotron frequency. This scenario is a possible candidate for recent AUG ECH scenario with tungsten wall to control impurity accumulation in plasma center.

Conclusions

We have shown that exact Maxwell-Vlasov boundary value problem solution for EC plasma heating by updated 3D STELEC code solidly confirms early [1] results on O- and X-mode coupling in tokamaks at fundamental harmonic leading to remarkable power deposition profile broadening important e.g. in NTM suppression concept for ITER. Weak coupling exists also at 2nd harmonic.

References

1. Vdovin V.L. Role of Upper Hybrid resonance and diffraction effects at Electron Cyclotron Heating in tokamaks , Proceedings of EC and ECE 14th Joint Workshop, invited lecture, p.323-333
2. Podoba Yu.. Radio frequency heating on the WEGA stellarator, PHD thesis, Ernst-Moritz-Arndt-Universität Greifswald. 2006

3. Prater R. et al, *Proc. of the EPS25* (Praha 1998) Vol. 22C p. 1410

V. Gyrotrons and EC Technology

SUMMARY OF TECHNOLOGY PRESENTATIONS AT EC-16

W. KASPAREK
IPF Stuttgart

In the technology sessions of EC-16, 24 talks and posters were presented, six on gyrotrons, seven on the system design, transmission lines, components, and the launcher for ITER, three on vacuum windows, seven on new ECRH systems. Additionally, a discussion of future requirements was held.

In the session on gyrotrons, G. Denisov et al. presented the development of tubes in Russia. The present status of the 170 GHz tube for ITER is characterized by powers of 0.65 MW for 800 s, and 1.05 MW for 200 s pulses. The gyrotron features low frequency drift, and up to 60% efficiency. Limitations in pulse length are due to isolator heating, and a window failure stopped the tests for some time. The experiments were resumed with the new test stand (80 kV, 50 A), and it is planned, to reach 1 MW power with 1000 s pulse length this year. In parallel, a 1.5 MW CW gyrotron operating in $TE_{28,12}$ has been designed, and on a prototype a power of 2 MW (short-pulse) was measured. The development of multi-frequency gyrotrons in the range 105 to 140 GHz continued, and 1 MW at 50-60% efficiency was reached. While two-frequency tubes work reliably, the design and reliability of 4-frequency gyrotrons is still and issue.

Gyrotron development at JAEA continued very successfully, as was shown by K. Sakamoto et al. A reliability test of the 170 GHz/1MW tube at 800 kW with 10-min shots every 30 min showed 72 successful shots out of 88, which is a good result; however, for ITER this would not be sufficient, therefore recover technologies are to be developed. Modulation experiments up to 5 kHz using the body power supply together with the acceleration power supply result in strong current modulation and thus in a strongly reduced collector loss, and increased power up to 1.2 MW. Work on a dual-frequency gyrotron 170 GHz/137 GHz, 1.5 MW design was presented, and short-pulse tests showed power in excess of 1 MW, and high-quality output beams with similar parameters. Moreover, a 110 GHz/1.5 MW gyrotron for JT-60 SA is now under test.

In Europe, a 170 GHz coaxial gyrotron with a power of 2 MW is developed for ITER, and three talks by F. Albajar, G. Gantenbein, J. Jin et al. dealt with this subject. The test of the first prototype yielded 1.4 MW maximum power limited by voltage stand-off problems as well as parasitic oscillations in the beam tunnel. In the meantime, the tube has been refurbished, and especially a new beam tunnel and a new q.o. converter have been installed. A test at KIT Karlsruhe with the corresponding pre-prototype tube and the new corrugated beam tunnel resulted in 2.2 MW (short-pulse). In parallel, the new launcher was tested, which had been optimized using a novel numerical method for the synthesis of launchers; a fundamental Gaussian mode content of 96.3% at the launcher aperture was obtained, and a further improvement to 99.1% mode purity (at the window plane) is expected with the use of new phase correcting mirrors. These results give good prospects for the refurbished tube, nevertheless the decision, if Europe will deliver 2 MW coaxial tubes or conventional 1 MW gyrotrons has to be made relatively soon.

At CPI, Palo Alto, the development of a $TE_{22,6}$ gyrotron, 95 GHz, 2 MW is ongoing, and results were presented by S. Cauffman. The work is done for the U.S. Air Force, and this customer has specified 2 MW output power at 55 % efficiency, and a maximum weight of 1.5 tons. This required the development of a high-power, lightweight collector made from Glidcop. The results obtained so far are 1.4 MW, at 75 A, limited by parasitic oscillations in the beam tunnel. After improvement of absorbers, 1.4 MW were obtained at 45 A with efficiency of 51%; however, at higher currents, saturation of the power at 1.7 MW occurred due to beam tunnel oscillations, which requires further improvement of this part.

Overviews on the ITER ECRH system and its transmission system where given by M. Henderson et al., and F. Gandini et al. Problems concerning administration of the project, time schedule, in-kind purchasing were discussed. The system allows now a broad range of EC applications, as the Equatorial Launcher and Upper Launchers have been continuously optimized. New scenarios are X2 mode operation at half field, and possibly X3 launch, as well as increase of current drive at mid-radius. The 127 GHz tubes for start-up are replaced with 70 GHz gyrotrons. The system design is still ongoing, and components are optimized. Space problems in the RF hall allow only for installation of 24 instead of 26 tubes. The transmission lines are now rated for 2 MW, the length is about 160 m, and the expected efficiency is 92%. The conceptual design of the Launchers is essentially frozen.

An issue is the deflection of ITER Upper Launcher due to its own weight and magnetic forces in case of disruptions, as was reported by D. Strauß et al. The optimization results show a maximum deflection of the port and the plug of 11.6 mm, which has to be compared with the 13 mm spacing around the plug. This value might not be sufficient for safe operation, and remote handling in view of tolerances.

The upgrades of the vacuum barrier window for ITER were presented by T. Scherer. This includes corrugated waveguide inserts up to the diamond disc, which shield the cuffs from stray radiation, and thus reduce the excessive loss measured in earlier tests. Material investigations are ongoing with the goal to reduce the surface loss of the diamond discs. If surfaces could be prepared such that only bulk loss ($\tan\delta = 2\cdot10^{-6}$) contributed to the window absorption, then powers well above 2 MW could be handled.

A talk by A. Vaccaro et al. dealt with corrosion problems for window designs, where the brazing is in contact with the cooling medium. He calculated that low viscosity silicone oil is a good alternative to water. The cooling performance is only slightly reduced, which still allows operating the windows for ECRH at W7-X with the rated power of 1 MW. J. Stober et al. discussed windows for multi-frequency gyrotrons. For the gyrotrons and the torus windows at ASDEX Upgrade, double-disc tuneable windows, single-disc Brewster windows have been used, not always successfully. At present, single discs with anti-reflection corrugations are under investigation, showing good prospects.

Components for the ITER transmission lines have been discussed by R. Olstad et al. and D. Rasmussen et al. A complete set of components for HE_{11} mode transmission is available, and many are approved for 1 MW. New designs include a mode analyzer based on a partly transparent mitre bend mirror, a robust directional coupler, and 3.5 m long waveguide segments with helical corrugations, which are cheaper. Present developments concentrate on the power upgrade of components for 2 MW; especially for polarizers and waveguide switches, cooling is a big issue. To qualify the parts, a 2 MW test facility is foreseen using a 170 GHz gyrotron and a ring resonator.

A component, which presently is not foreseen in the ITER transmission lines, but could be useful for power combination in case of power upgrades, as well as for fast and slow switching of the power between launchers, was discussed by W. Kasparek et al. A prototype of such a high-power diplexer was tested in the ECRH system of W7-X, and proof-of-principle experiments confirmed the calculated performance. An issue is the frequency stability of gyrotrons, therefore frequency tracking systems are under development. Application for NTM stabilization and in-line ECE at ASDEX Upgrade is foreseen, and results will help to decide if these diplexers may become reliable components in ECRH systems.

The ITER ECE transmission components have been presented by H. Pandya et al. A main requirement is the enormous frequency band of 70 GHz to 1000 GHz, where low-loss, stable transmission over 40 m should be guaranteed. HE_{11} mode is foreseen for transmission, and the question is open if corrugated waveguides (diameter 63.5 mm with groove period of 0.45 mm) can be used above the Bragg-limit of about 300 GHz, or if alternatives like dielectric or dielectrically coated waveguides have to be used. As vacuum barrier, windows with anti-reflection layers and/or reflection reducing corrugations are proposed.

Besides ITER, work for new ECRH systems was presented. H. Idei et al. described a phased array antenna for electron Bernstein wave heating and current drive on QUEST, operating at 8.2 GHz. Eight individually phased radiators, which are fed with arbitrary polarization allow beam steering over a large angular range without mechanics. Successful experiments at QUEST were reported, with a driven current of 10 kA at a launched power of 30 kW. The system is further improved to reach the nominal power of 200 kW. In addition, application of this antenna for diagnostics is underway.

The 110 GHz ECRH system for KSTAR became operational as well. M. Joung et al. described commissioning and operation of this system. After problems with the original gyrotron, a 110 GHz tube from DIII-D could be loaned, and after fast installation by Gycom and due to an efficient matching optics unit, 250 kW could be injected to KSTAR, showing pre-ionization in X2 and O2 mode, which was optimized with gas-puffing. The 2.8 MVA power supply system for ECRH on KSTAR was discussed by T. Seo. It consists of a main power supply 80 kV, 25 A with IGBT inverter, a transformer with rectifier, a filter deck, a HV switch with crowbar; via a snubber coil, the gyrotron is connected. The body power supply consists of a HV DC supply with an IGBT modulator.

A large project, still not approved, is the planned ECRH system at JET. The work on the technical feasibility of this 170 GHz system fed by 12 tubes at 1 MW each, for up to 20 s pulses was reported in several talks by C. Sozzi, M. Lennholm, S. Garaviglia, and H. Braune et al. A big challenge is the wish to have the system ready for operation already in 2015; therefore, an early decision on the project is needed to cope with time schedules. At present, the conceptual design is ongoing. It is envisaged that the gyrotrons are supplied by Gycom in Russia. The power supplies for the tubes will consist of 8 refurbished power supplies, where 4 old systems with 180 kV will be modified to 8 units with 90 kV each. In addition, 4 new pulse-step modulators will be purchased. The transmission system will be very similar to that of ITER; corrugated evacuated waveguides for HE_{11} transmission will be used. Therefore, the installation could act as kind of test facility for the ITER transmission lines. A special feature is the requirement of a 2nd barrier window for safety reasons. For the 12 beams, two launchers are foreseen which must fit into one port. The launchers will have two-axes steering with independent poloidal and toroidal movements. Both steering mechanisms could use the ITER Upper Launcher mirror assembly with pneumatic mirror drive. In addition to the steering mirrors, a fixed focusing mirror and a mirror at the side wall of the port will be used to obtain good optical performance, and a large steering range.

Finally, the technical requirements for ECRH on DEMO have been discussed by M. Henderson. Key issues are an overall efficiency of larger than 50% to keep the recirculating power low, and a high reliability for safe reactor operation. This means that power supplies with efficiencies of 97% are needed, and that all support systems like cooling pumps, auxiliary supplies, and control systems must have low power consumption. The gyrotrons, which probably operate at a frequency around 200 GHz, will possibly need multi-stage collector potential depression. Unit powers of 2 MW are expected; the question, if multi- or single-frequency tubes are needed, was left open. Transmission lines should be simple, nevertheless transmit microwaves with high mode purity, and an efficiency of > 92%. High reliability calls for a simple overall system design and corresponding control. The lifetime of all components must be high, techniques for restart after a problem must be developed, and automated operation is a must. Especially the launcher, which is the most critical component in the system, must be optimized. If possible, (remote) steering should be avoided; the most reliable launcher was defined as "hole" in the first wall.

DEVELOPMENT OF HIGH POWER LONG PULSE ITER GYROTRONS

K. SAKAMOTO, K. KAJIWARA, Y. ODA, K. TAKAHASHI, K.HAYASHI, N. KOBAYASHI

Japan Atomic Energy Agency (JAEA) 801-1 Mukoyama, Naka-shi, Ibaraki 311-0193 Japan

M.A. HENDERSON, C. DARBOS

ITER Organization, St. Paul-lez-Durance, 13067 France

Recent progress on the development of a high power 170 GHz gyrotron obtained in Japan Atomic Energy Agency (JAEA) is presented. Firstly, a repetitive operation of 800 kW/600 s with an interval of 20-30 min was performed. The electrical efficiency was 52-57 % with a depressed collector. The 72 shots of 88 shots was very stable 600 s oscillation. No damage was found on the gyrotron after the test. Secondly, 5 kHz-full power modulation was demonstrated with pulse duration of 60 s. For this purpose, a full beam current modulation was realized to minimize the collector heat load by adopting an anode voltage switching. Thirdly, in the dual frequency gyrotron development, the 1.3 MW oscillations at 170 GHz and 136.8 GHz were successfully demonstrated at short pulse operation by changing the cavity field and the anode voltage keeping other parameters constant.

1. Introduction

FOR ITER, a 20MW Electron Cyclotron (EC) system [1-3] is planned for central heating and current drive (H&CD) applications as well as off-axis CD to control magneto-hydrodynamic (MHD) instabilities. As a power source, Japanese Domestic Agency (JADA) will procure eight 1MW-170GHz gyrotrons. The tube should have an electrical efficiency of ≥50 %. In JADA, a basic specification for ITER was satisfied at 170 GHz gyrotron, i.e., 1MW, 800 s 55 % in 2006 [4]. Using this gyrotron, the extended experiments for ITER procurement is carried out, such as a demonstration of reliability test, high power generation, and high frequency power modulation up to 5 kHz in CW-mode. In addition, using the gyrotron as a power source, developments of the ITER transmission line components [5] and launcher mm wave components [6] have been conducted. These provide database for accomplishment of ITER EC H&CD system. In parallel, a development of dual-frequency gyrotron is underway for a future advanced EC H&CD system. This dual-frequency gyrotron will give a large flexibility to the EC H&CD experiments.

In this paper, present activities of the gyrotron development in JAEA are presented. In section 2, the result of the gyrotron reliability test for RF generation is described. In section 3, the result of the development of high frequency power modulation in CW operation is presented. In section 4, the development of the dual-frequency gyrotron is presented. The summary is given in section 5.

2. Reliability test for repetitive operation

First repetitive operation has been conducted in 2008. The 10 pulses of 0.8 MW-400 s have been generated in series at every 30 min [7]. Next campaign was carried out at the end of 2009 to obtain a statistical database. Here, the 0.8 MW-600 s shots were repeated with an interval of 20~30 min. The total shots are more than 100 shots. The electrical efficiency of each shot was in the 52 %-57 %.

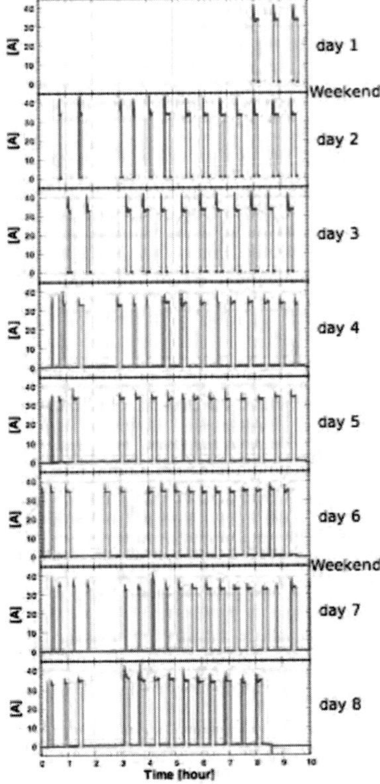

Fig.1 Shots of repetitive operation test. Vertical axis is the beam current of the gyrotron. Generation power is ~0.8MW.

In Fig.1, the time history of the repetitive operation represented by the beam curent of the gyrotron is shown. Fig.2 shows the summation of the pulse duration of shots.

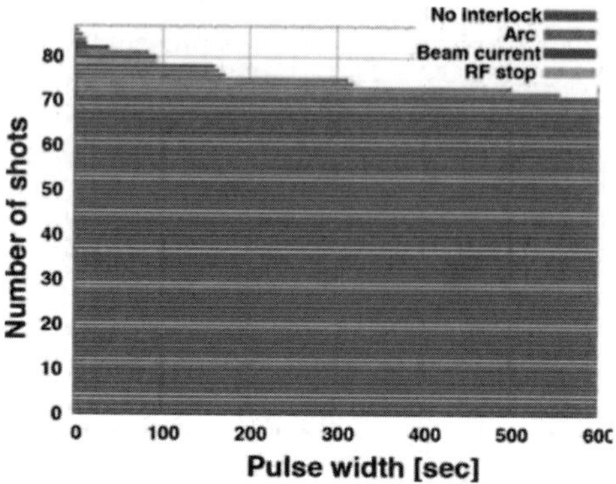

Fig.2 Summary of pulse duration of repetitive operation at 0.8 MW. The interval of each shot is 20-30 min.

Reasons of the pulse termination are a light intensity increase measured by a photo-multiplier through the viewing port, sudden increase of the beam current, etc. One cause of the light increase is the mode transition resulting in the RF scattering and the discharge in the tube. The initial shots of every day were spent for the conditioning. The 88 shots are accounted as a denomenator, and 72 shots was very stable 600 s operation, i.e., the success rate of 600 s shot was 82 %. A present operation sequence of the JAEA gyrotron test stand has no re-start system when the pulse is terminated by the interlock during the shot [8]. To increase the full pulse operation ratio, a quick recovery system should be adopted. Fig.3 is the transition of the ion pump current corresponding to the gyrotron vacuum pressure. The pressure increases by the power generation, however, the pressure returned to the original level soon after the shot and no influence gives to the following shots. The base pressure decerased shot by shot. It should be noted that the finite level of the base pressure, $\sim 1\text{x}10^{-5}$ Pa, existed througout the test. A pin hole leak was identified after the test at the ion pump section. This pin hole was fixed by the Vacseal® and the ion pump current turned to ~0 μA. It is considered that this level resisual gas does not cause a major problem on the gyrtorn operation. As a tentative conclusion, the gyrotron will be reliable for ITER experiment. To increase the operation ratio further, the development of the quick recovery system during the shot will be effective in addition to the effort to reduce the interlock-event frequency.

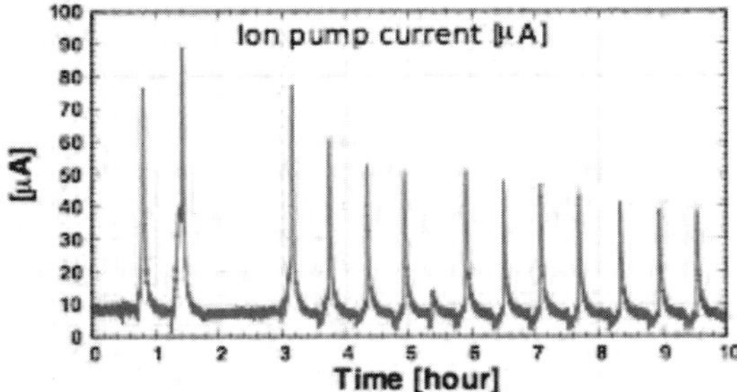

Fig.3 Transition of the ion pump current corresponding to the vacuum pressure in the gyrotron by the RF power generation. Output power is 0.8 MW.

3. High Frequency Power Modulation

In ITER, high frequency power modulation with the pulse duration of ~1min is required for the suppression of the neo-classical-tearing (NTM) mode, whereas the modulation period is limited by ~1 s at most of the present-day NTM suppression experiments. Generally, the conventional modulation method causes large power consumption at the gyrotron collector or at the power supply system. So, the extension of the pulse duration has been the issue for ITER application. A present reference design for ITER is that the full current modulation is expected up to 1 kHz and it is compromised to be a half power modulation at 5 kHz to avoid the large heat deposition to the collector. Here, we demonstrate the full power modulation is possible at 5 kHz by utilizing an anode voltage switch of the triode magnetron injection gun (MIG).

3.1 JAEA Gyrotron and Power Supplies

In Fig.4, the gyrotron and the power supply configuration of the JAEA test stand is shown. The JAEA gyrotron is featured by the triode MIG. The body voltage V_{body}, which is equivalent to the depressed collector voltage V_{dep}, is fed by the body power supply (BPS). The anode voltage V_a is controlled to change a pitch factor of the electron beam independently with the beam acceleration voltage V_b. In addition, as the beam extraction voltage V_{ak} ($=V_{cathode}-V_a$) decreases the beam current I_b does, and the I_b go to 0 A at $V_{ak}=0$. This indicates that the feature of the triode MIG is preferable for the power modulation with the beam current modulation [9], which results in lower heat load at the collector.

3.2 Power modulation with Body Power Supply

Fig.5 is an example of the power modulation at 300 Hz carried out by using the ITER-relevant BPS alone. Pulse duration is 9 sec, and 3 sec-300 Hz BPS voltage modulation V_{body}=0~23 kV was tried in the middle phase of the pulse.

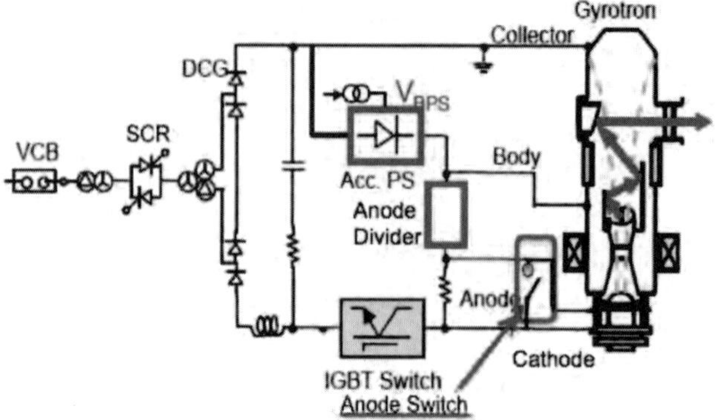

Fig.4 Configuration of power supply and gyrotron with triode magnetron injection gun.

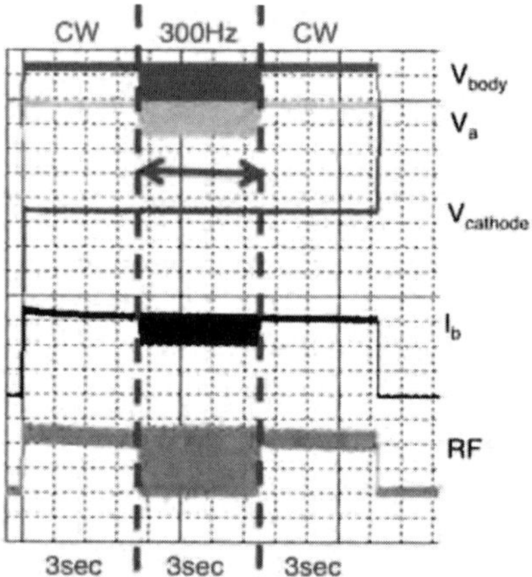

Fig.5 Power modulation at 300 Hz with BPS voltage modulation. Output power is ~0.6 MW.

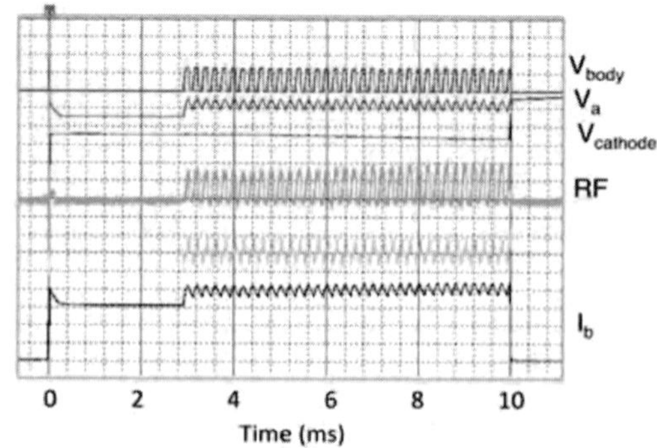

Fig.6 Power modulation at 5 kHz with BPS voltage modulation.

This is a demonstration of the gyrotron operation for the NTM control, where the power modulation is required suddenly during the shot. The V_{ak} changes with the V_{body} modulation. The beam current and the mm wave power are modulated I_b=20A-30A and P_{RF}=0-0.6 MW by V_{body}=0-23 kV, respectively. As the beam current drops 2/3 in the non-oscillation phase, the collector heat load is mitigated. However, when the modulation frequency increases, the V_{ak} could not follow because of the slow time constant of the circuit. Fig.6 is a result of 5 kHz BPS modulation. Although the full RF power modulation is realized because of a mismatch of the oscillation condition in the non-oscillation phase, the change of the beam current is small and consequently the collector heat load becomes large.

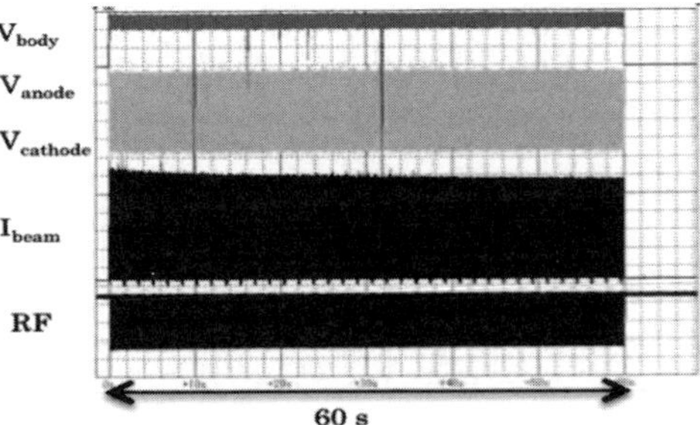

Fig.7(a) Experimental result of power modulation at 5 kHz. Pulse duration is 60 s. RF power is ~1.1 MW.

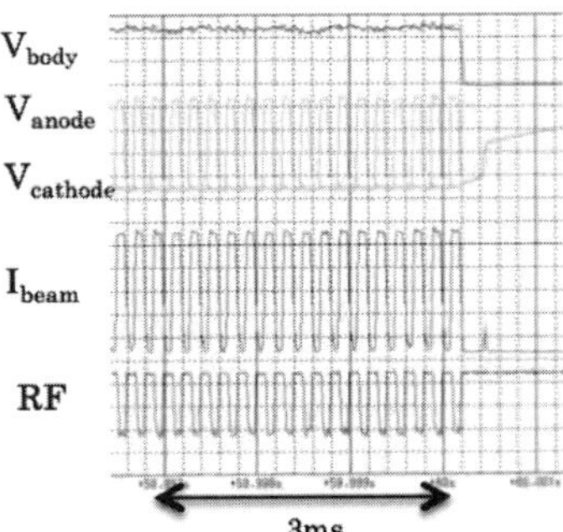

Fig.7(b) Expansion of end of the pulse. By the anode switching, the RF power changes between 0 and ~1.1 MW.

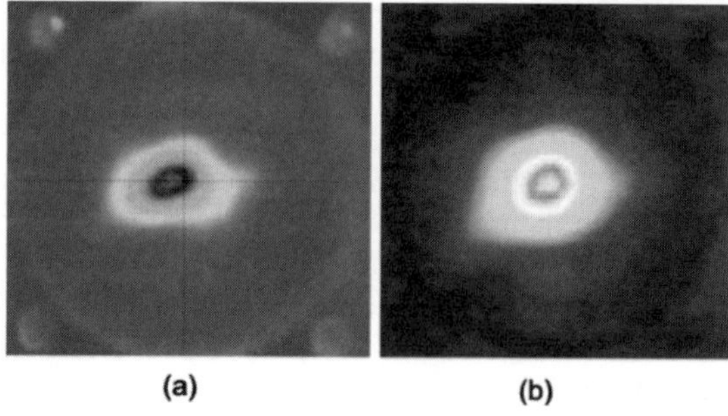

Fig.8 Power profiles of output RF beam at the window for 170 GHz (a) and 136.8 GHz (b).

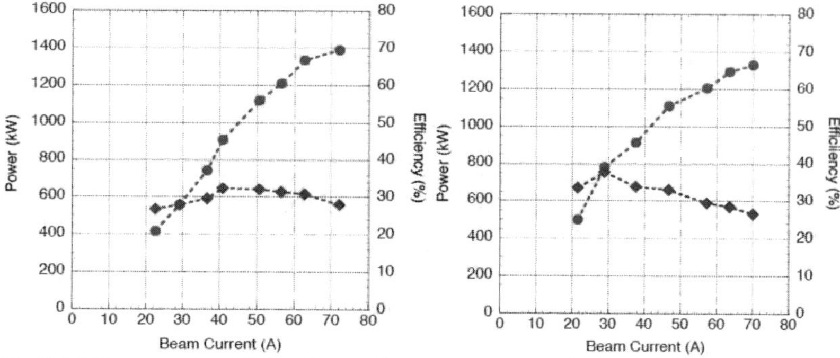

Fig.9 (a) Beam current dependence of output power and efficiency without depressed collector at 1ms.
(a) 170 GHz. V_b=71~72 kV (b) 136.8 GHz. V_b= 68~69 kV.

3.3 5 kHz power modulation with Anode switch

To realize the power modulation with full beam current modulation, the anode switch composed of IGBT's is proposed and tested. Fig.7 (a) is an example of the 5 kHz full beam current and power modulation for 60 s at the output power of ~1.1 MW. Here, V_{dep}~24.3 kV, V_{catode}~-45.2 kV (main power supply voltage), I_{beam}~50A. Fig.7 (b) is the expansion of the pulse end. By the anode voltage switching, the beam current I_{beam} is modulated between 0 A and ~50 A and the power changes between 0 and ~1.1 MW, respectively. As the beam current goes to zero, the heat load on the collector is significantly decreased. Furthermore, the main power supply and the body power supply are in CW operation; consequently, the beam energy keeps a high stability, which brings about stable RF oscillation. It is concluded that the anode switching is very simple and effective for CW power modulation.

4. Dual frequency gyrotron

A development of a dual-frequency gyrotron is underway for a future advanced EC H&CD system. The design was made to satisfy the matching condition at the output window and to have a similar radiation angle from the window for both frequencies, and the modes of $TE_{31,11}$ and $TE_{25,9}$ are selected, which have frequencies of 170 GHz and 136.8 GHz, respectively. The oscillation is obtained by setting a proper magnetic field at the cavity (cavity filed). The optimum beam radius at the cavity is selected by setting a proper mirror ratio between the cavity field and MIG field. A pitch factor of the beam can be optimized by setting an optimum V_{ak} keeping the beam voltage constant. In Fig.8 (a) and (b), the measured power profiles at the window are shown. Very similar RF beam pattern was obtained at both frequencies. In Fig.9, the beam current dependence of the output power and the oscillation efficiency is shown at short pulse operation (~0.5 ms). Up to now, the output power of 1.3 MW was obtained. And, 1 MW with efficiency of more than 30 % was also achieved for both frequencies.

5. Summary

A progress of the gyrotron development is described. The repetitive operation shows the robustness of the gyrotron. For further increase of the reliability for the EC H&CD system, a quick recovery system will be effective in addition to the efforts of the realization of the further stable oscillation. The 5 kHz CW-relevant power modulation was realized using anode switching of the triode MIG. This configuration offers a simple configuration of the power supply and preferable system for gyrotron, i.e., stable voltage and low heat load at the collector. For the advanced EC H&CD system, the dual frequency gyrotron with the triode MIG is proposed and the preliminary results are demonstrated. By controlling the anode voltage, the optimum oscillation condition is obtained by fixing the beam voltage. This will make the EC H&CD system more attractive for fusion application.

References

[1] How J. (ed) 2007 Project Integration Document PID ITER_D_2234RH Ver3.0 and https://user.iter.org/

[2] M. Henderson et al, "An Overview of the ITER EC H&CD System", IRMMW-THz Conference, 21-25 Sept. 2009, Busan, Korea.

[3] C. Darbos et al, "ECRH system for ITER", Proc. Of 18th Topical conf. on Radio Frequency Power in Plasmas, June 2009, Gent, Belgium, p.531 (2009).

[4] K. Sakamoto et al, Nat. Phys. **3** 411 (2007).

[5] K.Takahashi, et al., "Investigation of characteristic in corrugated waveguide transmission line for fusion application", Proc. 8th IEEE Int.Vacuum Electronics Conf., Kitakyushu, Japan, p.275 (2007).
[6] K.Takahashi, et al, Nucl. Fusion **48** 054014 (2008).
[7] K.Kajiwara, et al., J.Plasma and fusion research, vol.4, 6 (2009).
[8] K.Kajiwara, et al., submitted to J.Infrared and mm waves.
[9] K.Sakamoto, et al., Nuclear Fusion, 49, 9, 095019 (2009).

THE EUROPEAN 2 MW GYROTRON FOR ITER

FERRAN ALBAJAR, TULLIO BONICELLI

Fusion for Energy[*†], *Barcelona, 08019, Spain*

STEFANO ALBERTI[1], KONSTANTINOS A. AVRAMIDES[3], SANTE CIRANT[4], GERD GANTENBEIN[2], TIMOTHY P. GOODMAN[1], STEFAN ILLY[2], ZISIS IOANNIDIS[3], JEAN-PHILIPPE HOGGE[1], JIANBO JIN[2], STEFAN KERN[2], GEORGE LATSAS[3], IOANNIS GR. PAGONAKIS[1], BERNHARD PIOSCZYK[2], TOMASZ RZESNICKI[2], MANFRED THUMM[2], IOANNIS TIGELIS[3], MINH QUANG TRAN[1], JOHN VOMVORIDIS[3]

European Gyrotron Consortium (EGYC) among 1.EPFL-CRPP (Lausanne, Switzerland), 2.KIT- IHM (Karlsruhe, Germany), 3.HELLAS (Athens, Greece), 4.CNR-IFP (Milano, Italy), 5.ENEA-FPN (Frascati, Italy)

PATRICK BENIN, CHRISTOPHE LIEVIN

Thales Electron Devices, Vélizy -Villacoublay, 78140, France

CAROLINE DARBOS, THIBAULT GASSMANN, MARK HENDERSON

ITER Organization, St. Paul-lez-Durance, 13067, France

Europe is developing a gyrotron capable of delivering 2 MW RF power at 170 GHz in continuous operation (CW) for the ITER EC H&CD system. This paper discusses the advantages and risks of the 2 MW project, the status of the development, and the expected results from the tests planned to start during the 2nd half of 2010 on the industrial gyrotron prototype, which will provide essential elements for the decision between the 2 MW (coaxial cavity) and 1 MW (cylindrical cavity) gyrotron option placed along the track of the European project plan.

1. Background

Fusion for Energy (F4E) is responsible for the in-kind procurement of gyrotrons generating 8 MW for the ITER EC H&CD system. The specifications

[*] The views expressed in this publication are the sole responsibility of the author and do not necessarily reflect the views of Fusion for Energy. Neither Fusion for Energy nor any person acting on behalf of Fusion for Energy is responsible for the use which might be made of the information in this publication.

for ITER go beyond the present capabilities of the gyrotrons installed in any existing facilities. Therefore a new generation of gyrotrons is needed.

Europe is developing a gyrotron capable of delivering 2 MW RF power at 170 GHz in continuous operation (CW). This doubles the original ITER specification in terms of generated microwave power, offering the potential advantage for the EC H&CD system to reduce the number of gyrotron units and associated sub-systems, e.g. superconducting magnets, MOU, transmission lines and loads, or alternatively to enhance the total power without increasing the number of transmission lines and launcher entries [1]. In addition, the 2 MW tubes could bring substantial savings in the ITER costs to the end-of-life due to the limited lifetime of gyrotrons compared to the ITER operation time and thus the need of regular refurbishment and replacement.

The coaxial cavity gyrotron uses a coaxial resonator with a corrugated inner conductor, which relaxes the limits of the more conventional cylindrical waveguides due to mode competition and limiting current [2].

The development and technical validation of the higher power sources are especially important for future power upgrades of ITER and for DEMO. The coaxial cavity concept, in particular, brings diversity to the EC power source system of ITER mainly relying on the cylindrical cavity gyrotron under development in Japan and Russia.

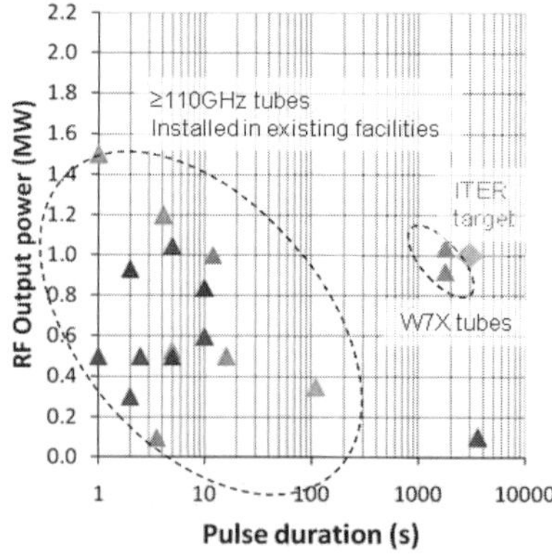

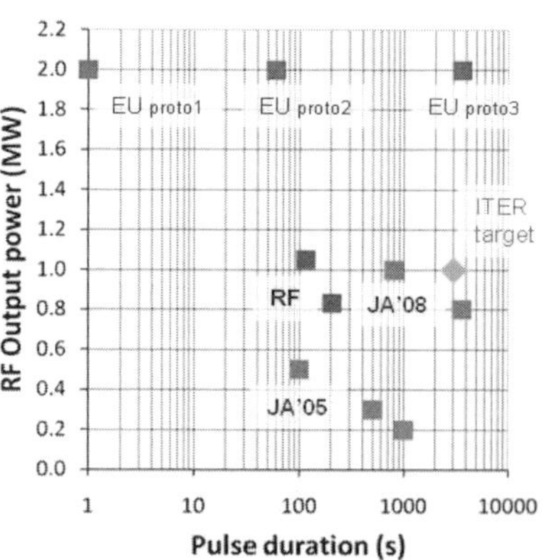

Figure 1. On the left-hand side, the capabilities in terms of RF output power and pulse duration of existing gyrotrons at frequencies equal or higher than 110 GHz installed in fusion devices; on the right-hand side, the best results obtained from the Japanese and Russian gyrotron prototypes for ITER, the ITER minimum target (1 MW, 3000 s) and the three foreseen European industrial prototypes of the 2 MW coaxial cavity gyrotron for ITER.

2. The development of the EU gyrotron for ITER

2.1. *Structure of the Gyrotron Development Programme*

The development programme of the EU gyrotron for ITER is run in strong collaboration with the EU Associates CRPP, KIT, CNR, ENEA, HELLAS (CEA, FOM, TEKES were also involved in a previous stage) and the industrial partner Thales Electron Devices (TED) [3-4]. A dedicated EC test facility was established at the CRPP, in Lausanne, for the full power testing of the 2 MW tube in steady-state conditions as well as for other EC system components. The test facility was equipped by fully solid-state power supplies with excellent dynamic performances, operational flexibility and efficiency, and with cooling and cryogenics systems, control and acquisition system, and other auxiliary systems [5-7].

A second gyrotron test facility at KIT, in Karlsruhe, hosts the short pulse pre-prototype coaxial cavity gyrotron [8], conceived as a modular tube enabling an easy replacement of components and with good diagnostics. The experiments with this tube are extremely useful to identify problems sufficiently in advance and verify the new designs. A very low power test stand is also available at KIT for the validation of quasi-optical gyrotron components [9].

The EU programme also benefits from the experience acquired during the recent development of high power, long pulse gyrotrons in Europe [10-12].

2.2. *Present status*

The first industrial gyrotron prototype was tested at the CRPP EC Test Facility in 2008. As a main result, the gyrotron operated stably in the desired $TE_{34,19}$ mode, thus the basic radio frequency design of the tube was proved. The level of output power and the excitation mode sequence observed in the experiments were in good agreement with the multi-mode simulations [13].

The tests showed the limitation of the operating parameters range to an efficiency domain lower than the predicted by interaction codes, and voltage stand-off problems. An output power of about 1.4 MW (the goal was 2 MW) was measured for short pulses, ca. 2 ms (the goal was 1 s). The maximum pulse length in the desired mode was 60 ms at a reduced power (ca. 0.5 MW).

The main reasons limiting the performance of the 1st gyrotron prototype have been identified from the analysis of the test results and the opening and inspection of the tube, and a very important effort has been made to improve the design of the internal components:

- new electron gun designed to enhance the quality of the electron beam at the cavity for a large range of operation parameters, avoid the presence of any trapped electrons, and reduce the electric field on the surfaces [14],
- new beam tunnel with an irregular azimuthal symmetry to suppress high frequency parasitic oscillations [15],
- new quasi-optical mm-wave output coupler designed to maximize the Gaussian output mode purity and reduce the level of stray radiation [16].

The new internal configuration has been validated in the KIT pre-prototype tube obtaining very remarkable experimental results: 2.2 MW of RF power at the right operating mode has been measured in a reproducible way with an efficiency of 30% (no depressed collector operation - equivalent to ~50% or higher with depressed collector). Moreover the quality of the RF beam at the gyrotron window is very high, ~97% of Gaussian mode content, which is higher than ITER specifications. The measurements of output power and efficiency, and the excitation mode sequence are in notable agreement with the mono and multi-mode wave-beam simulations [17], see also fig. 2.

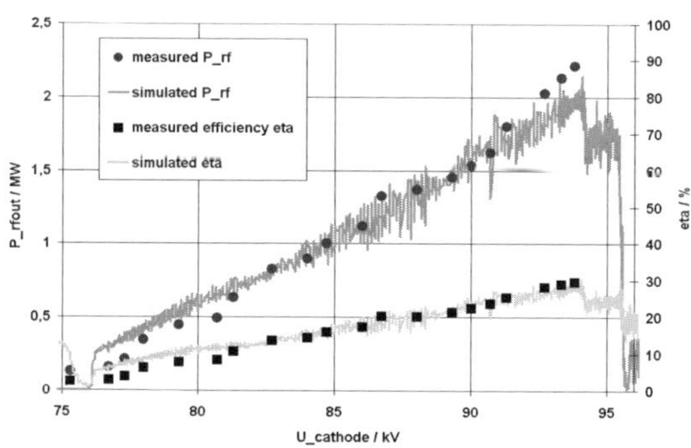

Figure 2. The measured RF output power and efficiency from the short pulse KIT pre-prototype coaxial cavity gyrotron incorporating the improved internal designs and comparison with theoretical simulations.

The 1st industrial gyrotron prototype is presently being refurbished with the improved internal components. The new configuration reduces substantially the technical risk for the 1st gyrotron prototype of not achieving the target results (2 MW – 1 second). New tests with this gyrotron are planned to start during the second half of 2010. The extension of the pulse duration to 1 min and 1 hr is the main objective of the 2nd and 3rd industrial prototypes, which will complete the development phase which is expected to last until 2016.

3. Risk assessment and mitigations approaches

3.1. *The European strategy: risk assessment and mitigations*

The main technical risk on the EU gyrotron package is associated to the development of the 2 MW gyrotron: mainly, the higher thermal loads to be handled by the internal components, and the relatively late start of the development programme involving industry. As a main mitigation measure, the design of a more conventional 1 MW cylindrical cavity gyrotron has been prepared making optimal use of the existing expertise in Europe from the development of the gyrotrons for the W7-X Stellarator.

The tests on the refurbished gyrotron are expected to provide essential data enabling F4E to take a decision between the continuation of the development of the 2 MW gyrotron and the 1 MW fall back option. The date of the decision is mainly driven by the EU schedule to develop and manufacture the coaxial cavity or cylindrical cavity gyrotrons on time for ITER. In case of shifting to the 1 MW back-up solution, it is foreseen to manufacture and test two industrial prototypes before the procurement of eight gyrotron tubes and superconducting magnets instead of four as required in the 2 MW strategy to fulfil the EU obligations towards ITER.

Other technical and commercial risks are e.g. the few number of suppliers worldwide with specific know-how about gyrotron technologies, the complexity of the quality control during manufacturing of tubes and superconducting magnets and the relatively long duration of the contracts, the very demanding technical specifications requested by the ITER Organization (IO), the complex F4E and IO contractual procedures and requirements, and the uncertainties on the sharing of responsibilities and liabilities and the actual level of local support during the on-site installation. If not addressed these risks could result in delays, reduced performances, lower reliability or lifetime, issues with the interfaces, and cost increase. Several actions have been identified to avoid, mitigate or reduce these risks such as the multi-prototyping strategy, following of several R&D paths, the structured collaboration involving EU Associates and industry, anticipation of long-lead components, developing tools to ease the project management, conducting cost and design reviews, establishing more detailed interface control with IO, etc. Some of these dispositions have been already launched and some others are in preparation.

3.2. *Cost versus benefit of higher power sources*

As a result of a Project Change Request recently accepted into the ITER baseline, all EC transmission line and launcher components are now specified to be compatible with the 2 MW sources [18]. As a consequence, the EC system could allow the operation of the Japanese and Russian gyrotrons with an output RF power higher than 1 MW [19], and could benefit from further developments in high power gyrotrons (up to the 2 MW power level) for the regular substitution of tubes needed during the operation time of ITER and for possible power upgrades of the heating systems. For making these enhancements possible the power supply system of the EC system should be specified to be flexible to accommodate these changes and not constrained by the space available in the RF building.

The technology associated with the EC sub-systems has advanced over the past years. For example, most of the transmission line components compatible with the 2 MW operation are commercially available [20]. Likewise, modifications have been implemented into the design of the upper launchers and proposed for the equatorial launcher to render them compatible with the 2 MW level [21]. Some specific components, e.g. line switches, mirrors, and bellows, require a further analysis and have been re-designed.

However the margin of safety between the operating point and the design limit of the transmission line and launcher components decreases with increasing power level. Therefore the technical risk of the sub-systems and of the overall EC system for ITER increases with the higher power sources, and with the 2 MW gyrotron in particular, with a potential drawback on system reliability. Some dispositions to optimise components and increase the design limit have been defined by IO and should be addressed independently of the European decision between the 2 MW and 1 MW gyrotron.

4. Conclusions

In strong collaboration with the EU Associates and industry, Europe is developing a gyrotron doubling the original ITER specification in terms of generated microwave power, offering attractive advantages for the ITER EC H&CD system and the potential power upgrades of the heating systems.

The 1st European gyrotron prototype for ITER is presently being refurbished with improved internal components, which are believed to significantly increase the chances of obtaining the target results of 2 MW – 1 second. The tests on the refurbished gyrotron will yield essential data for the decision point between the 2 MW/1 MW gyrotron placed along the track of the EU project plan as a risk mitigation measure. Other technical and commercial risks associated to the development and procurement gyrotron phases are identified and disposition actions launched or foreseen.

All ITER sub-systems are now specified to be compatible with the 2 MW sources. This is also central for the JA and RF gyrotrons which may be offering the possibility to operate at powers significantly higher than 1 MW, and for the new generation of gyrotrons for ITER which may be needed for power upgrades and for regular substitutions due to the limited lifetime of gyrotrons.

The development of the higher power gyrotrons is strategic for Europe but also for ITER, DEMO and future fusion devices. On the longer term, the coaxial cavity gyrotron could give the possibility to extend the output power to even higher levels [22]. It is therefore strongly recommended that the development of the 2 MW coaxial cavity tube continues regardless of the decision on the strategy to be followed for the procurement of the European gyrotrons for ITER.

References

1. ITER DDD 5.2, G 52 DDD 5 01-05-29 W 0.1, see in particular Sec. 1.2.7.2.
2. B. Piosczyk et al, *IEEE Trans. Plasma Science* **32**, 413 (2004).
3. T. Bonicelli et al, *Fusion Engineering and Design* **82**, 619 (2007).
4. F. Albajar et al, Proc. of the 15th Joint Workshop on ECE and ECRH, 415 (2009).
5. J.-P. Hogge et al, *Journal of Physics: Conference Series* **25**, 33 (2005).
6. T. Bonicelli et al, Proceedings of the13th European Conference on Power Electronics and Applications, 1 (2009).
7. D. Fasel et al, *Fusion Science and Technology* **53**, 246 (2008).
8. B. Piosczyk et al, *IEEE Trans. Plasma Science* **32**, 853 (2004).
9. T. Rzesnicki et al, *Int. Journal of IRmmW* **27**, 1 (2006).
10. G. Dammertz et al, *IEEE Transactions in Plasma Science* **30**, 808 (2002).
11. V. Erckmann et al, *IEEE Transactions in Plasma Science* **27**, 538 (1999).
12. S. Alberti et al, *Fusion Engineering and Design* **53**, 387 (2001).
13. J.-P. Hogge et al, *Fusion Science and Technology* **55**, 204 (2009.)
14. I. Pagonakis, 34th Internat. Conf on Infrared, Millimeter, and Terahertz Waves, Busan, Korea (2009).
15. S. Kern et al, 34th Internat. Conf. on Infrared, Millimeter and Terahertz Waves, Busan, Korea, (2009).
16. J. Jin et al, *IEEE Transactions on Microwave Theory and Techniques* **57**, 1661 (2009).
17. M. Thumm, *Terahertz Science and Technology* **3**, 1 (2010).
18. ITER Project Change Request 116, ITER_D_2V4PD9 (2009).
19. K. Sakamoto et al, *Fusion Science and Technology* **52**, 145 (2007).

20. R.A. Olstad et al, *Journal of Physics: Conference Series* **25**, 166 (2005).
21. M.A. Henderson et al, *Fusion Science and Technology* **53**, 139 (2008).
22. M. Thumm et al, Proc. 13th Joint Workshop on ECE and ECRH, Nizhny Novgorod, Russia, 365 (2004).

THE SIX GYROTRON SYSTEM ON THE DIII-D TOKAMAK

JOHN LOHR, M. CENGHER, J.C. DeBOO, J. DOANE, Y.A. GORELOV,
G.L. JACKSON, C.P. MOELLER, D. PONCE, and R. PRATER
General Atomics, PO Box 85608, San Diego, California 92186-5608, USA

The gyrotron complex on the DIII-D tokamak now comprises six gyrotrons in the 1 MW class with pulse lengths during plasma operations up to 5 s. The system performance, representative experiments, gyrotron related tests, diagnostics and long term plans are presented.

1. Introduction

The complete six-gyrotron system [1], Fig. 1, has now been operating at the DIII-D tokamak for two experimental campaigns. The frequency of all gyrotrons is 110 GHz, corresponding to the second harmonic electron cyclotron resonance for the maximum central magnetic field for DIII-D of 2.1 T. A flexible launcher system with ±20° scan range both poloidally and toroidally permits the injected rf beams to intersect the resonance at any point from the center to the plasma periphery in the upper half plane. Both co- and counter-current drive are readily available in addition to heating.

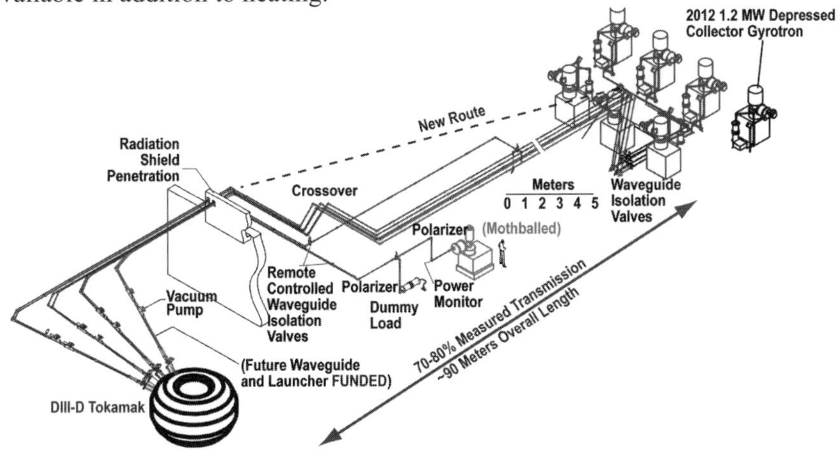

Figure 1. The DIII-D ECH system has 6 gyrotrons in regular service. The waveguide lines average 90 m in length and have as many as 14 miter bends. The number of miters will be reduced by elimination of the crossover gallery, which previously allowed any gyrotron to be connected to any launcher. The new routing will follow the dashed line directly from the gyrotron vault to the tokamak shield wall.

Gyrotron pulse lengths to 5.0 s are regularly used and typical maximum injected power is 3.4 MW with a peak-injected energy of 16.6 MJ. Although the gyrotron design is for 10 s maximum pulse length, typical DIII-D shots have current and field flattops of ~5 s, so the administratively limited pulse length provides a good match to the requirements for tokamak operations while extending the gyrotron collector fatigue life. Despite ongoing power supply and gyrotron development, the gyrotron reliability has been ≈85%, Fig. 2, for the past several years; and the system plays a part in nearly all of the plasma experiments on DIII-D. The transmission lines are evacuated circular corrugated waveguide with no isolation windows except the gyrotrons' own diamond disk windows. No arcing or other problems have been encountered with these lines. Full control of the elliptical polarization of the injected rf waves is provided by pairs of polarizers located in 2 miter bends near the gyrotrons. A plasma equilibrium is calculated and then trajectories of the rf beams, determined for the appropriate plasma density and temperature profiles, are calculated along with the polarizer mirror orientations to launch the desired elliptical polarization for the experiment. Usually the extraordinary mode will be launched for essentially 100% absorption at the Doppler shifted second harmonic resonance for densities below the cutoff density. Ray tracing using the injection geometry and the plasma parameters is used to calculate the heating and current drive profiles.

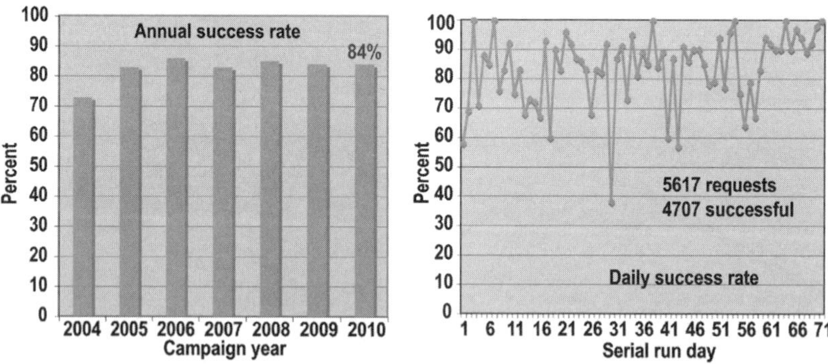

Figure 2. Individual gyrotron reliability has averaged about 85% for a number of years even though power supply development and gyrotron conditioning were in progress. The failures are not statistically independent, since in some cases two gyrotrons are fed from one power supply and more failures occur during daily startup than after stable operation has been achieved.

2. Representative Experiments

Three experiments will be described briefly to indicate the range of the experiments being performed using the system.

Real time poloidal scanning has recently been implemented on the DIII-D system by replacing air turbine motors on the launcher antenna poloidal drives with dc electric motors that can be controlled by the DIII-D plasma control system (PCS). In the first tests of this capability, a plasma with normalized plasma pressure, β_N, of about 2.5 was produced, which had a steady $m/n = 3/2$ neoclassical tearing mode [2]. The antennas were set for co-current drive and swept poloidally so that the rf current drive from three gyrotrons passed across the $q = 3/2$ resonant surface. The neoclassical tearing mode (NTM) amplitude decreased and then increased again as the current drive swept across the island location. The gyrotron power was ≈1.5 MW for this test and the restoration of the NTM after the sweep passed the resonant surface showed that the partial suppression was not due to some other plasma effect. The experiment is summarized in Fig. 3.

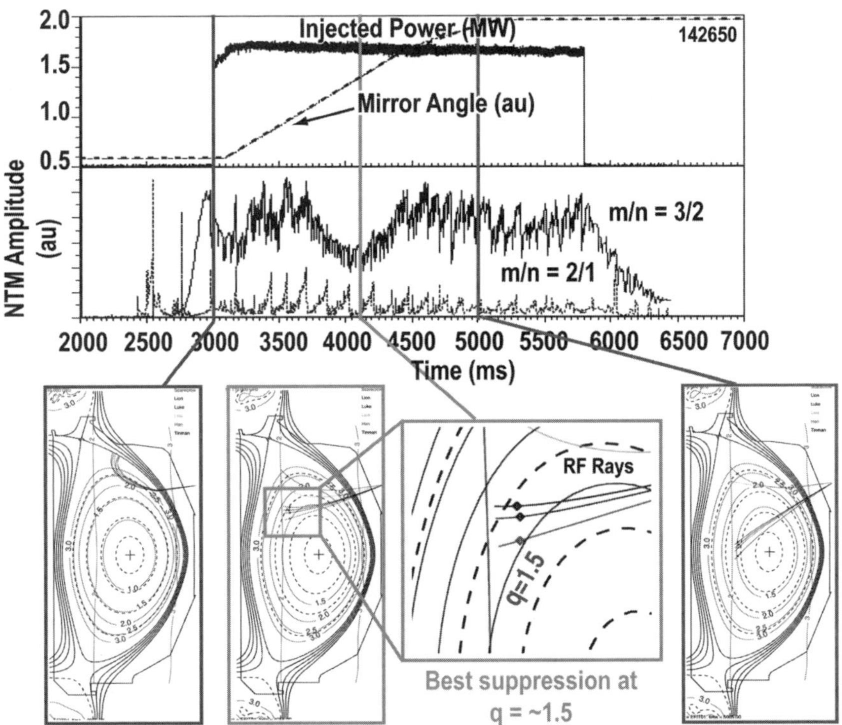

Figure 3. Scanning the rf beam poloidally across the $q = 1.5$ surface reduces the amplitude of the $m/n = 3/2$ NTM, which then grows after the current drive moves inside the resonant surface. The scan was under feed forward control of the plasma control system, injecting 1.5 MW of ECCD from 3 gyrotrons.

The flexibility and precision with which EC heating can be applied to a plasma was demonstrated in a modulated transport experiment designed to periodically vary the electron temperature gradient driving transport while keeping the total energy input to the plasma constant [3]. As shown in Fig. 4, two groups of three gyrotrons each with their antennas aimed to heat at normalized radii of 0.6 and 0.7 respectively were modulated 180 degrees out of phase. This rocked the $T_e(r)$ gradient slightly, allowing synchronous detection techniques to be used to ascertain the effect of only the gradient on transport.

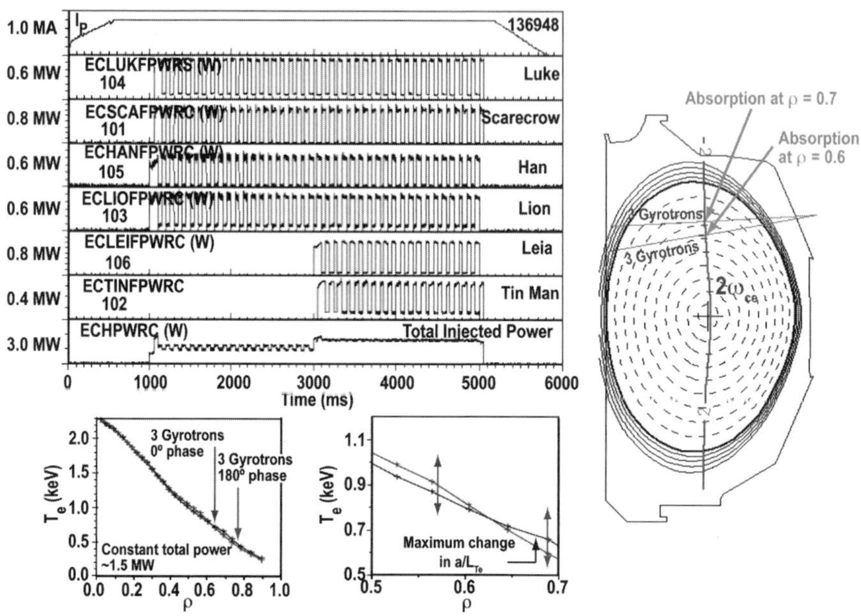

Figure 4. Transport studies using modulated ECH but constant injected power allowed simplified analysis as the $T_e(r)$ profile "rocked" from deposition at two different values of normalized radius.

A third series of experiments has involved studies of plasma breakdown at very low loop voltages using electron cyclotron heating (ECH) assist [4]. In these experiments, 1.0–1.5 MW rf power was injected into the tokamak with toroidal magnetic field present. As seen in Fig. 5 showing images from a fast framing camera with a C^{III} filter, initial breakdown along the second harmonic resonance preceded a clean transition to closed flux surfaces with reduced volt-second consumption and a loop voltage which never exceeded 3 V. This result is important for the new generation of superconducting tokamaks, which are limited by the superconducting coil set to developing low loop voltages during the breakdown phase.

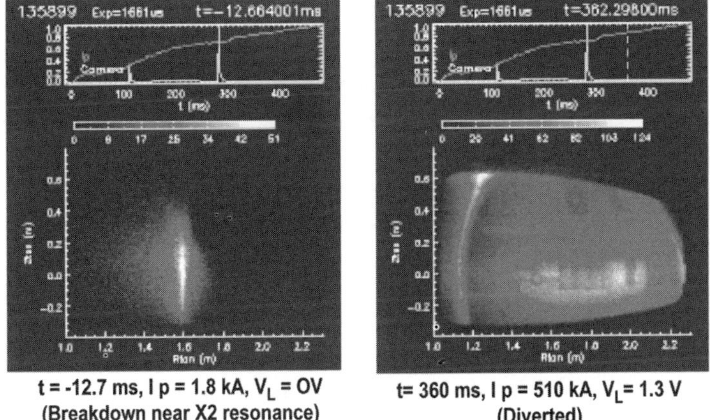

Figure 5. Fast camera views in C^{III} light of the resonance cylinder and the transition to closed flux surfaces show the success of an ECH assisted startup experiment with $V_{loop} \leq 3$ V.

3. Gyrotron Performance Experiments

A number of experiments designed to assess aspects of gyrotron performance have been carried out on the DIII-D system. Two of these will be described. The first of these is the direct spectroscopic measurement of the light passing through the gyrotron diamond output window during conditioning. The object of this measurement was to attempt to understand the conditioning process for gyrotrons being brought to full performance. In Fig. 6, two spectra are presented, one from a normal conditioning pulse and one from a short pulse which terminated on an overcurrent fault. The fault spectrum shows hydrogen lines, as expected from the hydrogen braze and clear evidence of titanium, which can only have come from the vacion pumps, suggesting that it may be possible to compound the problem of a gyrotron fault by release of gas from the pumps.

It has always been difficult to determine injected rf power in tokamaks. Various solutions have included switchable dummy loads at the vacuum vessel and transmission line efficiency measurements coupled with gyrotron performance data. At DIII-D we discovered that the diagnostic bolometer arrays can be used to infer the injected power under certain circumstances. An example is

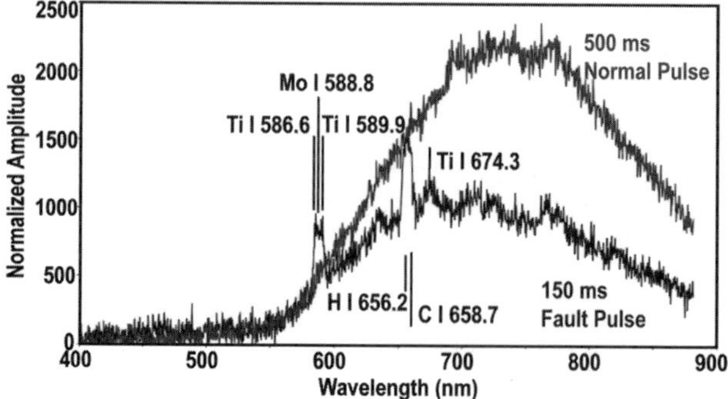

Figure 6. The visible spectrum of light exiting through the gyrotron diamond window can be used to diagnose breakdown location during conditioning. The presence of Ti lines indicates arcing in the vacion pumps.

shown in Fig. 7, in which several combinations of gyrotrons injecting known power into the empty vacuum vessel produce a calibrated response from the bolometers in agreement with the injected power. We also are developing a leaky mirror miter bend with -70 dB coupling for the final miter in the transmission line. This four-port device can measure the power in both orthogonal polarizations and in the contaminating modes resulting from misalignment or other sources of mode conversion.

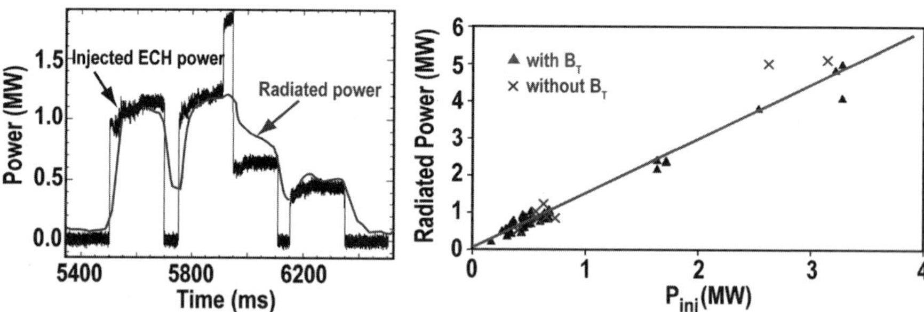

Figure 7. The injected rf power can be measured using the integrated signals from the bolometer arrays with no plasma. On the right is a check of the response linearity with and without resonance.

4. System Enhancements

The theoretical transmission line efficiency for a typical 31.75 mm diameter. line, 90 m in length and having 13 miter bends, is about 80%, or -1 dB [5]. This efficiency, which was verified using a pure $HE_{1,1}$ beam at low power, until recently was achieved on only one of the six transmission lines in the DIII-D system [6]. The typical efficiency measured in hot tests on the DIII-D system, however, is about 75%, or -1.25 dB, due, it is believed, to angular misalignment of the free space Gaussian rf beam where it enters the waveguide. This coupling should be ~98% efficient at exciting the $HE_{1,1}$ waveguide mode [7], however measurements of heating of the first few meters of the waveguide line and of rf power density profiles at different points in the waveguide line suggest that mode conversion at the injection point from misalignment is causing excitation of higher order lossy modes. Although the transverse alignment of the rf beam where it enters the waveguide is better than 1 mm, the tilt angle is difficult to measure. From free space propagation measurements of the rf beam after the matching optics unit (MOU) and without waveguide, it has been seen that the mechanical and laser based alignment techniques were not adequate for ensuring coaxiality of the rf beam and the waveguide. In the worst case, the injected beam was tilted by nearly 4° with respect to the waveguide axis. It was seen that the rf propagating in the waveguides was not a pure $HE_{1,1}$ beam except in the one system with -1 dB attenuation. Following careful realignment of the beams with attention to coaxiality, a good quality $HE_{1,1}$ beam was achieved in each line, and the rf launched into free space following various lengths of waveguide was close to a propagating Gaussian beam.

Once the theoretical efficiency is obtained, additional efficiency improvements could be realized by reducing the number of miter bends, each of which has about 1.5% mode conversion loss [8]. We are rerouting the existing waveguides on the DIII-D gyrotron installation as seen in Fig. 1 to eliminate 13 miters and plan to replace a large fraction of the remaining miters with low loss versions having matching sections at the input and output arms. Earlier versions of low loss miters [8] were demonstrated to have greatly reduced mode conversion losses of about 0.4% but with long arms which were incompatible with the present transmission lines on DIII-D. New miter designs [9] are now available that will permit improved miters to be virtually direct replacements for the presently installed miters. With the alignment improvements and decreased numbers of improved units, it is believed that the transmission line efficiencies can be increased to ~90%, yielding an additional ~700 kW injected at a cost of ~$0.4/W, which is much less than the cost of newly installed power.

5. System Upgrades

The DIII-D gyrotron system has received funding for an additional 1.2 MW depressed collector gyrotron and associated launcher, high voltage power supplies, controls and transmission line. It is proposed that an 8th gyrotron injecting 1.5 MW will follow. The launcher mirrors on all systems will be replaced with improved designs capable of handling pulses longer than 5 s and the toroidal mirror drives, as has already been done with the poloidal drives, will be upgraded to dc electric motors.

Acknowledgments

This work was supported by the U.S. Department of Energy under DE-FC02-04ER4698.

References

1. John Lohr, et al., *Fusion Sci. Tech.* **48**, 1226 (2005).
2. R.J. La Haye, *Phys. Plasmas* **13**, 055501 (2008).
3. J.C. DeBoo, et al., "Probing Plasma Turbulence by Modulating the Electron Temperature Gradient," accepted for publication in *Phys. Plasmas* (2010).
4. G. Jackson, et al., "Understanding and Predicting the Dynamics of Tokamak Discharges During Startup and Rampdown," submitted to *Phys. Plasmas* (2009).
5. John L. Doane, *J. Infrared and Millimeter Waves* **13**, 123 (1985).
6. M. Cengher, et al., *Fusion Sci. Technol.* **55**, 213 (2009).
7. K. Ohkubo, et al., Proc. 10th Joint Workshop on Electron Cyclotron Emission and Electron Cyclotron Heating, Tony Donne and Toon Verhoeven, Eds, World Scientific, Singapore, 597 (1997).
8. J.L. Doanc and C.P. Moeller, *Int. J. Electronics* **77**, 489 (1994).
9. A.V. Chirkov, et al., Proc. Strong Microwaves in Plasmas 271 (2006).

PROGRESS IN STABLE OPERATION OF HIGH POWER GYROTRONS *

G. GANTENBEIN, G. DAMMERTZ, S. KERN, G. LATSAS[2], B. PIOSCZYK, T. RZESNICKI, A. SAMARTSEV, A. SCHLAICH[1], M. THUMM AND I. TIGELIS[2]

Karlsruhe Institute of Technology (KIT), Association EURATOM-KIT, Institute for Pulsed Power and Microwave Technology (IHM)
[1]Institute of High Frequency Techniques and Electronics (IHE)
Kaiserstrasse 12, 76131 Karlsruhe, Germany
[2]National and Kapodistrian University of Athens, Faculty of Physics, 15784 Athens, Greece.

1. Introduction

Due to their ability to produce millimeter-wave power in the MW range in continuous wave operation gyrotrons play an important role in fusion energy research. With the increasing demand for overall millimeter-wave power the power per unit was increased in gyrotron tubes in recent years. Thus 1 MW output power is considered as a standard, prototype tubes with 2 MW output power have been verified successfully.

One of the issues coming along with the design and operation of a gyrotron with very high output power and high electron beam current is the possibility of parasitic oscillations which reduce the efficiency of the system and may result in a power deposition at unfavorable locations in the tube which may suffer from thermal overload.

In this contribution we report on recent results on stable operation of high-power gyrotrons by suppression of parasitic oscillations which originate from the beam tunnel area and which are excited via an electron-cyclotron-resonance interaction with a frequency close to but slightly lower than the design frequency. We describe experimental investigations with a dedicated gyrotron

* Part of his work, supported by the European Communities under the contract of Association between EURATOM and KIT, was carried out within the framework of the European Fusion Development Agreement and the European Gyrotron Consortium (EGYC). The views and opinions expressed herein do not necessarily reflect those of the European Commission. The European Gyrotron Consortium (EGYC) is a collaboration among CRPP, Switzerland; KIT, Germany; HELLAS, Greece; CNR, Italy; ENEA, Italy.

test setup, which was configured to observe parasitic RF oscillations in the beam tunnel, and to validate improved beam tunnel structures. This program was started to remove problems with parasitic RF oscillations which were found to limit the performance of the series gyrotrons for the W7-X project (1 MW, 140 GHz) [1]. Similar results were found in the coaxial 2 MW 170 GHz EU ITER gyrotron project carried out by the European Gyrotron Consortium (EGYC) [2].

2. Experimental Set-up and Measurement System

The gyrotrons are equipped with a beam tunnel structure for oscillation suppression in the compression zone, which is standard for many gyrotrons: The beam tunnel consists of an alternating stack of copper and attenuating ceramic rings. While the ceramics should attenuate possible RF fields, thereby lowering the quality factor in order to increase the starting currents of possible oscillations, the copper rings define the electrical potential and avoid static charges on the insulators.

In the case of the W7-X gyrotrons, the frequency of the parasitic oscillations was in the range 120 – 130 GHz. From calorimetric measurements of the cooling circuit of the beam tunnel the roughly estimated power is 1 – 5 kW compared to 1 MW total output power. One of the main limiting consequences during operation of the gyrotron was that the gyrotron parameters have to be modified in order to maintain the design mode and to avoid mode jumps after a certain pulse length. Thus the gyrotron must be operated in a regime with lower efficiency and output power.

The power limitations due to parasitic RF oscillations, which were found in experiments with the coaxial pre-prototype for the 2 MW 170 GHz ITER gyrotron, appeared less serious, but were still measurable [3]. In these measurements, this gyrotron had a beam tunnel structure similar to that of the 140 GHz gyrotron mentioned above.

For an experimental investigation, a test program was set up using the modular step-frequency tunable short pulse (~ms) D-band gyrotron at KIT. Furthermore, since this tube is equipped with a Brewster window and is designed for broadband operation with regard to all components, experiments over wide parameter ranges could be performed, yielding more information about the behaviour of the measured oscillations.

The step-frequency tunable gyrotron was designed for operation in the frequency range 105 – 143 GHz in different operating modes. Frequency-step tuning is performed by changing the magnetic field and exciting corresponding

TE modes in the cavity. Typical high power operating parameters of the gyrotron are an accelerating voltage of 80 kV and a beam current of 40 A. However, beam currents up to 60 A have been used in the experiments.

In the first experimental setup a standard azimuthally symmetric beam tunnel was installed as a reference. This reference beam tunnel was intended to be susceptible to parasitic oscillations. In a second experimental campaign, this beam tunnel was replaced by a modified version with broken azimuthal symmetry, to improve its attenuation for azimuthally symmetric modes, which was insufficient in the reference beam tunnel.

Since the frequency measurement of parasitic oscillations was problematic due to mixed signals and low power levels, the measurement system was enhanced by employing a spectrum analyzer in addition to the filter bank and the HP modulation domain analyzer used previously [4]. The minimal sweep time of a spectrum analyzer is of the order of the gyrotron pulse length (~ ms), while it should be much shorter than the measured pulse. A solution is to do repeated pulses, synchronize them with the analyzer's sweep and measure only a small part of the spectrum during a single gyrotron pulse. Keeping in mind that the frequency axis of the analyzer simultaneously becomes a time axis for the pulse when such a method is used, spurious frequencies appearing during startup and switch-off can easily be excluded. Figures 1a and b show measurement examples from such a system (details are given in [5]). It should be noted that in the experiments described in the following section, the appearance of a LF parasitic oscillation (48 MHz), as can be seen in Figures 1a and b, sometimes

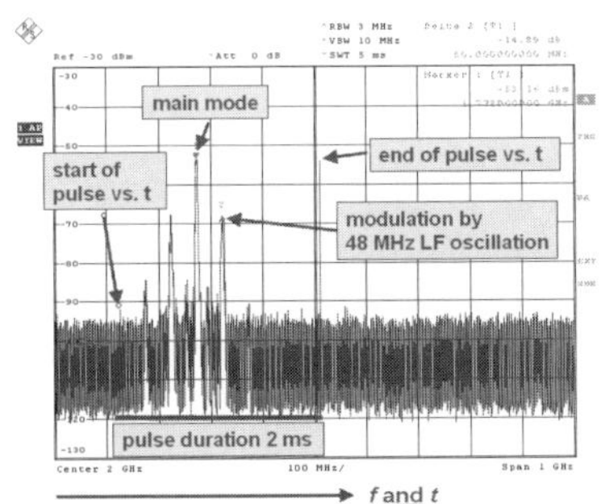

Fig. 1a: Measurement example: Spectrum of a main mode line ($TE_{22,8}$ downmixed, originally at 140 GHz), modulated by a 48 MHz LF oscillation.

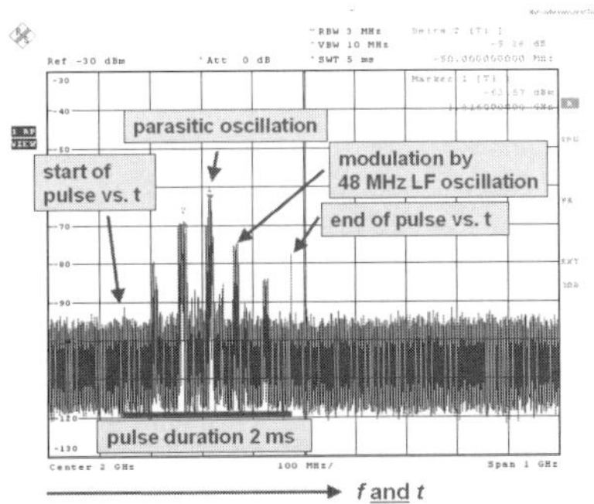

Fig. 1b: Spectrum of a parasitic oscillation (downmixed, originally at 134 GHz), modulated by a 48 MHz LF oscillation. Note the higher phase noise in comparison to Figure 1a.

suppressed parasitic oscillations in the beam tunnel.

3. Experimental Results with Reference Beam Tunnel

In the first experimental campaign, using the azimuthally symmetric reference beam tunnel, parasitic oscillations and their influence on gyrotron behaviour could clearly be seen. The first observation was a reduction in output power and efficiency, and increased stray radiation power as soon as frequencies of parasitic oscillations could be measured, as shown in Figure 2. The stray radiation power was measured through a release window at the mirror box of the gyrotron. It is also observed that the frequencies of the parasitic oscillations are much more sensitive to parameter changes than the frequencies of cavity modes (see Figure 3).

4. Analysis using the Brillouin-Diagram

A more detailed analysis of the measured frequencies can be done using the well known Brillouin-diagram. While a measured frequency constitutes a horizontal line in the Brillouin-diagram, it is helpful to determine a corresponding axial wave number k_z in order to draw a single point for each measurement. This is done by calculating the electron beam parameters at some chosen axial position z from the known operating parameters. Using the known local relativistic cyclotron frequency $\Omega_c \approx (2\pi\ [28\ GHz]\ B[T])/\gamma$, with the magnetic flux density B and the relativistic factor γ, and the axial electron velocity v_z, the beam line $\omega - k_z v_z = \Omega_c$ which indicates possible gyrotron interaction

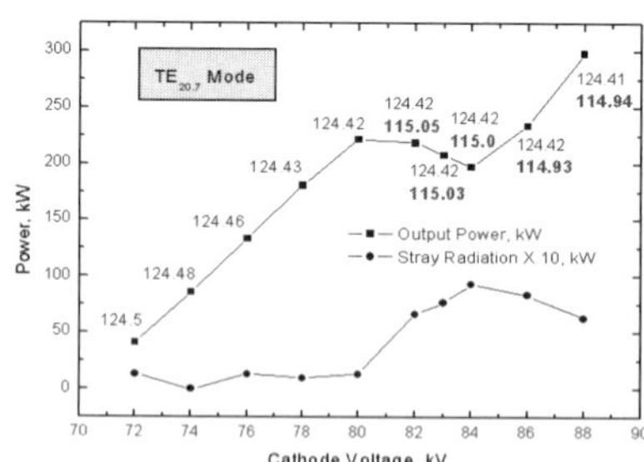

Fig. 2: Output power vs. cathode voltage for $TE_{20,7}$ operating mode in the first experimental setup (reference beam tunnel). Numbers are measured frequencies in GHz, lower bold numbers refer to parasitic oscillations.

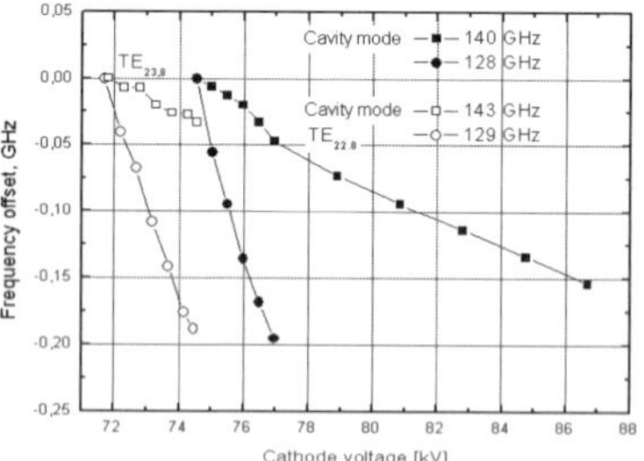

Fig. 3: Frequency variation of cavity modes and parasitic oscillations around 128 GHz vs. acceleration voltage.

points can be drawn. The crossing point of such a beam line and the measured frequency then determines the interaction point for the measured frequency. The wave number is then simply given by $k_z = (2\pi f - \Omega_c)/v_z$, with f being the measured frequency.

The calculated beam line is a function of the axial z coordinate, since the magnetic field changes along z. By calculating these interaction points for several measured frequencies and comparing those with a local waveguide mode hyperbola, which also must fit to the interaction points in the Brillouin-diagram, we found a method to determine the actual z-position of a parasitic oscillation.

In Figure 4 the method described above was applied to parasitic frequencies around 115 GHz, which were observed during operation in two different cavity modes ($TE_{20,7}$ and $TE_{22,7}$). A very good fit is obtained by setting z to 349 mm or 356 mm, as was done in Figure 4. This shows that the parasitic mode is a gyro-

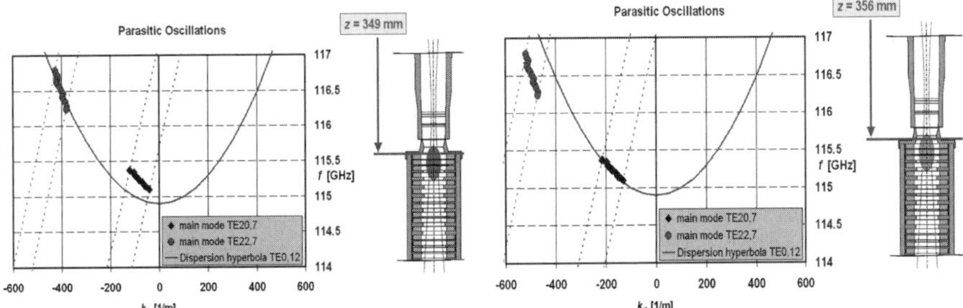

Fig. 4. Brillouin-diagram representation of parasitic oscillations around 115 GHz for z= 349 mm (left) and z= 356 mm (right).

backward wave (negative wave number) at the exit of the beam tunnel.

5. Experimental Results with Improved Corrugated Beam Tunnel

In the second experimental campaign the beam tunnel was modified by breaking the azimuthal symmetry. This was done by introducing irregular corrugations in the copper rings with a depth of about $\lambda/4$, in order to cut off azimuthal RF currents and thereby let the attenuation of the ceramics also have an effect on azimuthally symmetric modes. This complicates the constitution of a resonant field structure and effectively damps $TE_{m,p}$ modes, in particular symmetric TE modes (m = 0).

Experiments with this configuration did not show any beam tunnel parasitic oscillations as described above for the reference beam tunnel. Even at a beam current of 60 A, well above the design value (40 A), no beam tunnel RF oscillations in the high frequency range were observed.

It should be noted that parasitic oscillations similar to the observed ones have already been reported by a group from JAEA, Japan [6] and CPI, USA [7] with similar results as described above.

In addition to the experiments with the frequency-step tunable gyrotron, the coaxial 2 MW 170 GHz ITER pre-prototype gyrotron was also equipped with a similarly corrugated beam tunnel. This improvement led to complete suppression of any parasitic RF oscillation; the gyrotron now operates up to the limit given by mode competition in the cavity and produces 2.2 MW output power at an efficiency of 30 % without collector depression [3].

6. Conclusions

A dedicated experimental program on parasitic oscillations in megawatt class gyrotrons was conducted. Starting with a standard beam tunnel as reference, the appearance of parasitic oscillations was verified and their deteriorating effect on the cavity interaction was demonstrated. Careful analysis of the measurements verified the simulation model, such that these oscillations can now be considered well understood.

Based on these findings, an improved beam tunnel structure employing a breaking of the azimuthal symmetry by corrugations, was successfully tested in a second experimental campaign. The parasitic oscillations in the beam tunnel were completely removed, as proven by a frequency measurement setup with improved sensitivity, and the gyrotron behavior changed to the desired and usual one.

References

[1] G. Gantenbein et al., 22nd IAEA Fusion Energy Conference, Geneva, October 13-18, 2008. FT/P2-24.
[2] J.-P. Hogge et al., Fusion Science and Technology, **55**, pp. 204-212.
[3] T. Rzesnicki et al., accepted for publication in IEEE Trans. Plasma Science, DOI: 10.1109/TPS.2010.2040842.
[4] H. O. Prinz, IEEE Potentials, **26**, pp.25-30. DOI:10.1109/ P.2007.4280329.
[5] G. Gantenbein et al., accepted for publication in IEEE Trans. Plasma Science, DOI: 10.1109/TPS.2010.2041366
[6] K. Sakamoto, et al., Nuclear Fusion, **43**, 2003, pp. 729–737.
[7] K. Felch et al., IRMMW-THz 2008, Pasadena, CA, Paper T2A2.1608.

THE ITER EC H&CD SYSTEM

M.A. HENDERSON, B. BECKET, D. COX, C. DARBOS, F. GANDINI, T. GASSMAN, O. JEAN, C. NAZARE, T. OMORI, D. PUROHIT, A. TANGA, V.S. UDINTSEV

ITER Organization, St. Paul-lez-Durance, 13067 France

F. ALBAJAR, T. BONICELLI, R. HEIDINGER, G. SAIBENE

F4E, C/ Josep Pla 2, Torres Diagonal Litoral-B3,E-08019 Barcelona, Spain

S. ALBERTI, R. BERTIZZOLO, R. CHAVAN, A. COLLAZOS, T.P. GOODMAN, J.P. HOGGE, J.D. LANDIS, I. PAGANAKIS, L. PORTE, F. SANCHEZ, O. SAUTER, M.Q. TRAN, C. ZUCCA

CRPP, Association EURATOM-Confédération Suisse, EPFL Ecublens, CH-1015, Suisse

U. BARUAH, M. KUSHWAH, N.P. SINGH, S.L. RAO

Institute for Plasma Research, Near Indira Bridge, Bhat, Gandhinagar, 382428, India

T. BIGELOW, J. CAUGHMAN, D. RASMUSSEN

USIPO, ORNL, 055 Commerce Park, PO Box 2008, Oak Ridge, TN 37831, USA

A. BRUSCHI, S. CIRANT, D. FARINA, A. MORO, P. PLATANIA, G. RAMPONI, C. SOZZI

Istituto di Fisica del Plasma, Association EURATOM-ENEA-CNR, Milano, Italy

M. DEBAAR, D. RONDEN

Association EURATOM-FOM, 3430 BE Nieuwegein, The Netherlands

G. DENISOV

Institute of Applied Physics, 46 Ulyanov Street, Nizhny Novgorod, 603950 Russia

K. KAJIWARA, A. KASUGAI, N. KOBAYASHI, Y. ODA, K. SAKAMOTO, K. TAKAHASHI

JAEA 801-1 Mukoyama, Naka-shi, Ibaraki 311-0193 Japan

W. KASPAREK, H. KUMRIC, B. PLAUM
IPP, Universitat Stuttgart, Pfaffenwaldring 31, D-70569 Stuttgart, Germany

G. AIELLO, G. GANTENBEIN, S. ILLY, J. JIN, S. KERN, A. MEIER, B. PIOSCYZK, T. RZESNICKI, T. SCHERER, S. SCHRECK, A. SERIKOV, P. SPAEH, D. STRAUSS, M. THUMM, A. VACCARO
KIT, Association EURATOM-KIT, Kaiserstr. 12, D-76131 Karlsruhe, Germany

E. POLI, H. ZOHM
IPP-Garching, Association EURATOM-IPP,D-85748 Garching, Germany

M. SHAPIRO, R. TEMKIN
MIT Plasma Science and Fusion Center, Cambridge, MA 02139, USA

A 24MW CW Electron Cyclotron Heating and Current Drive (EC H&CD) system operating at 170GHz is to be installed for the ITER tokamak. The EC system will represent a large step forward in the use of microwave systems for plasma heating for fusion applications; present day systems are operating in relatively short pulses (≤10s) and installed power levels of ≤4.5MW. The magnitude of the ITER system necessitates a worldwide collaboration. This is also reflected in the EC system that is comprised of the power supplies, sources, transmission line and launchers. A partnership between Europe, India, Japan, Russia, United States and the ITER organization is formed to collaborate on design and R&D activities leading to the procurement, installation, commissioning and operation of this system.

1. Introduction

ITER is planning to install a 24MW Electron Cyclotron (EC) system[1,2] that will be used for central heating and current drive (H&CD) applications as well as off-axis control of magneto-hydrodynamic (MHD) instabilities. The EC system will be operating at 170GHz for pulse lengths up to 3,600 sec and is comprised of up to 26 gyrotrons, 13 matching high voltage power supplies (HVPS), 24 transmission lines (TL) connected to either one equatorial (EL) or four upper (UL) launchers. The layout of the EC system is illustrated in figure 1. The power supplies (PS), gyrotrons and part of the TLs are located in the RF building. The TL then passes through the Assembly hall building to the tokamak building, which houses the tokamak and the EC launchers. An estimated 20MW will be injected into the plasma from the original 24MW of generated microwave power resulting in a 83% transmission efficiency (from gyrotron window to plasma) or roughly 40% electrical efficiency from the power pulled from the grid.

The EC system is being designed and procured based on a partnership between the ITER organization (IO) and five participating Domestic Agencies (DAs) involved in the EC system. The partitioning of the responsibilities is outlined in Table 1. In general, the IO is responsible for the functional specifications of all components as well as the detailed design of the launchers, the integration management, part of the installation, commissioning and operation. Note that the detailed design of the EL and UL are being performed by Japan and Europe (respectively) via ITER Task Agreements.

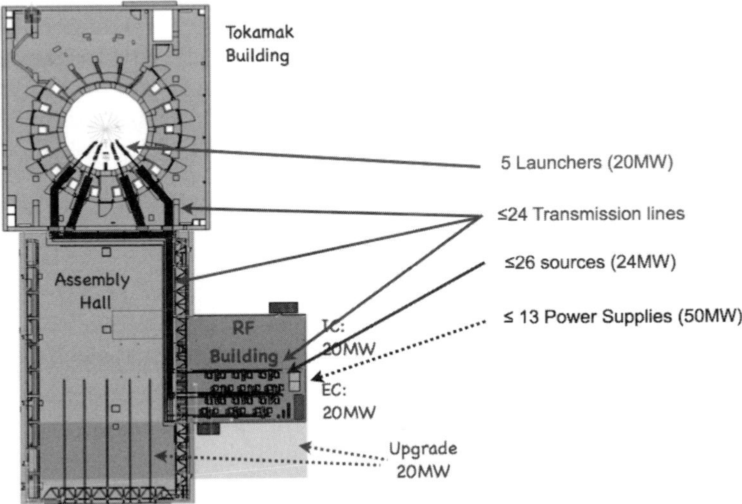

Figure 1. The ITER EC system comprised of power supplies, gyrotrons, transmission lines and launchers.

The procurement of all subsystems is performed by the DAs. The HVPS designed, procured and installed by Europe and India. Four DAs (Europe, India, Japan and Russia) develop the gyrotrons, procure, install and then commission them at the ITER site. The US develops and procures the TLs, which the IO installs and commissions. The launchers are procured by Japan (EL) and Europe (UL) and then installed, commissioned and operated by IO. The IO is responsible for the commissioning of the integrated system with the aim of delivering 6.7MW to the first plasma in 2019 and the full 20MW in 2021.

Table 1 Present sharing of the EC subsystems between the five domestic agencies involved in the EC procurements.

	Europe	**India**	**Japan**	**Russia**	**USA**
Power Supplies	12 sets	1 set			
Gyrotrons	8MW	2MW	8MW	8MW	
Transmission Lines					24
Launchers	4 UL		1 EL		

This paper provides an overview of the planned EC system, with a technical description described in the second chapter followed by the envisioned functional capabilities in Chapter 3 and an outline of the envisioned schedule in chapter 4.

2. Technical Description

The in-kind procurement strategy taken to build the ITER system requires careful management of the interfaces between each EC sub-system and with interfacing services (such as coolants, vacuum, buildings, etc.)[3]. The EC sub-systems are to come as a complete package including cabling, control systems, water manifolds, etc, such that all packages fit together without the IO procuring any components. This is illustrated in figure 2 in which the procurement interfaces are defined. The interface between the HVPS and the gyrotrons is at the cable connection near the gyrotron cathode with the HVPS adapting to the voltage and current requirements specified by the gyrotron procurements. Similarly, the interface between the gyrotron and TL is formed at the output of the gyrotron matching optics unit (MOU), with the gyrotron providing a HE_{11} mode purity of >95% and ≥0.96MW at the waveguide aperture. The end of the TL is at the waveguide connection prior to the diamond window located in either the upper or equatorial port cell. All waveguide components from the diamond window to the plasma are associated with the EL or UL as it forms the primary vacuum and tritium confinement barriers.

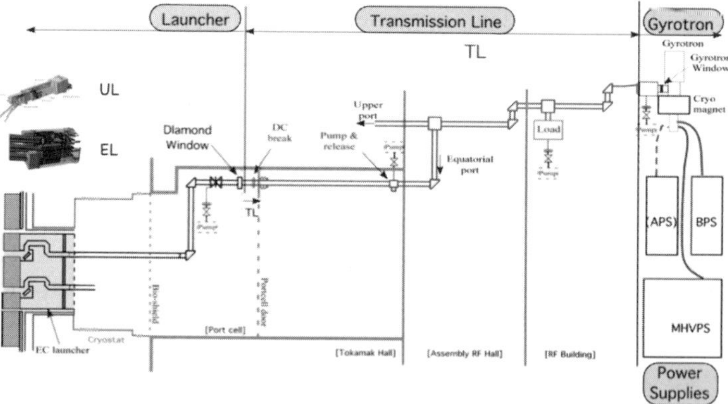

Figure 2. The interface boundaries of the four EC subsystem procurements are at the HV connectors at the gyrotron, the MOU output flange and the waveguide flange on the gyrotron side of the diamond window unit.

The division between the EC sub-systems described in figure 2 is different from conventional EC plants and should be kept in mind by the reader in reviewing the technical description of these sub-systems in the following sub-chapters.

2.1. *Power Supplies*

The power supplies[4,5] consist of a Main (MHVPS), body (BPS) and (in the case of the triode gyrotrons) anode (APS) HV power supplies. The main parameters of the three HVPS are provided in table 2. Each MHVPS feeds two 1MW (or one 2MW) gyrotron, with a separate BPS (and APS) for each gyrotron. The number of MHVPS was increased from 2 to 12 as a result of the 2007 ITER design review, offering a greater control of the delivered power and a more modular design to avoid significant interruption in operation in the event a PS failed.

Table 2 The general parameters of the three HVPS (main, body and anode) used to power the EC gyrotrons.

	Voltage	Current	Modulation	Stability
Main HVPS	55kV	90A	≤1kHz	1%
Body HVPS	~35kV	~200mA	≤5kHz	1%
Anode HVPS	~40kV	~50mA	≤5kHz	1%

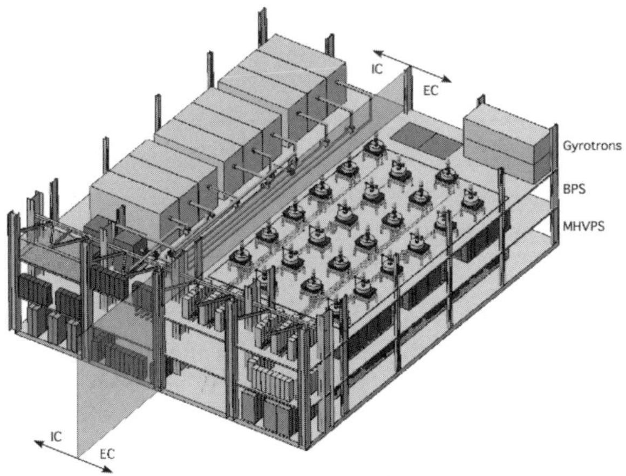

Figure 3. The RF building (B15) houses the IC and EC systems, which are each distributed on three floors. The first two floors are for PS and the top floor for the sources.

All power supplies are envisioned to use the Pulse Step Modulated (PSM) design that offers fine control and high frequency modulation (≤5kHz) of the applied voltage. The HVPS are configured on the bottom two floors of the RF building as shown in figure 3. The MHVPSs are on the first level, followed by the BPS (and APS) on the second level, all of which are directly below the gyrotrons on the third level. The vertical configuration minimizes the cable length between the supplies and gyrotrons, thus reducing the stored energy in the cable. The configuration also simplifies the building design keeping the heavier items (in particular the MHVPS transformers) on the ground floor and leaving the ceiling of the third floor relatively high for lifting the gyrotrons with an overhead crane. Note that the present size of the RF building is limited to house only 12 HVPS sets and 24 gyrotrons consistent with the 24 TL to be procured.

2.2. *Gyrotrons*

The gyrotrons[6-10] are to be provided by four DAs as outlined in Table 3, the gyrotron type varies with each procurement. The general gyrotron specifications are to generate ≥0.96MW at the MOU output (≥1.0MW at the window) and >95% coupling to the HE_{11} mode. The gyrotron procurement includes a Helium free cryomagnet, MOU, cooling manifold, local control system, support structure and ancillary systems. The tube should have an electrical efficiency of ≥50%.

Table 3 The four gyrotron procurements and the envisioned type and characteristics of each gyrpotron.

	Power	Cavity	Gun	Reference
EU-DA	2.0MW	Co-axial	diode	[9,10]
IN-DA	1.0MW	cylindrical	TBD	[8]
JA-DA	1.0MW	cylindrical	triode	[6]
RF-DA	1.0MW	cylindrical	diode	[7]

The RF building size is compatible with a 24MW EC system, which is predominately limited by the space required for the power supplies. This limits the number of gyrotrons to 24 (assuming all 1MW sources) or 20 (assuming four 2MW sources).

The gyrotrons will be distributed on a grid with typically ≥4m spacing and an equivalent distance between building iron structures (see figure 3). The spacing is maximized with in the finite size of the building to minimize impact on each gyrotron's magnetic field. All gyrotrons are positioned >90m from the tokamak center limiting the horizontal component of the stray field to ≤3mT along the electron beam trajectory.

2.3. *Transmission Lines*

The transmission line[11,12] has the primary function of transmitting the EC power from the gyrotrons up to the launcher diamond windows in the equatorial and upper port cells. The overall transmission efficiency should be ≥90% with a maximum degradation of the HE_{11} content of 97% relative to the mode purity at the output of the MOU. The transmission line performs additional functions, which include:

- Monitor forward and reflected power (using power monitor mitre bend)
- Deviate power to dummy load (via an in-line switch)
- Modify beam polarization for optimum plasma coupling
- Deviate power to either upper or equatorial launchers
- Provide pumping access for maintaining low pressure in line
- Electrically isolate line from gyrotron and launchers

The components achieving these functionalities are distributed in a typical TL illustrated in figure 4. Additional functionality and more detailed description of the transmission lines are available in references [11] and [12].

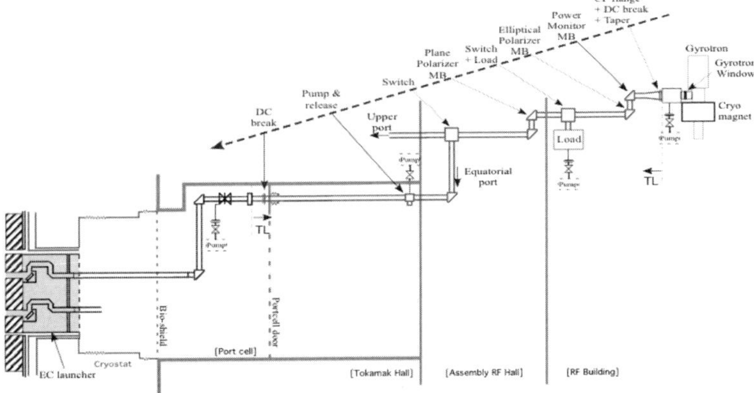

Figure 4. The principle components are shown between the MOU and diamond window for a generic TL.

2.4. *Launchers*

ITER requires two types of launchers to achieve the required functionality of central H&CD and the off-axis CD for stabilizing MHD activity. These functions are distributed between the equatorial (EL) [13,14] and upper (UL)[15-17] launchers, respectively, which are shown in figure 5 and 6. Both launchers are to be compatible with 1.8MW beams (losses from MOU and TL will be >10%). As mentioned above, the launchers include the diamond window, isolation (or gate) valve and roughly 10m of ex-vessel waveguide. The window and valve form the primary confinement barrier preventing tritium and other material from traveling up the TL. The 10m of waveguide compensates for the torus displacements (≤±30mm) due to temperature variations, plasma disruptions, thermal quenches and loss of coolant. Note that the waveguides are aligned during nominal operating conditions (to minimize mode conversion losses) and then deflect as the torus displaces.

The equatorial launcher has a total of 24 beams at the entry, which are grouped into three sets of eight beams. Each beam set is projected to a focusing then flat steering beam before passing through the blanket shield module (BSM) and into the plasma. The steering mirror directs the beam set in the toroidal direction from 20 to 41°, providing access from near on-axis to $\rho_T \leq 0.45$ (where ρ_T is the square root of the normalized toroidal flux).

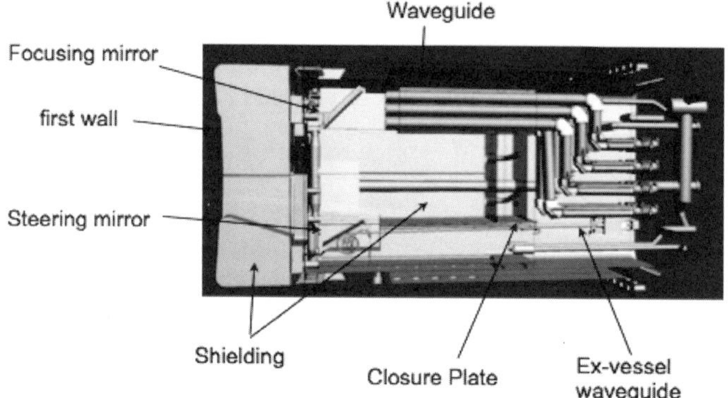

Figure 5. The equatorial launcher cut in a horizontal plane showing the waveguide orientation, focusing mirror, steering mirror, port plug structure and blanket modules.

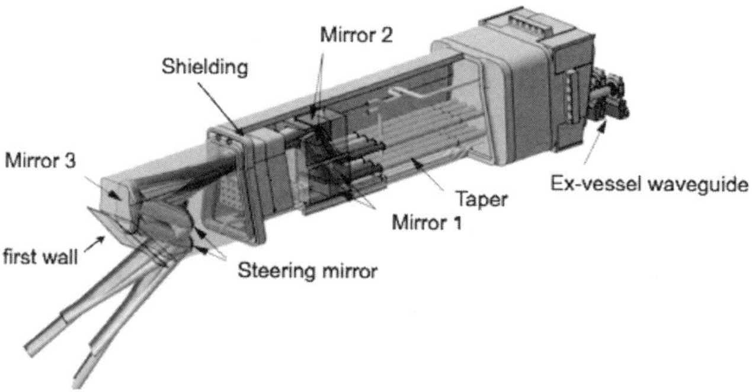

Figure 6. The upper launcher cut in a vertical plane showing the waveguide orientation, three fixed mirror sets, two steering mirrors, port plug structure and blanket module.

There are a total of four upper launchers with each having eight waveguide entries grouped into two sets of four beams. The beams propagate in free space through a set of four mirrors that regroup and focus the beams to provide a narrow and peaked current density profile for control of the MHD instabilities. The last mirror is a steering mirror that injects the beam over a poloidal range of ±12° to access the relevant q=2 and 3/2 rational flux surfaces. The beams have a nearly fixed toroidal injection angle of about 20° that provides a maximized peak current density profile over the desired access range.

3. Functional Description

The functional capabilities of the EC system strongly depend on the focusing and access range provided by the two launchers. The original baseline design (prior to 2007) partitioned the physics applications between the two launchers based on regions in the plasma. The EL accessing inside of $\rho_T \leq \sim 0.45$ for applications associated with central heating, current drive and sawtooth control and the UL accessing only the region where NTMs are expected to occur between $\sim 0.55 \leq \rho_T \leq \sim 0.85$. The combination of two launchers could not achieve the desired accessibility.

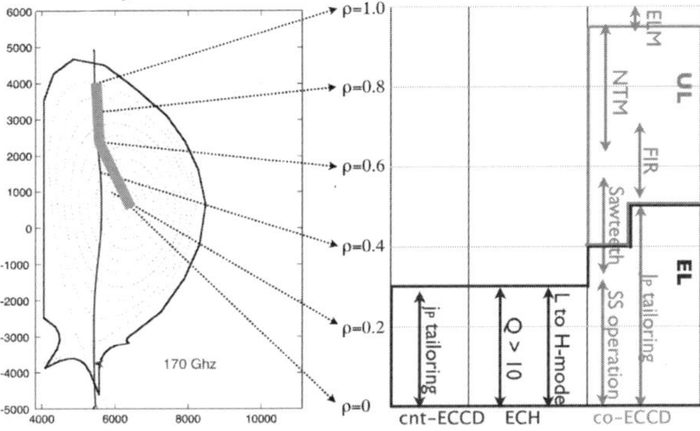

Figure 7. The revised equatorial and upper launchers access nearly the entire plasma cross section and is capable of injecting 20MW in the co-ECCD direction, 6.7MW in the counter-ECCD, or a balanced injection of the full 20MW for effectively pure heating.

The design teams have modified the launcher designs increasing the access range and avoiding regions of non-accessibility (as shown in figure 7 and 8) by increasing the access range of the UL. The physics applications have been repartitioned between the two launchers[18], dividing the functional requirements based on the need for broad (EL) or narrow (UL) deposition profiles. The EL would still be used for central heating and current drive applications[19], while the UL used for the control of both the NTM and sawtooth instabilities. In addition, one third of the EL beams are to provide counter current drive to decouple the heating and current drive contributions when depositing power centrally.

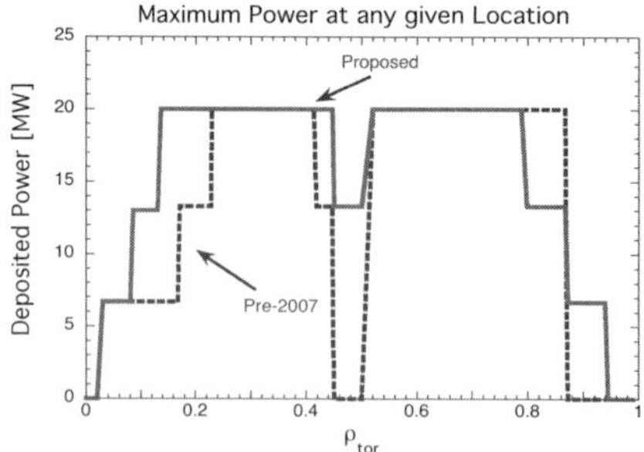

Figure 8. The maximum power that can be deposited at any given location in the plasma cross section based on the pre-2007 design (blue dashed line) and today's configuration (green solid line).

4. Schedule

The EC system is envisioned to provide 8MW (installed) for the first ITER plasma scheduled in late 2019 and the full 24MW (installed) for the second operating period in late 2021. The design, manufacturing, installation and commissioning activities are configured according to these milestones with some margin of float (typically ≥one year) prior to the planned plasma initiation date. The installation process will begin in early 2016 when the RF, assembly and Tokamak buildings are available for EC equipment installation.

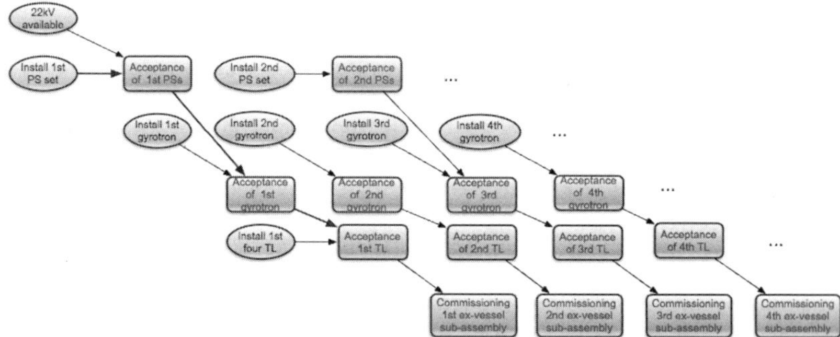

Figure 9. The installation strategy planned for the EC system with parallel activities on-going for the power supplies, gyrotrons and transmission lines.

The installation strategy is to install and commission a power supply in parallel with the corresponding gyrotron installation. Once the power supply is accepted, the gyrotron commissioning and acceptance will follow. A second PS will be installed and commissioned along with the installation of a second set of gyrotrons, while the first set of gyrotrons are being accepted as illustrated in figure 9. This procedure of parallel activity will continue into all PS and gyrotrons are installed (in 2019).

The TL will be installed independently of the gyrotrons and PS, but will be commissioned as soon as a given gyrotron has been accepted. The Launchers will be installed for the second plasma operating period, note that a simplified launcher is being considered for the first plasma operation in 2019 to minimize risk and relax resource profiles.

References

1. M. Henderson et al, *An Overview of the ITER EC H&CD System, IRMMW-THz Conference*, 21-25 Sept. 2009, Busan, Korea.
2. C. Darbos et al, *ECRH system for ITER*, RF conference, June 2009, Gent, Belgium.
3. M. Henderson and G. Saibene, Nucl. Fusion **48** (2008) 054017.
4. D. Fasel *et al*, *Fusion Sci. Technol.* **53** (2008) 246.
5. T. Gassmann et al, *Integration of IC /EC systems in ITER* ISFNT-09 conference, October 11-16, 2009, Dalian, China.
6. K. Sakamoto *et al Nature* **3** (2007) 411–4; *Development of High Power Long Pulse ITER Gyrotrons,* this Conf.
7. G. Denisov *et al Recent Development Results in Russia of Megawatt Power gyrotrons for Plasma Fusion Installations*, this Conf.
8. S.L. Rao et al, "Design Change, its Implications & Alternate Options for the EC Source System for ITER Plasma Start-Up", 5th IAEA TM on "ECRH Physics & Technology for Large Fusion Devices", 18-20 Feb. 2009, Gandhinagar, INDIA.
9. F. Albajar et al, *The European 2MW gyrotron for ITER*, this Conf.
10. J.-P. Hogge et al, "First experimental results from the European Union 2-MW coaxial cavity ITER gyrotron prototype," Fus. Sci. Technol., vol. 55, no. 2, pp. 204-212, Feb. 2009..
11. D. Rasmussen *et al R&D Progress on the ITER EC Transmission Line*, this Conf.
12. F. Gandini et al, *An Overview of the ITER EC Transmission Line,* this Conf.
13. Takahashi K. *et al* Nucl. Fusion **48** (2008) 054014.
14. K. Kajiwara et al, Fusion Engineering and Design 84 (2009) 72–77
15. D. Strauss et al, *Deflections and Vibrations of the ITER ECRH Upper Launcher*, this Conf.
16. T. Scherer et al, *Recent Upgrades of the ITER ECRH CVD Torus Diamond Window Design and Investigation of Dielectric Diamond Properties*, this Conf.
17. M. Henderson et al Nucl. Fusion **48** (2008) 054013.
18. G. Ramponi et al, Nucl. Fusion **48** (2008) 054012.
19. C. Zucca et al, Theory of Fusion Plasmas AIP Conf. Proc. **1069** (2008) 361

AN OVERVIEW OF THE ITER EC TRANSMISSION LINE

F. GANDINI, B. BECKET, C. DARBOS, T. GASSMAN, M. HENDERSON, O. JEAN, C. NAZARE, T. OMORI, D. PUROHIT

ITER Organization, St. Paul-lez-Durance, 13067 France

F. ALBAJAR, T. BONICELLI, G. SAIBENE

Fusion for Energy, C/ Josep Pla 2, Torres Diagonal Litoral-B3,E-08019 Barcelona, Spain

T. BIGELOW, J. CAUGHMAN, D. RASMUSSEN

US ITER Project Office, ORNL, 055 Commerce Park, PO Box 2008, Oak Ridge, TN 37831, USA

G. DENISOV

Institute of Applied Physics, 46 Ulyanov Street, Nizhny Novgorod, 603950 Russia

K. KAJIWARA, N. KOBAYASHI, Y. ODA, K. SAKAMOTO, K. TAKAHASHI

Japan Atomic Energy Agency (JAEA) 801-1 Mukoyama, Naka-shi, Ibaraki 311-0193 Japan

S. L. RAO

Institute for Plasma Research, Near Indira Bridge, Bhat, Gandhinagar, 382428, India

D. RONDEN

Association EURATOM-FOM, 3430 BE Nieuwegein, The Netherlands

M. SHAPIRO, R. TEMKIN

MIT Plasma Science and Fusion Center, Cambridge, MA 02139, USA

The ITER ECH&CD system will have an installed power of 24 MW at 170 GHz delivered to the launchers at the plasma side through a transmission line sub-system constituted by evacuated HE_{11} waveguides, DC breaks, power monitors, mitre bends, polarizers, switches, loads and pumping sections. Each line will be typically about 160 m in length and will connect the RF power sources alternatively to the equatorial launcher or to one of the upper launchers in order to accomplish the various physics requirements: heating, current drive and instability control. Two different source configurations are at the moment under study, pending the final decision about the European 2 MW coaxial gyrotron development. Each configuration will imply a dedicated transmission line layout. An overview of the actual design is presented and the technical requirements are discussed.

1. Introduction

The transmission line (TL) sub-system associated to the ITER electron cyclotron (EC) heating and current drive (H&CD) system [1] has reached the conceptual design maturity and is now going to be handled to US Domestic Agency for further design development and for procurement. Scope of TL is to transmit the microwaves generated by the 170 GHz gyrotrons installed in the RF Building to the launchers located in one equatorial (EL) and four upper (UL) Tokamak ports. It consists of evacuated HE_{11} corrugated waveguide [2] (63.5 mm inner diameter) and various sub-components distributed along the line to provide electrical isolation, polarization rotation and change of polarization ellipticity, switching between EL and ULs or towards local dummy load for gyrotrons daily conditioning, power monitoring, seismic isolation and TL pumping. The overall design has been recently optimized following changes in the interface with launchers, interface with RF power sources and size reduction of the RF Building where gyrotrons will be located.

2. TL's components location

Figure 1 shows the EC H&CD system layout and location of TL components for the case of 24 units of 1 MW generated power. It shows also the relative position of the three buildings crossed by the TL sub-system.

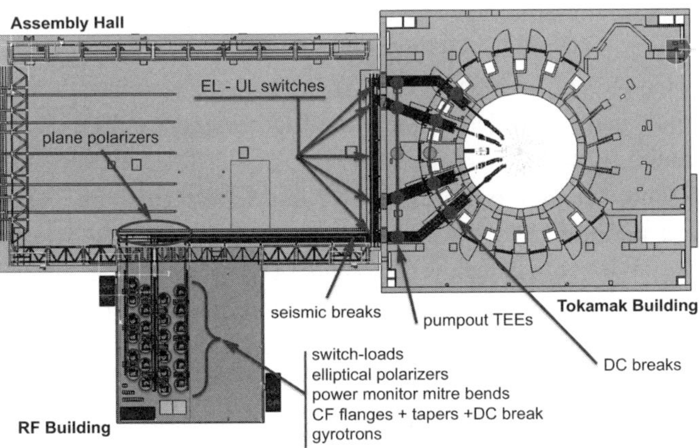

Figure 1. EC H&CD system layout and TL component's location.

Note that the RF Building size has been reduced since the original proposal and now its dimension is sufficient only to house the gyrotrons and power supply for the initial 20 MW injected power of both EC and Ion Cyclotron H&CD systems (IC equipments are not showed in the Fig.1). Each TL connects a gyrotron to the EL and to one of the ULs. ITER port numbering increases clockwise and EC power is delivered to ports upper 12 and 13, equatorial 14, upper 15 and 16. Gyrotrons are arranged on four staggered lines of six units each and TLs originating from the same line of six are bunched together in a two by three fashion (top to bottom). Each bunch of six TLs runs under a false floor at the 3rd level of the RF Building, penetrates the east Assembly Hall wall and bends north towards the Tokamak Building. Along the east Assembly Hall wall the TLs are rearranged to form three rows of eight TLs each: the top row comes from the eight gyrotrons closer to Assembly wall, the bottom row comes from the eight furthest gyrotrons and the middle row from the eight gyrotrons in between. At the corner between east Assembly Hall and south Tokamak Building wall, top and middle rows comprise eight mitre bends (MBs) each to deviate TLs west, while the bottom row comprises a set of eight switches to route the power alternatively straight to UL of port 16 or westbound parallel to the other two rows and to EL of port 14. In front of port 15 the middle row includes a set of eight switches to deviate alternatively the power to the corresponding UL or towards EL of port 14. In front of port 14, bottom and middle rows are connected to the EL through a couple of MBs for each TL while the top row includes a set of eight switches to route the power to the EL of port 14 or further west towards ULs of ports 13 and 12. Four TLs of the top row are routed to UL of port 13 and four to UL of port 12: each TL comprises a switch in order to deviate the power alternatively to the upper steering mirror or to the lower steering mirror of those ULs. Pumpout TEEs have been moved from the port cell area to the gallery, immediately after the penetration through the Tokamak Building south wall, in order to facilitate maintenance and connection to the service vacuum system. To cope with the non-orthogonal direction change required to route the TLs to UL of port 12 and 16 a special 140° MB has been designed and preliminary tests showed transmission losses comparable to the alternative solution realized using two standard MBs coupled in a periscope like fashion. At the TLs input gyrotron side and just inside port cells, in line DC breaks insure 5 kV electrical isolation between TL, source and launchers. TLs are mutually electrically isolated and grounded at a single point. The eight European 1 MW units will be installed in the furthest area from Assembly Hall and therefore they will be connected to the EL (port 14) and to the port 16 UL.

In the same area the eight 1 MW units can be substituted by four 2 MW units with the single TL connected to each 2 MW unit replacing the middle row pair of each TLs bunch. The TLs layout in this case is the following: the top row comes from the eight gyrotrons closer to Assembly wall, the middle row comes from the four (2 MW units) furthest gyrotrons and the bottom row from the eight gyrotrons in between. In this way the 2 MW units are going to be routed to the UL of port 15 ensuring the greatest flexibility of the overall system: the eight switches used to divert the power to the UL are to be rearranged in pair to allow the alternative connection of each 2 MW unit to the upper or to the lower UL steering mirror following the same scheme used for upper ports 12 and 13.

3. Technical requirements for the TL components

Beside the general scope described in the introduction, the various components of the TL system have to satisfy a number of technical requirements. The overall transmission efficiency has to be equal or greater than 90% and the power rate has to be 2.0 MW for at least 3000 sec with a 25% duty cycle. The various TLs have to switch independently between EL and ULs in less than 2 seconds and each TL has to be switchable to a CW load (1.5 MW power rate) for gyrotrons daily conditioning. The European 2 MW gyrotrons will have a dedicated CW load with 2 MW power rate. As mentioned in the previous section, a DC isolation of at least 5 kV is to be provided between TL, source and launchers. Power monitoring of forward and reflected power, full polarization control (rotation range: $-90^\circ \leq \alpha \leq +90^\circ$; ellipticity range: $-45^\circ \leq \beta \leq +45^\circ$) and arc detection capability (intervention time ~5 μsec) have to be integrated into MBs. Tokamak Building sits on seismic pads and is detached from Assembly Hall, therefore the TL section that bridges across those buildings has to include a seismic break. It is also required a pumping access in the vicinity of the Tokamak Hall wall penetration and at each load sufficient to maintain a pressure lower than 10^{-2} Pa throughout the TL.

A schematic of a generic TL is illustrated in Fig.2 together with identification of interfaces with the other EC sub-systems and with the affected buildings. Some recent Project Change Requests (PCRs) have modified the TL boundary changing the interface location between TL and sources and between TL and launchers. The RF Conditioning Unit (RFCU), originally included in the TL scope, has been split in its functional components: a Matching Optic Unit (MOU), an in-line switch plus a dummy load and two polarizers (elliptical plus plane). The MOU has been moved to the gyrotron scope while the polarizers have been integrated in the second and third mitre bend. The diamond window, originally placed at the port plug closure plate and interface between TL and launcher, has been moved to the port cell where the larger inter-space between adjacent TLs allows placement of the isolation valve and safe spacing between critical components (about 2 m in between MB, isolation valve, diamond window and DC break) reducing risks associated to higher order modes generation.

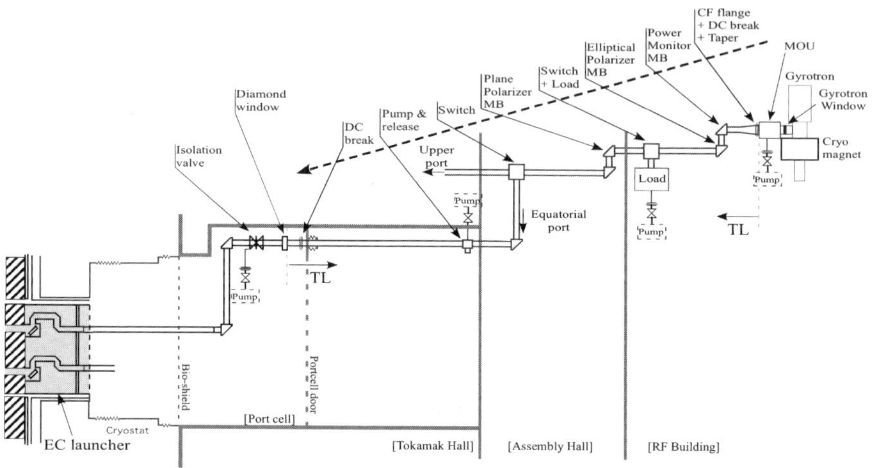

Figure 2. Schematic of a generic transmission line.

As consequence of this interface change, about 15 m of waveguides and 2 MBs have been moved from each TL to the interfacing launchers. A brief description of all the in-line components, from gyrotron to launcher, depicted in Fig.2, is given below.

CF-adapter. Small waveguide piece that has a CF flange on one side and standard coupling on the other, used to connect to the MOU.

In-line DC break. Provides DC isolation of the TL from the MOU and Tokamak. Note that the MOU DC Break may be combined with the CF-adapter and taper.

TEM_{00} to HE_{11} taper. Taper to minimize mode conversion loss at the transition from free space to guided beam propagation. Note that this component may be combined with the CF-adapter and DC-break.

Standard HE_{11} waveguide. Straight waveguide $\geq$ 2 m length (length to be optimized based on functional, cost, assembly and manufacturing criteria) or specially cut length to account for shorter distances to another component.

Power monitor MB. A mitre bend with coupling holes in the mirror surface that couples forward and reflected power into a set of WR-6 waveguides (~80 dB) for monitoring forward and reflected power in the TL. Note that this may include in-line arc detectors viewing both MB arms.

Elliptical polarizer MB. A mitre bend with $\lambda/8$ grooves machined in the surface to provide control of the elliptical polarization by rotating (90° in $\leq$ 2sec) the grating (in real time) about the normal surface of the mirror.

In-line switch. In-line switch that directs power to one of two output arms. Switching occurs in $\leq$ 2 sec and when RF power is off.

Dummy load (long pulse or calorimetric). Long pulse load (CW) used for daily gyrotron conditioning and during commissioning phase. A short pulse (~1.0 sec) calorimetric load can be installed in place of the CW load for gyrotron installation and power characterization. Note that the long pulse load could be assessed to potentially provide the equivalent functionality.

Plane polarizer MB. A mitre bend with $\lambda/4$ grooves machined in the surface to provide control of the plane polarization by rotating (90° in $\leq$ 2 sec) the grating (in real time) about the normal surface of the mirror.

Standard MB. A mitre bend providing a 90° change of propagation direction. The MB may include arc detectors that are embedded in the mirror and capable of viewing along both arms of the MB.

In-line pumpout TEE. A straight section of waveguide that provides pumping access to the internal volume of the TL. This component also serves as a releaser.

140° MB. A mitre bend that provides a 140° change of propagation direction. The MB may include arc detectors that are embedded in the mirror and capable of viewing along both arms of the MB.

Coupling. The coupling system that connects two neighboring waveguide components together and ensures the vacuum and alignment integrity (includes vacuum seals).

References

1. M. A Henderson *et al.*, A revised ITER EC system Baseline design proposal, *Proceedings 15th Joint Workshop on Electron Cyclotron Emission and Resonance Heating,* Yosemite N.P., California, USA, 2008.
2. D. Rasmussen *et al.*, Design of the ITER electron heating and current drive waveguide transmission line, *Proceedings 4th IAEA ECRH Technical Meeting*, Vienna, Austria, 2007.

Deflections and Vibrations of the ITER ECRH Upper Launcher

D. Strauss, T. Scherer, G. Aiello, A. Meier, S. Schreck, P. Spaeh, A. Vaccaro

Karlsruhe Institute of Technology, Association KIT-EURATOM,
D-76021 Karlsruhe, Germany
Institute for Materials Research I

The ECRH system at ITER is composed of gyrotrons, transmission lines to the torus and two types of launchers injecting the mm-wave power into the plasma. While the Equatorial Launcher heats the plasma center the four Upper Launchers can focus the power to localized current drive allowing the stabilization of neoclassical tearing modes and the sawtooth instability. The cantilevered design of upper port being fixed to the vacuum vessel and the correspondent plugs gives a spring like system with a length of 10 m and a very small gap to the neighboring components close to 10 mm. Vibrations and deflections during regular operation and disruptions are therefore a key design driver for the structural system.

During the last design upgrade the structural system has been improved to lower the plug deflections. The natural frequencies of the system upper port and plug have been analyzed. For disruptions the static deflection has been simulated and put into context with the effect of dynamic amplification. During operation vibrations might compromise the ability of targeting correctly the magnetic islands which is also discussed in this context.

1. Introduction

The electron cyclotron upper port launching system for ITER consists of four Upper Launchers which stabilize the plasma by local current drive 12. The design being developed by a group of European associations, KIT, IPP, IPF (D), CRPP (Ch), CNR (I) and ITER-NL (NL), has just been successfully revised in the “Preliminary Design Review” at ITER. The mm-wave beams are generated in gyrotrons and propagate in transmission lines to the Launchers. Ultra low loss CVD diamond torus windows serve as a first tritium barrier, in the rear part of the Upper Launchers a transition to a quasi-optical beam propagation is realized in order to reduce critical heat loads on the mirrors and to optimize efficiency. After passing through a dog-leg shaped beam path through shielding blocks - suppressing neutron streaming to acceptable values – the beams can be targeted by friction-free front steering mirror units to the points of interest in the plasma. The main targets are neoclassical tearing modes and the sawtooth instability.

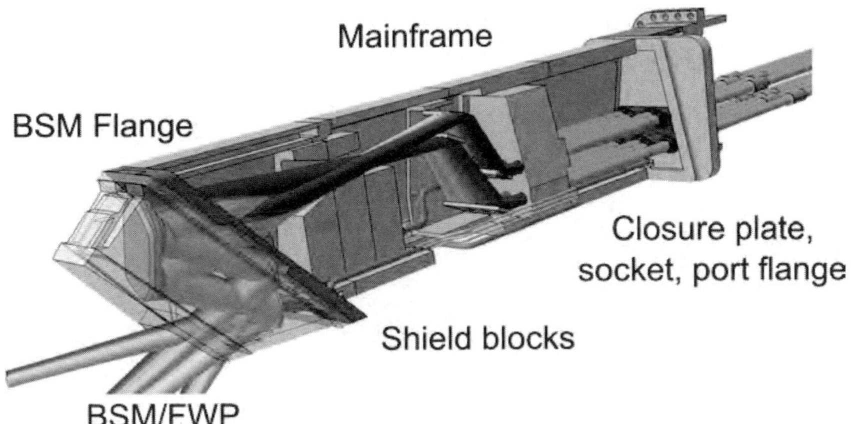

Fig. 1: The ITER ECRH Upper Launcher.

For the development of the Upper Launchers four major design drivers can be defined: (i) a mm-wave system capable of fulfilling the main physics requirements, (ii) a structural design to withstand the severe loads during disruptions within the minimal gap to the neighboring components, (iii) a robust design of the plasma facing components - first wall panel and blanket shield module – and (iv) neutron shielding to fulfill nuclear requirements.

While in the earlier design stages the different aspects could be optimized more or less independently, in the last remaining development steps an optimization of one parameter usually affects all the other aspects of the design. The design goal therefore is to get an optimized balanced system with acceptable compromises for each key design aspect.

The structural system is designed for lowest deflection during disruptions. As the mm-wave system requires as much space as possible in the front section the rear massive single wall part is especially enforced. The Upper Launchers are cantilevered systems, the plug with a length of over 5 m is attached on its rear side to the port extension being itself welded at its front to the vacuum vessel. This sums up to a length of around 10 m compared to a gap of 25 mm between the plug and the neighboring components such as first wall panels and of cooling tubes fed through the port extension. Due to manufacturing issues and assembly tolerances only 13 of the 25 mm remain for deflections.

2. Numerical analysis

During disruptions highest loads are applied to the structural components. The fast current quench leads to eddy current loops interacting with the magnetic field. The resulting Lorenz force then deforms the port plug. For the EM simulation a 20° sector model of ITER was used. The load scenario was the upward vertical displacement event followed by a fast linear current quench in front of the upper port plug as defined in 3. As a result the Lorenz forces on all elements of the plug are calculated as a function of time.

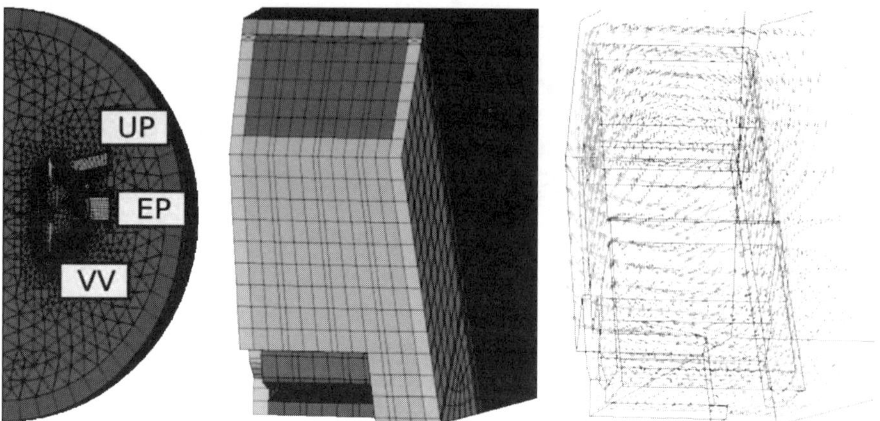

Fig. 2: EM sector model, Blanket Shield Module and eddy current induction.

For the deflection analysis and the optimization of the design the peak load values where taken as the input for a static deflection 4. The analysis shows a total deflection of the plug itself of 9 mm and for the entire ensemble plug + port 11.6 mm for the given design. Roughly 80% of this value comes from a toroidal deflection, the remaining 20% from a tilt around the radial axis. The gap reserved for deflections in the design is 13 mm. Taking into account the last upgrade of the ITER load specifications the linear current quench duration is reduced by 10%.

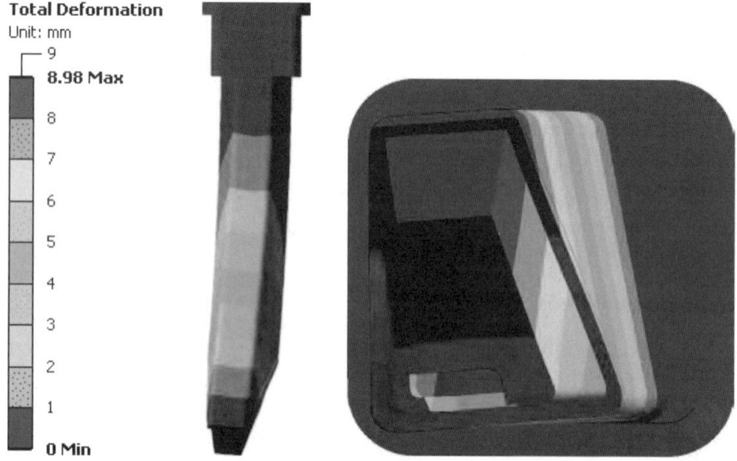

Fig. 3: 9 mm static deflection of the Upper Launcher.

Including this 10% effect in the deflection the gap criterion is still fulfilled. The deflection still can be reduced by different measures such as introducing slits into the First Wall Panel. Another option is to shift shielding 5 from the front, where maximum induction occurs, to the rear part of the Launcher, reducing the loads by up to 20%. Further it is possible to increase the wall thickness of the rear single wall main frame decreasing deflections by another 10% 5.A. As the maximum deflections only occur at singular points of the geometry dedicated machining can increase locally the gap. These optimizations will be part of the design work to be done until final design.

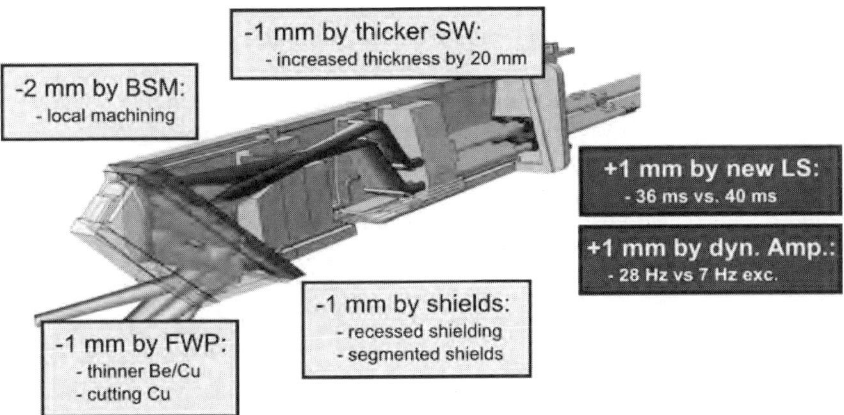

Fig. 4: Effects on max. deflection to be considered towards final design.

As the static analysis of the structure solves exclusively for potential energy, dynamic effects have to be taken into account. Therefore a Dynamic Amplification Factor DAF was introduced to represent the deviations to be expected in transient simulations 6. The dynamic system can be described by a linear harmonic oscillator with a set of natural frequencies and an excitation frequency given by the disruption duration. The excitation frequency is fixed and defined by the disruption from the load scenario, while the natural frequency of the upper port plug depends on design parameters such as wall thicknesses and other geometric peculiarities. A central goal of the structural design development is therefore to increase the natural frequency to values significantly above the excitation frequency of 6.25 Hz in order to minimize the DAF. In an analysis of a dummy port plug with a natural frequency of 14 Hz a dynamic amplification of up to 30% (toroidal) was calculated. The actual Upper Launcher design shows a significantly higher natural frequency of 25 Hz. In a coarse first approach the amplification term $(1-(\omega/\omega_0)^2)^{-1}$ from a linear harmonic oscillator can be taken to evaluate the effect of dynamic amplification. For the

14 Hz plug the DAF results to 25%, for the 25 Hz system the DAF is expected to be at 7%.

Conclusions

The preliminary design of the ITER ECRH Upper Launcher is capable to fulfill the gap criteria of 13 mm during the critical disruptions. While the upgrade of the load specification increases the deflections by around 10% and dynamic amplification leads to a further increase close to 10% a series of countermeasures can be taken to lower the deflections even close to 10 mm: A reduction of up to 20% is possible by shifting the shielding to the rear part, considering also the shielding requirements for nuclear radiation damage a reduction around 10% is realistic. Thicker walls in the rear single wall section can easily reduce the deflection by 10%, slitting of the highly conductive Be/CuCrZr layers in front of the First Wall Panel further reduces induction and Lorenz forces. Another option is to locally allow a higher gap. Considering the needs and the wide set of options it can be concluded that the design is flexible enough to introduce cost aspects as the dominating design driver for the final steps in the Upper Launcher structural design development.

Acknowledgements

This work was supported by Fusion for Energy under the contract No. F4E-2009-OPE-051. The views and opinions expressed herein reflect only the author's views. Fusion for Energy is not liable for any use that may be made of the information contained therein.

References

1. R. Heidinger, D. Strauß, T. Scherer et al., Conceptual design of the ECH upper launcher system for ITER, SOFT 2008.

2. M.A. Henderson, R. Heidinger, D. Strauß et al., Overview of the ITER EC upper launcher, Nucl. Fusion 48 (2008).

3. G. Sannazaro et al., Load Specifications, ITER_D_222QGL_v3.0 (2005).

4. D. Strauß et al., Electromagnetic and structural analyses of the UPP for the ECRH in ITER, Proc. EC-15 Yosemite/USA (2008)

5. Serikov et al., Progress in neutronics for the ITER ECRH launcher, Fusion Eng. and Design (2008), doi:10.1016/j.fusengdes.2008.06.044.

A. Vaccaro, D. Strauss et al., Mechanical Analysis of the EC upper launcher with respect to electromagnetic loads, SOFT 2008.
6. G. Sannazaro, Upper and Equatorial Port Plug Dynamics, ITER_D_22F7M7_v1.0 (2004).

TRANSMISSION LINE AND ITS COMPONENTS FOR ITER ECE DIAGNOSTIC

HITESH KUMAR B. PANDYA, SUMAN DANANI
ITER-INDIA, Institute for Plasma Research, GIDC, Electronic Estate, Sec.-25, Gandhinagar-380 025 India

KAUSHAL PATEL
Charotar Institute of Technology, Changa, Anand, India

The paper describes a possible scheme of the transmission line for the ITER ECE diagnostic. This includes wire grid polarisers, straight jointed sections of waveguide, mitre bends and other optical components. The brief description of each component is given. Information on the transmission efficiencies of various components available in the literature is summarised, which indicates that the number of mitre bends may be a concern. The paper also describes our study of the mitre bend through simulation and the results of the simulation are presented.

1. Introduction

The Electron Cyclotron Emission (ECE) diagnostic is one of the important plasma diagnostic on ITER [1] for the measurement of electron temperature. This diagnostic is also used to study many plasma physics phenomenon like temperature fluctuation, non-thermal electrons population and the power loss due to ECE.

In ITER for the ECE measurement frequency range of interest is 70 to 1000 GHz [2]. The intensity of ECE signal is low, at high frequencies. Moreover, during calibration the thermal black body calibration sources have very low power output signal at any frequency. These very low power signal with wide band frequency rang need to be transmitted over a long distance (~ 30 - 40 m) from tokamak to the diagnostics hall. Therefore a very low loss transmission line is needed. A circular corrugated waveguide which transmits power in the lowest order hybrid mode (HE_{11} mode) has a very low transmission loss for wide frequency band and is therefore a very good choice. Also the HE_{11} mode inside the corrugated waveguide couples efficiently with other optical components.

2. Transmission Line Lay Out

The transmission line lay out from the port plug to the ECE diagnostic hall is depicted in figure 1. There are two waveguides emerging from the port plug carry ECE radiation from two different locations of the ITER plasma. There is a polarizer splitter unit in the port cell (i.e. outside the cryostat and bio shield) , which splits each waveguide line into two lines corresponding to the X and O-modes Thus, four waveguide lines, each of 34 meter length transmit the EC radiation to the diagnostics hall. There is a pump out unit in each transmission line for evacuation and thus avoid atmospheric line absorption.

3. Transmission Line Components and Estimation of Transmission Loss

A brief description of the transmission line components follows. The transmission efficiency of the components available in the literature is summarised and an estimate of the transmission loss of total transmission line is presented

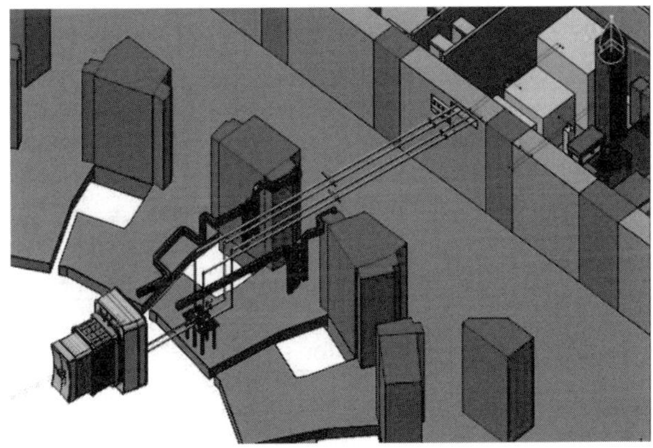

Figure 1 Transmission line lay out

3.1 Circular Corrugated Waveguide

A circular corrugated waveguide of diameter 88.9 mm is proposed for ECE transmission in ITER. There are two types of losses in the corrugated waveguide [3]. These are: (1) Ohmic loss and (2) mode conversion loss.

The theoretically estimate of Ohmic attenuation for lowest order hybrid mode in the corrugated waveguide is generally low throughout the ITER ECE frequency range of 70-1000 GHz. Only at certain frequencies, when the corrugation period becomes a multiple of half wavelength, the attenuation becomes very high due to Bragg reflection (fig.6 of ref.3). It appears that long (i.e. ~ 2 meters) section of corrugated waveguide with a pitch less than 0.4 mm is not presently available. So the 'stop bands' (discreet spikes of high attenuation) are unavoidable above 357 GHz ($\lambda = 0.84$ mm). The effect of these narrow width stop bands on the radiometry and Michelson interferometry measurements needs to be ascertained. In addition, the power transmission above 350 GHz is not properly measured and studied. A design of the transmission line for whole required frequency range (i.e. 70 – 1000 GHz) is a challenging task.

As already discussed, for the case of corrugated waveguide, the ohmic loss is negligible but the mode conversion loss may be dominant. Mode conversion losses at high frequencies are due to curvature and tilt. The curvature and tilt are due to axial variations in the diameter or deviations from circularity in the waveguide, misalignments at waveguide supports, imperfect waveguide couplings. Therefore, care must be taken in the fabrication and installation of the corrugated waveguide. More detailed description is given in Ref. 3.

In the past, several experiments have been carried out to measure transmission loss of the circular corrugated waveguides of different diameters at millimetre wave frequencies. The results are summarised in table 1.

Table 1 Experimental measurements of the insertion loss of a corrugated waveguide section

Corrugated waveguide dimension		Frequency (GHz)	Measured insertion loss (dB/meter)	Reference
Diameter (mm)	Corrugation period (mm)			
63.5	0.254*	60/140	0.08/<<0.04	4
88.9	1.7	84	0.002	5
63.5	0.45	100 to 300	0.0002	6
63.5	0.66	50 to 220	1.6 for 80 m.	7
31.75		110	0.003	8

*Waveguide section length ~ 0.25 meter

3.2 Mitre Bend

Mitre bends are widely used in the millimetre wave transmission line to bend the path of radiation in the required direction. It uses a mirror to deflect the path by 90° and can be fabricated in various diameters for different frequencies. The loss in a mitre bend can be calculated by using a simple theory in which the mitre bend geometry is considered as a gap of length of one waveguide diameter between the two waveguides. For a corrugated waveguide mitre bend, the waveguide wall is corrugated up to the mirror surface, and therefore the theoretical derived formula (i.e. equation 22 of Ref. 9) for the loss calculation is halved to get the following mathematical relation [9]

$$HE_{11}\ loss\ in\ a\ mitre\ bend \cong 2.4\left(\lambda/_{D}\right)^{1.5} dB \quad (1)$$

Where λ is a wavelength of the radiation and D is a diameter of the waveguide.

The loss calculated by using above equation is consistent with measurements. In table 2 summarised a few data on the mitre bend insertion losses.

Table 2. The calculated and measured mitre bends insertion losses for HE_{11} mode

Waveguide diameter (mm)	Frequency (GHz)	Calculated loss (dB)	Measured loss (dB)	References
12.7	140	0.17	0.22 /0.3 ±0.1	10
63.5	170	0.011	0.05±0.02	10
63.5	100 - 300	0.025 – 0.0047	0.06	6
63.5	100 to 350	0.025 – 0.0038	0.25	11

There would be in-waveguide fused silica widows in the transmission line. The window is sealed into the waveguide with Helicoflex seal. The window would be set a appropriate Brewster angle to minimize reflection. The transmission loss of the window should be worked out.

3.5 Polarizer Splitter Unit

There is a polarizer splitter unit in each of the two waveguides at outside bio shield. The arrangement inside the splitter unit consists of an ellipsoidal mirror followed by a wire grid polarizer. This enables the O and X-mode radiations to be coupled to separate transmission lines. A holder of the ellipsoidal mirror is such that the mirror can be rotated and collected the radiation from calibration source [7].

By assuming 98% coupling between polarizer splitter input corrugated waveguide and the ellipsoidal mirror and again 98% power coupling from the mirror to the output corrugated waveguide, the transmission loss of the polarizer splitter unit is estimated to be ~ 4 %.

The splitter unit is in the secondary vacuum part. A pumping unit is connected at the splitter unit to create vacuum between the primary vacuum window and the secondary vacuum window. The optical components inside the splitter unit should be vacuum compatible. Any loss due to the radiation or temperature effect at the splitter is unknown at present

3.6 Estimated Transmission Line Loss

The total transmission line loss for proposed transmission line is estimated on basis of previous measured published data listed in above sections.

Table 3. Total estimated transmission line loss

Component name	Quantities	Loss (dB)
Corrugated waveguide ($\varphi = 88.9$ mm)	34 meter	0.07
Mitre bend	5/6	1.1/1.32
Pump out unit	2	0.44
Polarizer splitter unit	1	0.18
	Total loss	**2.0**

4. Study of the Waveguide Gap and Mitre Bend through Gap Theory and Simulation

As seen above, the insertion loss of the mitre bend is higher than all other components of the transmission line. Prior theoretical work that has been carried out for determining the loss in the mitre bend is based on mode matching technique [9]. A numerical simulation using FEM based code can also be used to determine the loss [12]. We have used both above methods for the 70 – 110 GHz range.

The ITER ECE transmission line has a corrugated mitre bend that transmits the HE_{11} mode. The available numerical solver HFSS does not able to solve hybrid HE_{11} mode. However, this problem can be solved by using a smooth wall waveguide propagating the TE_{01} mode since the gap theory is equally valid for the TE_{01} smooth wall waveguide case and the HE_{11} corrugated waveguide and both modes retain the same symmetry.

4.1 Modelling of the smooth waveguide gap and mitre bend with TE_{01} mode

The HFSS model to simulate the gap and mitre bend problems are depicted in figure 2. First we kept radiation frequency 70 GHz (λ = 4.28) and both waveguides radius, a = 8.56 mm (i.e. a/ λ = 2) with gap of 2a.The project is simulated. More detail about the model is described elsewhere [12].

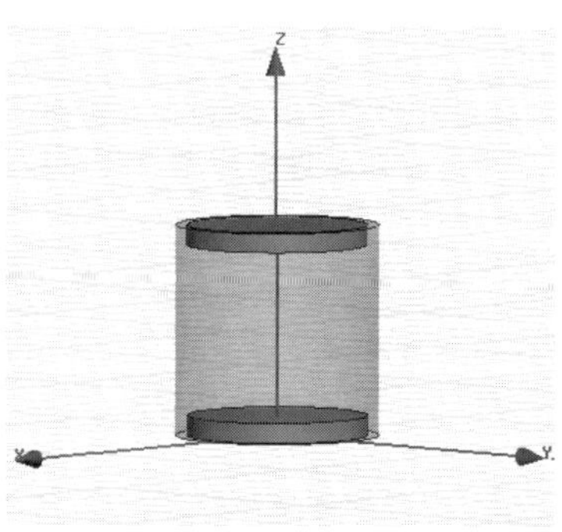

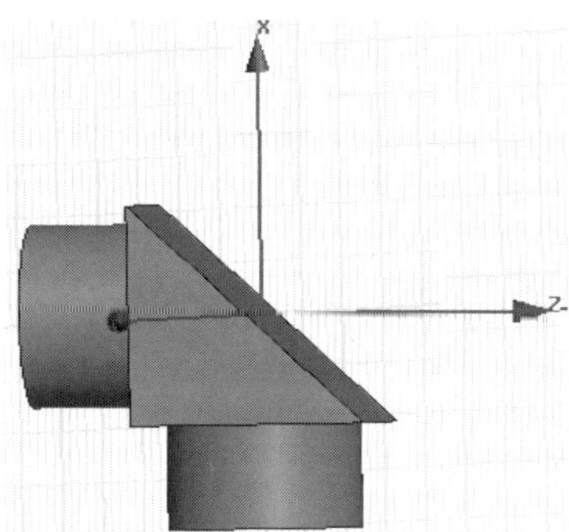

Figure 2 showing the model used to simulate the TE_{01} waveguide gap and Mitre Bend

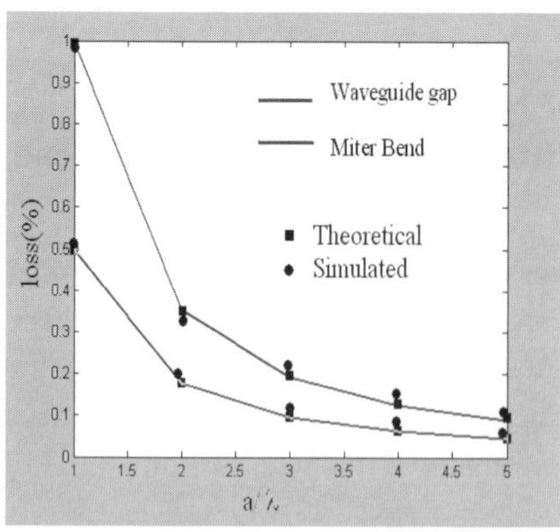

Figure 3 Loss (%) for waveguide gap and mitre bend for TE_{01} mode

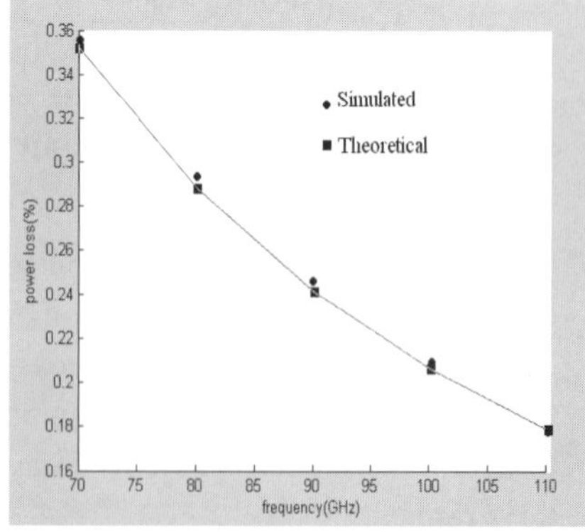

Figure 4 Power losses (%) for Waveguide gap for the radius of 8.56mm

Above figure 3 shows the loss for the waveguide gap and the mitre bend for TE_{01} mode in smooth waveguide. It is infer that the theoretical calculated values are approximately same as the simulation values in both cases and the loss in the gap is double the loss in the mitre bend. Our numerical simulations conclude that the waveguide gap theory is valid for the bend of the frequency (figure 4). Calculation is useful to estimate the loss of gap in the pump out unit and other gap in the transmission line.

Acknowledgments

The authors are thankful to Space Application Center, Ahmedabad for giving permission to use HFSS simulation. The authors also thankful to ITER design office and ITER ECE RO. The authors are grateful to ITER India design office and Dr. P. Vasu for their technical support and discussion.

References

1. DDD_55f_microwave_1.4
2. M.E. Austin, et al. "*Receiver front-end optics and performance assessment for ITER ECE system,*" S006937-F ITER ECE Report (U. Texas).pdf, February 2007
3. J. L. Doane, Fusion Scie. Technology, **53**, 159 (2008)
4. A. Cavallo, J. Doane and R. Cutler, Rev. Sci. Instrum., **61**, 2396 (1990)
5. K. Ohkubo et al. Int. J. Infrared Millim. Waves, **15**, 1507 (1994)
6. N. Isei, A. Isayama et al. Fusion Eng. and Design, **53**, 213 (2001)
7. Nagayama Yoshio, et al. J. Plasma Fusion Res., **78**, 601 (2003)
8. Shinichi Moriyama et al. Fusion Eng. and Design, **74**, 343 (2005)
9. J. L. Doane and C. P. Moeller, Int. J. Electron., **77**, 489 (1994)
10. S. T. Han, et al. Int. J. Infrared Millim. Waves, **29**, 1011 (2008)
11. Max Austin, Private communication
12. David S. Tex,"Mode conversion losses in overmoded millimetre wave transmission lines", M.S. Thesis, MIT-2008

DEVELOPMENT OF COMPACT RESONANT DIPLEXERS FOR ECRH: DESIGN, RECENT RESULTS, AND PLANS

W. KASPAREK, B. PLAUM, P. BRAND, C. LECHTE, M. SALIBA, Y. WANG,
Institut für Plasmaforschung, Universität Stuttgart, D-70569 Stuttgart, Germany
kasparek@ipf.uni-stuttgart.de

V. ERCKMANN, F. HOLLMANN, G. MICHEL, F. NOKE, F. PURPS,
J. STOBER, D. WAGNER
Max-Planck-Institut für Plasmaphysik, EURATOM-Association,
D-17491 Greifswald, and D-85748,Garching, Germany

M. PETELIN, E. KOPOSOVA, L. LUBYAKO,
Inst. of Applied Physics, Russian Academy of Science, 603950 Nizhny Novgorod, Russia

N. DOELMAN, R. VAN DEN BRABER,
Dept. of Mech. Equipment, TNO Science and Industry, NL-2600 Delft, The Netherlands

A. BRUSCHI
Istituto di Fisica del Plasma, EURATOM-ENEA-CNR Ass., I-20125 Milano, Italy

W. BONGERS, D.J. THOEN,
FOM Institute for Plasma Physics "Rijnhuizen", NL-3439 Nieuwegein, The Netherlands

High-power diplexers can be used in ECRH systems as power or beam combiners (BC), slow and fast directional switches (FADIS) to toggle the power from continuously operating gyrotrons between two launchers, and discriminators of low-power ECE signals from high-power ECRH in launchers used for in-line ECE. In the paper, design options for resonant diplexers are presented, and detailed low-power investigations on transmission characteristics and insertion losses are discussed. Two types, a purely quasi-optical and a compact waveguide-compatible diplexer, respectively, have been tested with high-power in the 140 GHz ECRH system for the stellarator W7-X. Fast switching, arbitrary distribution of the gyrotron power to two outputs, as well as power combination could be demonstrated with the first prototype. In recent experiments with a compact design, a mirror drive for tracking of the resonator to the gyrotron frequency was implemented; first long-pulse tests on transmission characteristics are shown. Finally, plans for implementation of this diplexer in the ECRH at ASDEX Upgrade are described.

1. Introduction

In the past years, high-power diplexers for millimeter waves have gained increasing attraction owing to their potential use in electron cyclotron resonance heating (ECRH) systems as well as for plasma diagnostics [1]. Firstly, oversized four-port diplexers can be used for power or beam combination (BC), the sources being implied to generate slightly different frequencies. Combining two (or more) conventional gyrotrons [2] would essentially reduce the number of transmission lines and launchers for ECRH systems of large tokamaks, e.g. ITER. Secondly, diplexers can be used as non-mechanical, fast directional switches (FADIS) to toggle the power from continuously operating gyrotrons between two transmission lines or launchers. In any gyrotron, modulation of a voltage results in a small (some tens of MHz) shift of the radiation frequency f. By using a diplexer with a steep slope at the transition frequency, this shift is translated into switching of the (combined) power between outputs. In particular, such electronically controlled switching between two antennas - synchronous to the rotation of the magnetic islands in the tokamak plasma - would maximize the efficiency for stabilization of these neo-classical tearing modes (NTM) [3]. Thirdly, especially resonant diplexers are of interest for plasma diagnostics employing a combination of high-power sources and sensitive receivers. For the measurement of the ECRH power deposition zone in NTM stabilization experiments, the “Line-of-Sight” scheme [4,5] exploits the gyrotron transmission line in the reverse direction as an electron cyclotron emission (ECE) antenna, thus ensuring that the ECE observed originates from the same location as where the ECRH is deposited. The potential of these diplexers led to the development of several prototypes [1,6,7,8] with different technical approaches. A promising version is based on the ring resonator with gratings as beam splitters, which can be used in quasi-optical as well as in corrugated waveguide systems. In the following, the design of these diplexers are sketched, and detailed low- and high-power investigations of resonant diplexers are discussed. Plans for applications at ASDEX Upgrade are presented.

2. Design of high-power ring resonator diplexers

The resonant diplexers shown in Fig. 1a) -c) are 4-mirror ring resonators [1,6], consisting of two focusing mirrors and two plane phase gratings for coupling of the incident beams to the resonator via the -1st diffraction order. Additional mirrors provide matching to the input and output beam or the waveguide. The power transmission coefficients [1] from the input 1 to output 1 and 2, respectively, are determined by the grating efficiencies, internal loss of the resonator, the round-trip resonator length L, and frequency f. Special features are the resonances in the output 2 ("resonant channel"), being periodic with $\Delta f_F = c/L$, and broad transmission bands with periodic notches in output 1 ("non-resonant channel"), as given in the example in Fig.1d.

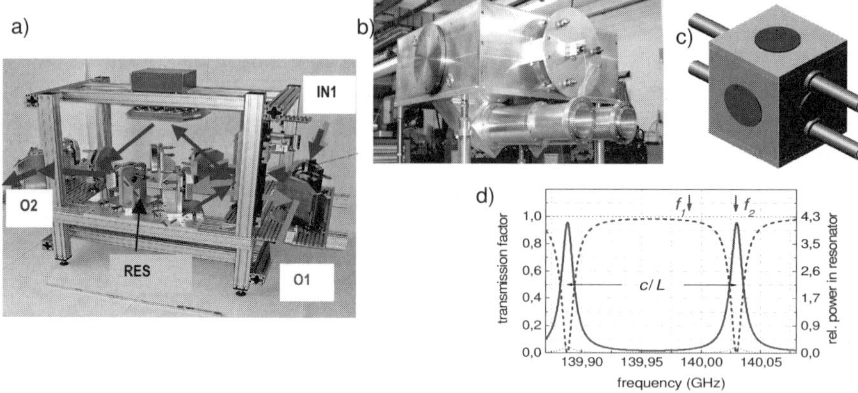

Fig. 1: Various designs of high-power diplexers based on a quasi-optical ring resonator (round-trip length L) with coupling gratings. a) photograph of the quasi-optical mock-up Mk I (L = 2.4 m), b) photograph of the closed diplexer Mk IIa (L = 2.1 m) featuring HE_{11}-TEM_{00} converters at in- and outputs, c) CAD model of the compact Mk IIIb (L = 0.9 m) with HE11 resonator. d) calculated transmission functions for input IN1 and the non-resonant output O1 (blue, dashed) and the resonant output O2 (red, solid). In the calculation around f = 140 GHz, L = 2.121m was assumed.

The detailed design can be adapted to quasi-optical transmission of TEM_{00} beams as well as corrugated waveguides transmitting HE_{11} mode. Up to now, 4 different prototypes have been built. Fig. 1 a) shows the first prototype Mk I for the TEM_{00} mode. In Fig. 1b), the prototype Mk IIa is presented, which features a compact design within a rigid box providing stable alignment as well as microwave shielding. Corrugated waveguide inputs and outputs are matched to the gaussian resonator mode by HE_{11}-TEM_{00} mode transitions. In Fig.1c), the compact device Mk IIIb is depicted, which contains 4 matching and 4 resonator mirrors within the casing and allows direct connection of corrugated waveguides owing to an internal HE_{11} resonator.

3. Experimental results

3.1. Low-power measurements

Measurements have been performed with low as well with high power (Mk I, Mk IIa). Low-power measurements on Mk IIa, which has a resonator length L = 2121 mm, grating efficiency R_1 = 0.22 were performed with a matched receiver; i.e. the transmission of the device in the nominal (HE_{11}) mode is obtained as seen in Fig.2. One can see, that the non-resonant output 1 provides narrow notches down to < -20 dB near the design frequency, as expected from calculations. The resonant output 2 shows the resonances at a distance corresponding to c/L = 141 MHz. From broadband measurements, the envelope of the peaks (determined by

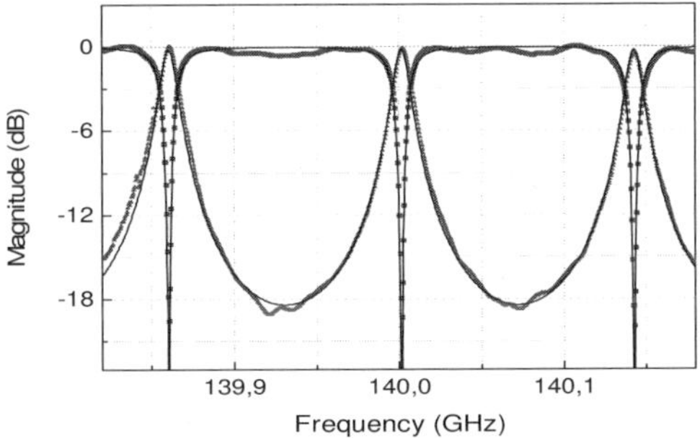

Fig. 2: Transmission functions of Mk IIa measured with matched HE11 receivers. Red triangles: resonant, blue squares: non-resonant channel. Theoretical curves (brown and black lines) are plotted for comparison.

the dispersion of the coupling gratings) yields a 97-% bandwidth of at least 500 MHz, making the device tolerant to typical frequency uncertainties of gyrotrons.

Precise data on the power transmission efficiency of the diplexer can be obtained from calorimetry [9]. In the non-resonant channel, the loss is mainly due to cross-talk (typically 2.2 %, i.e. near to theory), as well as absorption and mode conversion in the elements for coupling to the input and the output (0.8 %; about 0.4 % absorption in the matching mirrors, and 0.4 % stray radiation).

In the resonant channel, the absorption and scattering of two mirrors and two gratings with a power enhancement factor of 4.5 in the resonator results in an absorptive loss of 4.4%. If one assumes an average ohmic loss of 0.13 % per copper mirror and per reflection, the total ohmic loss sums up to 2.6 % (4 mirrors in the resonator with power enhancement [1] of 4.5, plus 2 matching mirrors). The remaining 1.8 % are absorbed in the water-filled Teflon hoses at the walls. The quasi-optical diplexer is a very efficient mode filter (especially in the resonant channel), as wrong modes entering the diplexer will not excite the resonator at the resonance frequency, but are mainly transmitted through the non-resonant channel; with the mode purity of about 97% for the HE_{11} generators used, this results in a cross-talk of 3.9% higher-order modes. The total loss averaged over both channels is therefore 5.7%. In case of a pure HE_{11} input mode with perfect alignment, an average insertion loss of 4% would be expected, about one half as

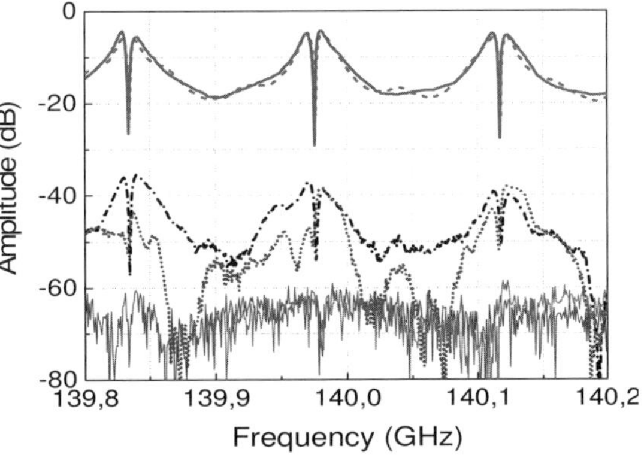

Fig. 3: Isolation of diplexer Mk IIb, i.e. HE11 power received in the isolated port for open outputs with absorbers at the resonant (thin solid line near noise level, blue) and the non-resonant (thin solid, green) output; for 100 % HE11 reflection from a plane mirror at the resonant (yellowish, solid,) and the non-resonant (pink, dashed) output; and for 100% multi-mode reflection from the resonant (black, dash-dotted) and the non-resonant (red, dotted) outputs.

absorptive power loss, the other half as cross-talk to the other channel.

The isolation of the diplexer was tested by measuring the power in the HE11 mode (i.e. with a matched receiver) in the isolated port 4 for different loads at the outputs. The result is shown in Fig. 3. For open (non-reflecting) outputs, a very high isolation of > 60 dB is obtained. If a 100 % reflection from a plane mirror, i.e. in HE_{11} mode, occurs, the power received in port 4 is given by the square of the transmission function for the corresponding output. In the more practical case, where a reflection (e.g. from the plasma vessel) occurs in multi-mode (simulated by a 100 % power reflection from a crumpled aluminium foil) reflector, the isolation is typically 40 dB.

3.2 High-power experiments with the ECRH system at W7-X

The Mk I as well as the MK IIa prototype have been tested with high power in the 140 GHz ECRH system for the stellarator W7-X [10]. For this purpose, the diplexers were integrated into the beam duct with the help of matching optics, such that one or two gyrotrons (B1 and B5) could be fed into the inputs, and the outputs were connected to cw calorimetric loads. Experiments were per-formed with power around 500 kW per gyrotron, which, for beam combination experiments, is close to the power limitation of the dummy loads (< 1 MW). The pulse length was limited to several tens of seconds by the un-cooled mirrors in the diplexers.

By modulating the body voltage U_B (square wave, $\Delta U_B \lesssim 5$ kV, $f_{mod} \lesssim 20$ kHz), frequency-shift keying of the gyrotron B1 with $\Delta f_{gyr} \leq 30$ MHz could be obtained. This allowed toggling the power in the rhythm of the frequency shift between the two outputs with a contrast of > 90 % [1].

Power combination was demonstrated by feeding two gyrotrons into the diplexer [2]. It was tuned such that a resonance was coincident with the frequency of gyrotron B1 after thermalization of the cavity. The frequency of gyrotron B5 was set about 40 MHz below the frequency of B1 using the dependence of the cavity temperature and thus the frequency from the generated power. Power combination with efficiency of about 90% was

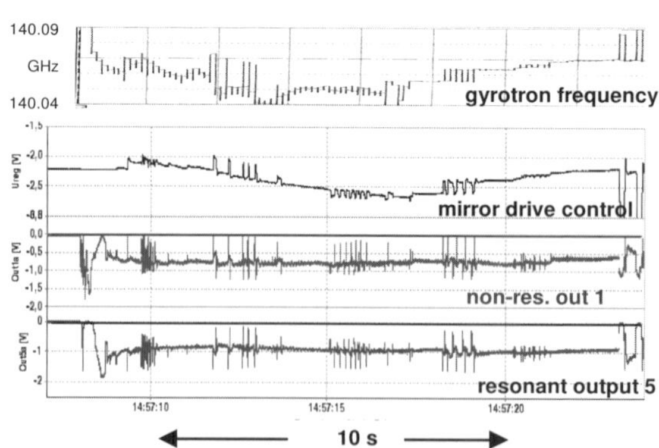

Fig. 4: Gyrotron frequency, position of the resonator mirror, and output signal for the non-resonant and resonant outputs showing the tracking of the slope of the diplexer resonance with respect to the gyrotron frequency.

demonstrated for 10 s pulses. The weak frequency stability of the resonant gyrotron B1 was setting the limitation of this experiment. To cope with this problem, a mirror drive for automatic tracking of the resonator to the gyrotron frequency was developed by TNO and implemented in the resonant diplexer Mk IIa.

Long-pulse tests on transmission characteristics and fast switching are presently underway. First results include the high-power (typ. 500 kW) confirmation of the transmission functions, long-pulse experiments up to 75 s, and a successful test of the automatic tracking of the diplexer resonance to the gyrotron frequency. In the 20-s pulse in Fig. 4, the gyrotron frequency was varied by variation of the power as shown in the top trace; the tracking was set to keep the slope of the diplexer transmission at the gyrotron frequency. As can be seen, the resonator mirror position (2nd trace) follows the gyrotron frequency on a fast timescale, leading to a regulation to constant powers in both outputs.

Further experiments with feedback-controlled tracking to the gyrotron frequency were performed with preprogrammed power sharing between the two outputs of the diplexer in the linear section of the steep resonance slope. The data as shown in Fig. 5 were obtained in a sequence of stationary pulses of 10 s duration by calorimetric measurements of the power in the two output channels. In the non-resonant channel a power of 355 kW is derived (for zero resonant power) as compared to 340 kW in the resonant channel (for zero non-resonant power). Assuming 2 % losses from the input to the non-resonant output, i.e. 362 kW input power, one obtains about 6 % losses in the resonant channel, which agrees well with the low-power measurements. The experiments showed that a very accurate and stationary power splitting between the two output channels can be realized by frequency tracking.

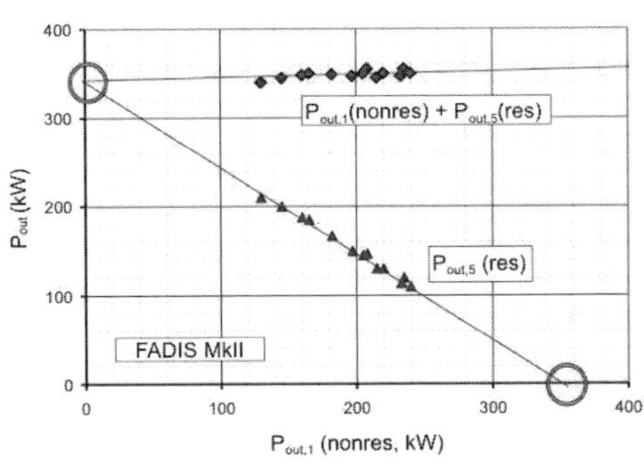

Fig. 5: Feedback controlled tracking of the gyrotron frequency with preprogrammed power sharing of one input beam (IN1) between both output channels, $P_{out,1}$ and $P_{out,5}$, respectively, The power is measured calorimetrically in 10 s pulses. The sum of both outputs (diamonds) and the output power in the resonant channel, $P_{out,5}$, (triangles) is plotted vs. the power in the non-resonant channel, $P_{out,1}$ for different power ratios.

4. Applications

For the main applications of high-power diplexers, namely power combination, fast and slow switching, various concepts exist. A proposal for use of diplexers in the ECRH systems of ITER was made [9], and first investigations on integration of diplexers into large ECRH systems have been performed. For the FTU tokamak, a diplexer is developed for fast switching between two launchers. At ASDEX Upgrade, proof-of-principle experiments on synchronous NTM stabilization are in preparation: The Mk IIa device (Fig. 1b) will be integrated into the new ECRH system [11] in this autumn. Owing to the favourable antenna design with symmetric, two-axes steerable launchers (Fig. 6, left), switching of the power in the rhythm of the rotating NTM islands is possible; else, one launcher can be used for synchronous NTM stabilization, the other for heating at an arbitrary position.

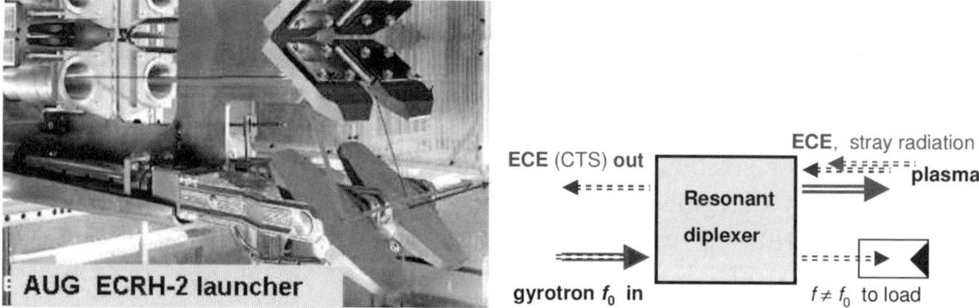

Fig. 6: Left: Launcher of ECRH-2 system at ASDEX Upgrade. Right: diplexer set-up for in-line ECE. The ECE receiver is connected to the port which is isolated with respect to the gyrotron input.

A second application, probably using the more compact diplexer Mk IIIb, will be the measurement of the power deposition zone in NTM stabilization experiments using the "Line-of-Sight" scheme [4, 5] exploiting the gyrotron transmission line in the reverse direction as an electron cyclotron emission (ECE) antenna, thus ensuring that the ECE observed originates from the same location as where the ECRH is deposited, see Fig. 6, right. The high-power beam can be fed via the resonant channel (corresponding to f_2 in figure 1d) of the diplexer to the launcher, while the ECE signal is coupled out in reverse direction via the non-resonant channel (the isolated channel with respect to the input) to the detection system. This scheme decouples the source from the detection system, and the notch-filter function of the non-resonant channel (*cf.* f_1 in figure 1d) suppresses the gyrotron stray radiation in the sensitive ECE receiver with high efficiency. As the notches are relatively narrow compared to typical filter bandwidths of the ECE receiver, they will not disturb the measurement. Moreover, with increasing distance from the centre frequency, the depth of the notches is reduced.

5. Summary and outlook

It has been shown, that high-power diplexers are a valuable component, which can strongly increase the performance and flexibility of ECRH as well as diagnostic systems. The results obtained up to now in the development of several prototypes including the high-power demonstration of fast switching and slow switching, and power combination from two gyrotrons including feed-back resonator control confirm the applicability of these devices. A proof-of principle experiment on synchronous NTM stabilization on ASDEX Upgrade is intended to demonstrate the applicability of diplexers in fusion experiments. An in-line ECE experiment on TEXTOR was successful, and its continuation at ASDEX Upgrade using a ring resonator diplexer aims at demonstrating the applicability of this technique in CW systems. In conclusion, the results motivate the further development of power combiners and fast switches until maturity, especially in view to applications in ITER and other next step devices. This includes especially compact designs for direct connection of corrugated waveguide, which use an eigenmode in the resonator corresponding to the HE_{11} in waveguide, thus avoiding transition losses from HE_{11} to TEM_{00} and back. Low-power measurements on the prototypes Mk IIb and Mk IIIb with HE_{11} resonators clearly confirm the feasibility of this concept. Another issue is the optimization of coupling gratings with equal efficiency and phase shifts for both polarizations, which are useful for the development of diplexers with polarisation-independent characteristics.

Further research is aimed to study other multiplexing schemes and arrays of phase-controlled gyrotrons. Gyrotron developers are encouraged to continue the development of frequency or even phase controlled gyrotrons, the availability of which would further extend the application palette of high-power diplexers.

Acknowledgments

This work is carried out in the frame of the virtual institute "Advanced ECRH for ITER" (collaboration between IPP Garching and Greifswald, FZK Karlsruhe, IHE Karlsruhe, IPF Stuttgart, IAP Nizhny Novgorod, and IFP Milano), which is supported by the Helmholtz-Gemeinschaft deutscher Forschungszentren.

References

1. W. Kasparek et al., Nucl. Fusion 48 (2008) 054010.
2. V. Erckmann et al., Fusion Sci. Technol. 55 (2009) 23-30.
3. M. Maraschek, et al., Phys. Rev. Lett. 98, 025005 (2007).
4. B. Hennen et al., this conference.
5. W.A. Bongers et al., Fusion Sci. Technol. 55, (2009) 188-203
6. M. I. Petelin, AIP Conference Proceedings 691 (2003), 251-262.
7. W. Wubie et al., Int. Conf. on Infrared a. Millimeter Waves, Pasadena, 2008.
8. A. Bruschi et al., Fusion Sci. Technol. 53 (2008), 97 – 103.
9. A. Bruschi et al., IEEE Trans. Plasma Science (2010), DOI: 10.1109/TPS.2010.2047658
10. V. Erckmann et al., Fusion Sci. Technol., 52 (2007) 291.
11. D. Wagner et al., Fusion Sci. Technol. 52 (2007), 313 – 320.

CONTROL ORIENTED ANALYSIS AND FEEDBACK CONTROL OF A SAWTOOTH INSTABILITY MODEL

G. WITVOET,[1,2,3] E. WESTERHOF,[2] M. STEINBUCH,[1]
M. R. DE BAAR[1,2] and N. J. DOELMAN[3]

[1] *Eindhoven University of Technology, Dept. of Mechanical Engineering, Control Systems Technology group, P.O. Box 513, 5600 MB Eindhoven, The Netherlands*
E-mail: g.witvoet@tue.nl, m.steinbuch@tue.nl

[2] *FOM – Institute for Plasma Physics Rijnhuizen, Association EURATOM-FOM, Trilateral Euregio Cluster, P.O. Box 1207, 3430 BE Nieuwegein, The Netherlands*
E-mail: e.westerhof@rijnh.nl, m.debaar@rijnh.nl

[3] *TNO Science and Industry, BU Mechatronic Equipment, Precision Motion Systems department, P.O. Box 155, 2600 AD Delft, The Netherlands*
E-mail: niek.doelman@tno.nl

A combined Porcelli-Kadomtsev numerical sawtooth instability model is analyzed using control oriented identification techniques. The resulting discrete time linear models describe the system's behavior from crash to crash and is used in the design of a simple discrete time feedback controller, which is successfully applied in simulation.

Keywords: Sawtooth instability; system identification; feedback control

1. Introduction

Sawtooth instabilities can trigger secondary instabilities like NTMs [1], and are associated with the mixing of fusion products and premature losses of alpha-particles in the plasma core. Feedback control of the sawtooth period enables optimization between these effects, which is essential for future fusion reactors. It has been shown that localized ECCD can alter the sawtooth period [2,3], and can therefore act as an actuator for this feedback control problem [4,5]. In order to derive a suitable control strategy for this type of system, this paper uses a numerical sawtooth instability model [6,7] as a case study to carry out control oriented system identification and analysis, with which a simple feedback controller for the sawtooth period using the ECCD mirror angle is designed. Simulations show that this controller is able to achieve a wide range of desired sawtooth periods within reasonable settling times.

2. Sawtooth instability model

To study the behavior of the sawtooth instability in a circular cross section tokamak, an *infinite dimensional impulsive dynamical system* is used:

$$\frac{\partial}{\partial t}B_\theta = \frac{\partial}{\partial r}\left(\frac{\eta}{\mu_0 r}\left(B_\theta + r\frac{\partial}{\partial r}B_\theta\right) - \eta J_{\mathrm{CD}}\right) \qquad \text{if } s_{q=1} \leq s_{\mathrm{crit}} \quad \text{(1a)}$$

$$B_\theta(r,t^+) = \begin{cases} B_\theta(r,t^-) & \text{for } r \geq r_{\mathrm{mix}} \\ \frac{1}{R} r B_\phi & \text{for } r < r_{\mathrm{mix}} \end{cases} \qquad \text{if } s_{q=1} > s_{\mathrm{crit}} \quad \text{(1b)}$$

where Eq. (1a) denotes the magnetic field evolution between crashes, and Eq. (1b) is a combined Porcelli-Kadomtsev full reconnection description of a sawtooth crash, where a critical value s_{crit} for the magnetic shear at the $q = 1$ surface denotes the crash condition. Moreover $r_{\mathrm{mix}} = 1.2 \cdot a/q_a$ and

$$\eta(r) \propto T_e(r)^{-3/2}, \qquad T_e(r) = T_0\left(1 + q_a\left(\frac{r}{a}\right)^2\right)^{-4/3} \qquad (2)$$

$$J_{\mathrm{CD}}(r) = J_0 \cdot \exp\left(-\frac{(r - r_{\mathrm{CD}})^2}{w^2}\right), \qquad r_{\mathrm{CD}} = r_{\mathrm{CD}}(\vartheta). \qquad (3)$$

The system's input is the ECCD mirror angle ϑ, and its output is the time between subsequent crashes τ_s (i.e. the sawtooth period). This model has been implemented in a Simulink® S-function, and yields the steady-state results shown on the left of Fig. 1. The figure on the right shows the derived DC-gain (steady-state amplification factor) of the system, i.e. $\partial\tau_s/\partial\vartheta$.

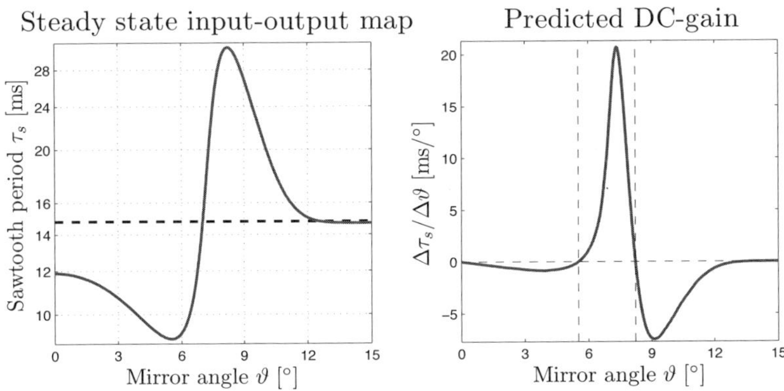

Fig. 1. Steady state dependency of τ_s on ϑ (left) and the derived DC-gain of the system as a function of ϑ (right)

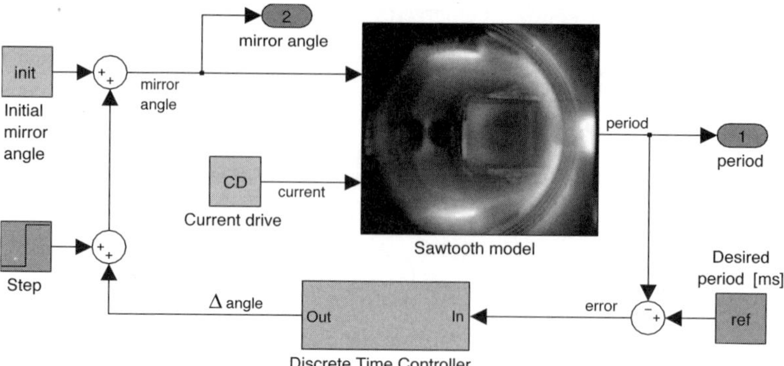

Fig. 2. Sawtooth model in a Simulink® closed loop scheme

3. Control oriented system identification

This paper focusses on controlling the blue region in Fig. 1, i.e. all operating points $\tau_s^i \in [9.25, 11.75]$. Controller design for this region is based on linear discrete time approximations $H_i(z)$ of the dynamic sawtooth period response of the system represented in Eq. (1). These $H_i(z)$ are obtained using a closed loop identification, illustrated in the feedback scheme of Fig. 2. The 'Sawtooth model' block is the numerical implementation of Eq. (1), controlled by any arbitrary stabilizing controller $C(z)$. Noting that k is the crash number index, the procedure is as follows:

(i) Let the closed loop stabilize on a specific τ_s^i and corresponding ϑ^i.
(ii) Apply a small stepwise perturbation on ϑ^i and set $k = 0$.
(iii) Monitor the responses of the resulting deviations $\Delta\vartheta_k^i$ and $\Delta\tau_{s,k}^i$ for the sequence $k = 0, 1, 2, \ldots$, i.e. after each crash.
(iv) Use approximate realizations techniques [8] to estimate the closed loop transfer functions from the stepwise input to:
   (a) $\Delta\vartheta_k^i$, i.e. the Sensitivity $S_i(z) = \frac{1}{1+C(z)H_i(z)}$, and
   (b) $\Delta\tau_{s,k}^i$, i.e. the Process Sensitivity $PS_i(z) = \frac{H_i(z)}{1+C(z)H_i(z)}$.
(v) Calculate the system's linearized dynamics $H_i(z) = PS_i(z)/S_i(z)$.

These linear crash-to-crash approximations, the $H_i(z)$, have the form

$$H_i(z) = \frac{n_1 z^{-1} + n_2 z^{-2} + \ldots + n_N z^{-N}}{1 + d_1 z^{-1} + d_2 z^{-2} + \ldots + d_N z^{-N}}, \tag{4}$$

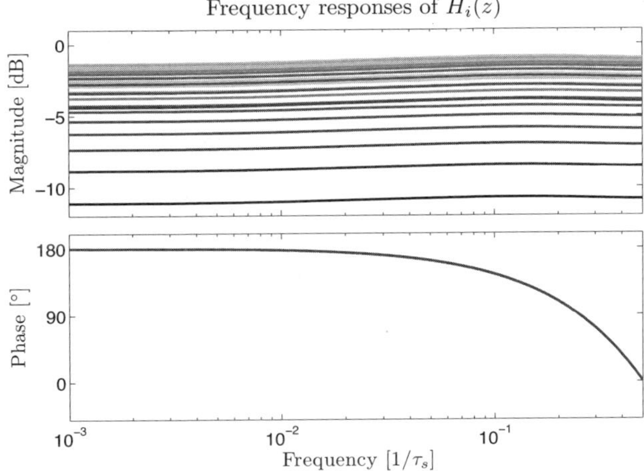

Fig. 3. Frequency responses of the identified linear approximations $H_i(z)$.

where N is the order of the approximation (here $N = 4$) and z is the *shift operator*, i.e. $\Delta\vartheta^i_{k-1} = z^{-1}\Delta\vartheta^i_k$, so that Eq. (4) can also be written as

$$\begin{aligned}\Delta\tau^i_{s,k} &= n_1\Delta\vartheta^i_{k-1} + n_2\Delta\vartheta^i_{k-2} + \ldots + n_N\Delta\vartheta^i_{k-N} \\ &\quad -d_1\Delta\tau^i_{s,k-1} - d_2\Delta\tau^i_{s,k-2} - \ldots - d_N\Delta\tau^i_{s,k-N}.\end{aligned} \tag{5}$$

Hence, $H_i(z)$ predicts the output $\Delta\tau^i_s$ at the current crash k, given inputs $\Delta\vartheta^i$ and outputs at N previous crashes. The frequency responses of a large number of $H_i(z)$ in the operating region of interest is shown in Fig. 3. All low frequent (steady-state) gains match the DC-gains of Fig. 1 perfectly.

4. Controller design and closed loop results

Given the linear models shown in Fig. 3, a number of specifications for the controller $C(z)$ can now be determined:

- to guarantee closed loop stability, the gain of $C(z)$ should be negative (due to the positive phase in Fig. 3) and should not be too large (due to the decrease in phase for increasing frequency);
- to obtain a good performance, $C(z)$ should be integrative (to guarantee zero steady-state error) and its gain should be as large as possible.

Based on these specifications a Tustin integrative controller was then designed using standard discrete time loopshaping techniques [9], guarantee-

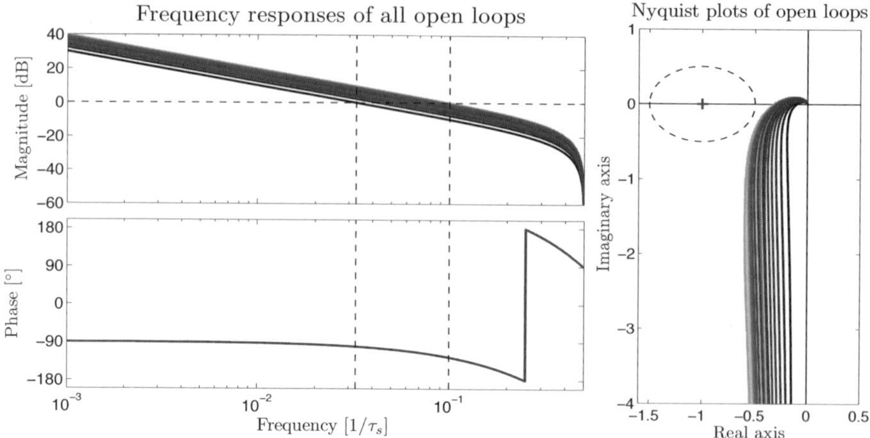

(a) Frequency responses; the dashed lines indicate the minimum and maximum crossover frequencies (i.e. bandwidths)

(b) Nyquist diagrams

Fig. 4. Representations of the open loop transfer function $C(z) \cdot H_i(z)$.

ing stability and performance for all H_i:

$$C(z) = -0.721 \cdot \frac{z+1}{2(z-1)}, \tag{6}$$

or equivalently, using the closed loop error $e_k = \tau_{s,k}^i - \tau_{s,\text{ref}}^i$

$$\Delta\vartheta_k^i = \Delta\vartheta_{k-1}^i - \tfrac{0.721}{2}\left(e_k + e_{k-1}\right). \tag{7}$$

As can be seen in the open loop Bode diagram of Fig. 4(a), this controller achieves bandwidths between 0.033 and 0.1 $\frac{1}{\tau_s}$, corresponding to settling times roughly between 10 and 30 sawtooth periods, while the Nyquist plot of Fig. 4(b) proves robust closed loop stability for all $H_i(z)$.

The controller in Eq. (6) is connected with the original numerical sawtooth model in a Simulink® feedback loop. A specific simulation result with stepwise changes in the desired period $\tau_{s,\text{ref}}$ is shown in Fig. 5. Indeed, the resulting closed loop is stable, and tracks the desired periods without any steady state error and with settling times in the order of 10 to 30 periods.

5. Conclusions

In this paper linear discrete time approximations of a combined Porcelli-Kadomtsev sawtooth model were obtained, using step responses and approximate realization techniques. Standard discrete time methods were then

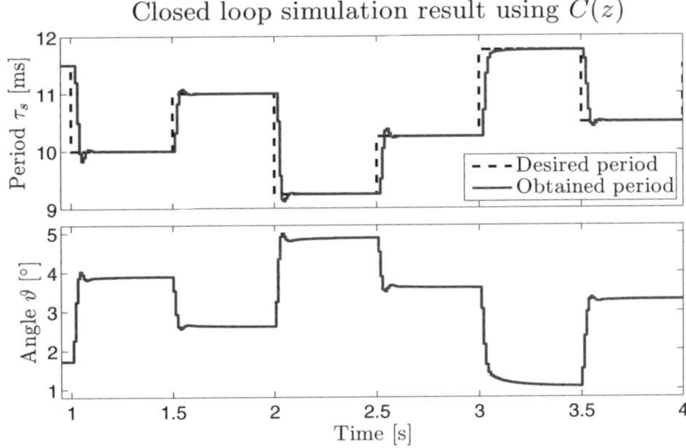

Fig. 5. Closed loop simulation result for a stepwise changing setpoint $\tau_{s,\mathrm{ref}}$.

used to design a controller for a specific operating region. This controller achieved perfect tracking with reasonable settling times in the whole region.

Future research includes the extension to other operating regions, and the application of automated controller design methods.

References

1. O. Sauter, E. Westerhof, M. Mayoral, B. Alper, P. Belo, R. Buttery, A. Gondhalekar, T. Hellsten, T. Hender, D. Howell, T. Johnson, P. Lamalle, M. Mantsinen, F. Milani, M. Nave, F. Nguyen, A. Pecquet, S. Pinches, S. Podda and J. Rapp, *Phys. Rev. Lett.* **88**, p. 105001 (2002).
2. C. Angioni, T. P. Goodman, M. A. Henderson and O. Sauter, *Nucl. Fusion* **43**, 455 (2003).
3. A. Mück, T. P. Goodman, M. Maraschek, G. Pereverzev, F. Ryter, H. Zohm and ASDEX Upgrade team, *Plasma Phys. Control. Fusion* **47**, 1633 (2005).
4. J. I. Paley, F. Felici, S. Coda, T. P. Goodman, F. Piras and the TCV Team, *Plasma Phys. Control. Fusion* **51**, p. 055010 (2009).
5. M. Lennholm, L.-G. Eriksson, F. Turco, F. Bouquey, C. Darbos, R. Dumont, G. Giruzzi, M. Jung, R. Lambert, R. Magne, D. Molina, P. Moreau, F. Rimini, J.-L. Segui, S. Song and E. Traisnel, *Fusion Sci. Technol.* **55**, 45 (2009).
6. G. Witvoet, E. Westerhof, M. Steinbuch, N. J. Doelman and M. R. de Baar, Control oriented modeling and simulation of the sawtooth instability in nuclear fusion tokamak plasmas, in *Proc. 48th Conf. Decis. Contr.*, (Shanghai, 2009).
7. G. Witvoet, M. Steinbuch, E. Westerhof, N. J. Doelman, M. R. de Baar and the TEXTOR team, Closed loop control of the sawtooth instability in nuclear fusion, in *Proc. Amer. Contr. Conf.*, (Baltimore, MD, 2010).
8. B. De Schutter, *J. Comput. Appl. Math.* **121**, 331 (2000).
9. K. J. Åström and B. Wittenmark, *Computer-Controlled Systems: Theory and Design*, 3rd edn. (Prentice Hall, 1997).

RECENT UPGRADES OF THE ITER ECRH CVD TORUS DIAMOND WINDOW DESIGN AND INVESTIGATION OF DIELECTRIC DIAMOND PROPERTIES*

T.A. SCHERER, D. STRAUSS, A. VACCARO, G. AIELLO, S. SCHRECK, A. MEIER, P. SPÄH

Karlsruhe Institute of Technology, Association KIT-EURATOM, D-76021 Karlsruhe Institute for Material Research I. Materials Technology Department

1. Introduction

The ITER ECRH system requires the use of ultra low loss CVD diamond windows mounted in a system of metallic parts (copper/steel). These components, being part of the primary vacuum boundary, have the function of tritium confinement in the vacuum vessel of ITER. Experiments of the first prototype with a real non-Gaussian input beam profile (ca. 90% Gauss content) show the influence of high frequency higher order modes (parasitic modes) introduced by excitation of oscillations in construction-conditioned small cavities in the window assembly [1,2]. A new design with inserted waveguide structures is able to avoid the parasitic excitations by covering gaps in the metallic structure of the window housing. A matchable distance of the waveguides related to the window disk position allows in addition to reduce amplitudes of such modes. A continuation of the corrugation up to the ends of the inserted waveguides is a supplemental improvement in the window design. The new optimized window assembly design will be shown.

A further reduction of the losses in the diamond window could qualify this material also for higher power fusion applications (e.g. DEMO) with more than 2MW/beamline. To study the influence of the CVD diamond surface chemistry to the high frequency properties (loss tangent: tanδ) different surface finishing methods for the diamond are being performed (e.g. hydrogen termination, oxidation, annealing and passivation). The evaluation of contact angles, surface conductivities and the corresponding loss tangent measurements will be

This work was partially supported by Fusion for Energy under contract F4E-2009-OPE-051. The views and opinions expressed herein reflect only the author's views. Fusion for Energy is not liable for any use that may be made of the information contained therein.

discussed due to the reduction of dielectric losses at higher ECRH beam power levels.

2. EU CVD Diamond Window (Prototype I)

In close interaction with the mm-wave design development performed at CRPP, it was decided to use the indirect cooling concept for the final design of the prototype version of the FS torus window. In this design, the CVD diamond disk with a diameter of d = 75 mm and a thickness of t = 1.11 mm (grown by Element 6, Cuijk, Netherlands) is integrated into a metallic housing with the cooling circuit separated from the copper disk by a copper cuff. The special configuration of the welding lip ("Γ shaped") allows to form the connection to waveguide which can inhibit significant contact of the inner surfaces of the welding zone with Be and tritium during ITER operation.

Fig. 1 Design of the CVD diamond torus window prototype I with the CVD diamond disk separating the torus vacuum section from the transmission line vacuum section. The red section is formed by a copper cuff with two symmetric cooling chambers separated from a vacuum segment at the outer diamond edge.

The time-temperature dependence and temperature distribution (IR imaging) during a long pulse RF experiment at 520 kW/30 s is shown in Fig. 2. The IR pictures are taken with an AVIO TVS-8500 IR camera (spectral range λ: 3.5 to 4.1 μm/4.5 to 5.1 μm) and indicate the temperature distribution at the end of the RF shot (JAEA, Naka) [1]. The main part of the diamond area shows an exponential saturated temperature behaviour. Some higher temperature spots correspond to the brazing zone. The temperature distribution of the diamond window is not radially symmetric, indicated by the warm tone colours in Fig. 2. At power levels up to 600 kW no arcing has been observed. The gap between copper and steel is about 100 μm. In given case of a mode mixing in the microwave beam additional parasitic cavity heating can be expected and is an obvious source for additional heating of the housing. An improvement of the beam purity up to 99% could avoid the parasitic heating effects in the window housing.

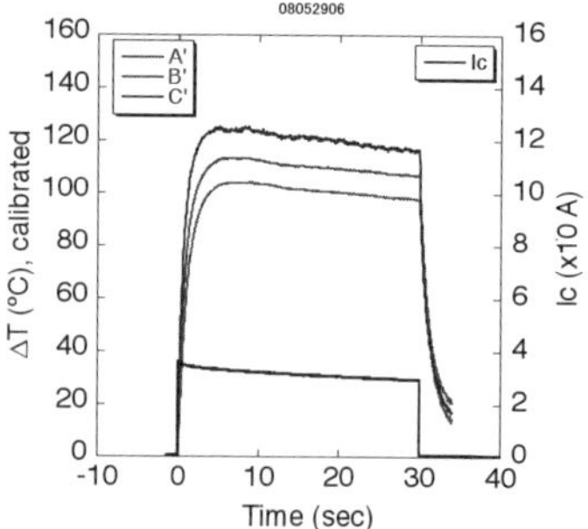

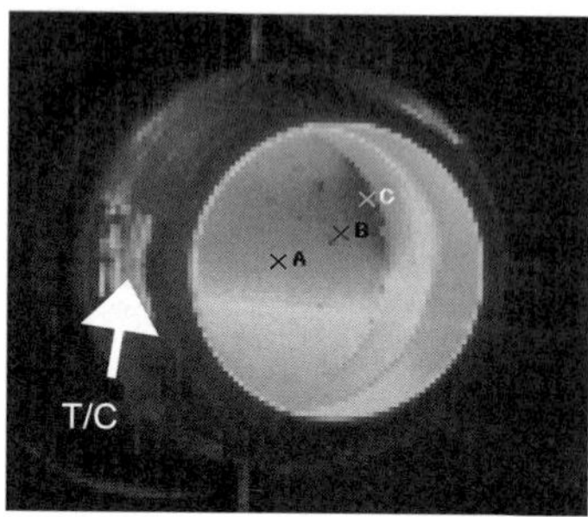

Fig. 2. IR measurements of the temperature of the CVD diamond window for 520 kW/30 s RF power at 170 GHz.

3. EU CVD Diamond Window (Prototype II)

To avoid the problem of parasitic higher order modes, a design change of the window unit is useful. By using a waveguide insert the small gaps of 50–100 μm between the copper cuffs and the SS waveguide vanish. An additional coating of the diamond disk edge with a thickness of about 2-3 times of the RF penetration depth with copper reduces the space for parasitic heating in the housing as well. The combination of both possibilities should allow an appropriate window operation for high power (1-2 MW) and long pulses.

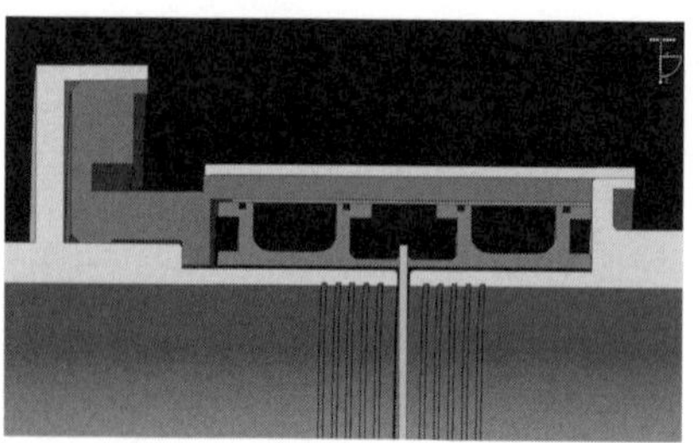

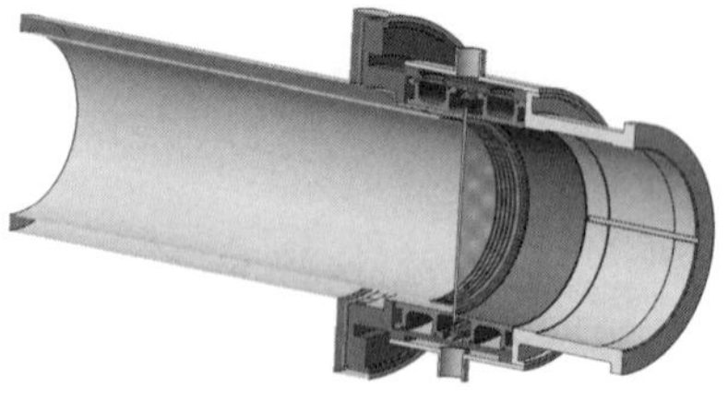

Fig. 3. Prototype II with inserted waveguides.

4. Evaluation of H-terminated CVD Diamond Disk Quality

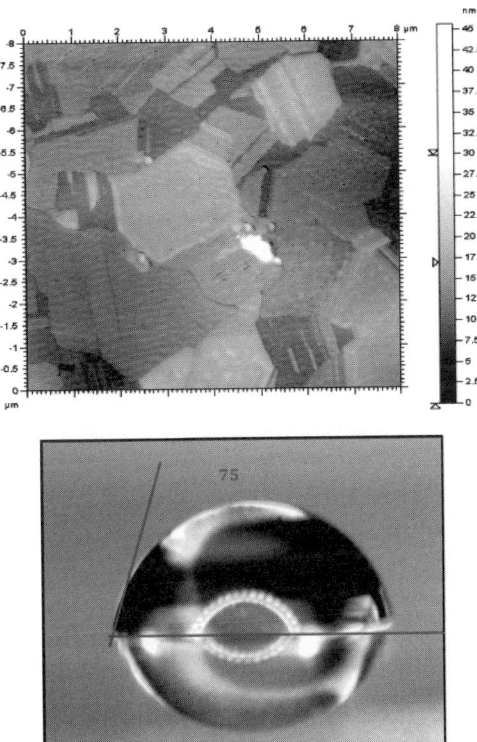

Fig. 4. AFM picture and contact angle measurement of hydrogen termination of a CVD diamond surface: 3 kW/150 mbar/850°C/-R (R: ramp down H_2 -pressure) [3].

To understand the behaviour of the bonding properties of a CVD diamond surface different chemical atomar/molecular surface finishing is foreseen for a series of experiments. In a first step a treatment of diamond with hydrogen in a plasma process shows a contact angle of about 75° with water. The measurement of the electrical conductivity with the Van-der-Pauw method (4 circular point electrodes on top) shows a higher resistivity of the order of several 10^4 Ω. This resistivity is also responsible for a higher microwave loss tangent at the diamond surface ($10^{-3} < \tan\delta < 10^{-4}$; f =170 GHz). In a second step the influence of an oxygen terminated or OH-terminated surface will be investigated. Fig. 5 shows the dependence of the Van-der-Pauw resitivity from the applied current. The back site of the diamond disk is less conductive in comparison to the front site.

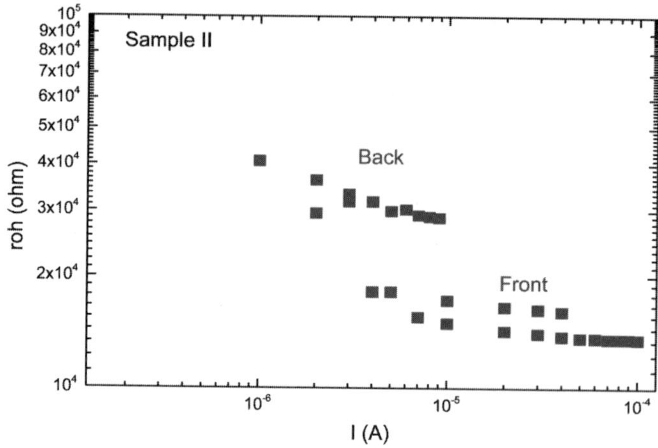

Fig. 5. Van-der-Pauw surface conductivity of the front and back site of the CVD diamond disk [3].

References

1. T. A. Scherer, R. Heidinger, A. Meier, K. Sakamoto, K. Takahashi, K. Kajiwara, M. Henderson, R. Chavan, "Design aspects and RF characterization of ITER-RF-CVD-Diamond windows," EC-15, Yosemite (USA), 2008, pp. 502-508.
2. T. A. Scherer, D. Strauss, M. Torge, A. Meier, "Investigations of dielectric RF properties of ultra low loss CVD diamond disks for fusion applications," IRMMW/THz conference, Busan (Korea) 2009.
3. Project Collaboration C. Nebel (Fraunhofer IAF Freiburg, Germany).

SILICON OIL DC200(R)5CST AS AN ALTERNATIVE COOLANT FOR CVD DIAMOND WINDOWS

A. VACCARO, G. AIELLO, A. MEIER, T. SCHERE, S. SCHRECK, P.SPAEH, D. STRAUSS

Karlsruhe Institute of Technology, Association KIT-EURATOM, D-76021 Karlsruhe, Germany, Institute for Materials Research I

G. GANTENBEIN

Karlsruhe Institute of Technology, Association KIT-EURATOM, D-76021 Karlsruhe, Germany, Institute for Pulsed Power and Microwave Technology

The production of high power mm-wave radiation is a key technology in large fusion devices, since it is required for localized plasma heating and current drive. Transmission windows are necessary to keep the vacuum in the gyrotron system and also act as tritium barriers. With its excellent optical, thermal and mechanical properties, synthetic CVD (Chemical Vapor Deposition) diamond is the state of the art material for the cw transmission of the mm-wave beams produced by high power gyrotrons.

The gyrotrons foreseen for the W7-X stellarator are designed for cw operation with 1 MW output power at 140 GHz. The output window unit is designed by TED (Thales Electron Devices, France) using a single edge circumferentially cooled CVD-diamond disc with an aperture of 88 mm. The window unit is cooled by de-ionized water which is considered as chemical aggressive and might cause corrosion in particular at the brazing. The use of a different coolant such as silicon oil could prevent this issue.

The cooling circuit has been simulated by steady-state CFD analysis. A total power generation of 1 kW (RF transmission losses) with pure Gaussian distribution has been assumed for the diamond disc. The performance of both water and the industrial silicon oil DC200(R) have been investigated and compared with a focus on the temperature distribution on the disc, the pressure drop across the cooling path and the heat flux distribution. Although the silicon oil has a higher viscosity (~x5), lower heat capacity (~x1/2) and lower thermal conductivity (~x1/3), it has proven to be a good candidate as alternative to water.

1. Introduction

The main heating system of the W7-X stellarator is electron cyclotron resonance heating (ECRH) with a total power of 10 MW in continuous wave (CW) operation at a frequency of 140 GHz. It consists of 10 gyrotrons with an output power of 1 MW each. The power that is dissipated by each gyrotron has to be removed by powerful cooling modules which are part of the gyrotron system. Because of the extreme corrosion sensitivity of diamond windows, the cooling systems have to operate very reliably [1].

The window unit consists of a diamond disc with copper cuffs brazed at each side. De-ionized water is used as coolant, which can lead to corrosion of the brazing material. Such issue could be prevented by replacing the water with alternative coolants, like Fluorinert^{TM}, developed by 3M and used for cooling of electronic parts, or the Dow Corning 200(R) silicone oils. This last option, in the version with kinematic viscosity 5cSt, is considered in the present work as alternative to water.

The DC200(R)5cSt is a linear polydimethyl siloxane with the general chemical formula $(CH_3)_3SiO\text{-}[(CH_3)_2SiO]_n\text{-}(CH_3)_3Si$. The siloxane chain is chemically very stable, and thus this liquid is chemically inert. Furthermore, it is characterized by high boiling point (well above 100°C) and low melting point. Most physical properties, such as viscosity, are little affected by temperature changes [2].

On the other hand, compared to water, the DC200(R)5cSt has a larger viscosity and smaller heat capacity and thermal conductivity. This means that the pressure drop will be higher, while the heat exchange will be to some extent degraded. CFD simulations are a required tool for a first assessment of the actual performance of this fluid for the cooling of the W-7X output window unit.

The cooling circuit has been simulated by static CFD analysis (continuous wave operation) and a comparison between the performance of water and DC200(R) is here presented. The following requirements must be met by the cooling system:

- The copper cuffs cannot withstand a pressure higher than 2 bar.
- The temperature of the coolant must be kept below the boiling temperature.

As a consequence, the following parameters are studied:

- The pressure distribution through the circuit.
- The maximum temperature reached by the fluid.

The behaviour of the silicone oil has been studied at different values of the mass flow rates.

2. CFD Model

2.1. *Geometry*

The W7-X output CVD diamond window has a thickness of 1.8 mm and a diameter of 106 mm. It is brazed to two copper cuffs 1 mm thick and with an internal diameter of 88 mm (the *aperture* of the window). A rim of 8 mm is available all around the window for direct cooling. The assembly window+cuffs is then mounted on a steel enclosing that provides the flow path for axial cooling.

Fig. 1 and Fig. 2 show the configuration of the assembly (volume occupied by the coolant, the copper cuffs and the CVD diamond disc) and the path of the coolant on the symmetry plane. After entering the steel structure through a pipe of 8 mm diameter, the coolant first flows around the outer ring – which has a higher hydraulic diameter – and then enters the inner ring through an array of small holes (36 holes, angular step 10°, ø 5mm). The outer ring acts like a buffer, while the holes are small enough to provide a relatively high resistance to the flow. This configuration provides a constant pressure distribution along the entire circumference of the unit. The coolant flows then in axial direction, removing the heat both indirectly – by contact with the copper cuffs – and by direct contact with the window rim. The unit is symmetric with respect to the window mid-plane (if the inlet and outlet connections are ignored); this means that the coolant passes to another external buffer through a second array of small holes and leaves the window unit by a pipe with 8 mm diameter.

The CFD model used in the present work consists of the fluid, CVD diamond and copper cuffs domains. The heat transfer at the interfaces with the steel structure can be neglected because of the lower thermal conductivity of such material (compared to copper and diamond). Because of the complexity of the flow path, the solution of such model requires a mesh with several millions of elements. The simulations have been run using ANSYS CFX on hexa meshes with $2{\times}10^6$ elements and tetra/prism meshes with up to $8.6{\times}10^6$ elements. The meshes have been generated using ANSYS IcemCFD.

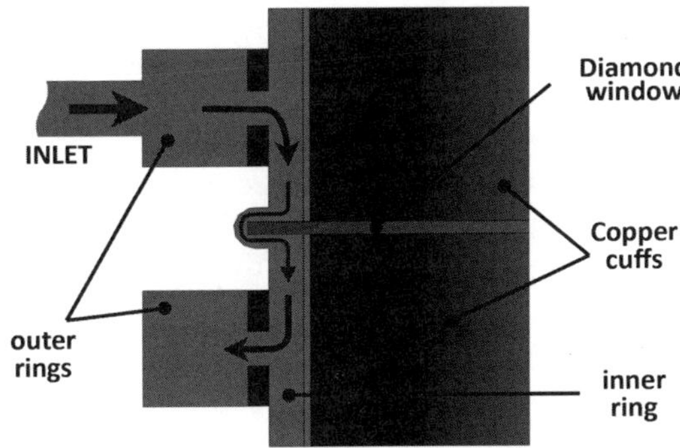

Fig. 1. Flow path on the symmetry plane of the window assembly.

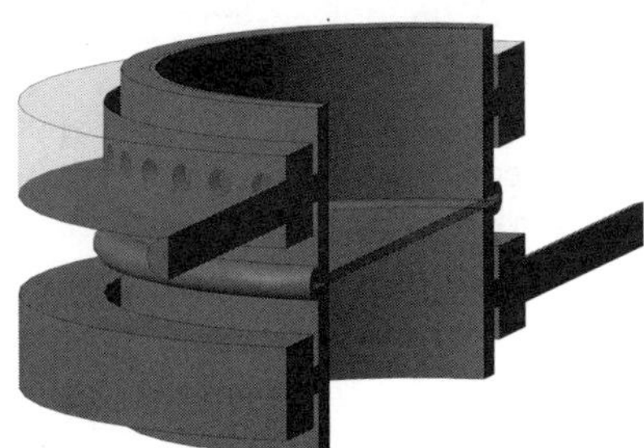

Fig. 2. Section of the cooling assembly. The picture shows the holes array in the upper half (inlet side) of the model.

2.2. *Materials Properties*

As a reference, the simulation has been run first considering water as coolant. The standard properties defined in CFX's library were used for water and copper.

The material properties for the silicon oil and the CVD diamond were obtained by Refs. [2] and [3] respectively. The actual values are summarized in Table 1.

Table 1. Material properties used in the simulations.

	Density [kg m^{-3}]	Dynamic Viscosity [N s m^{-2}]	Specific Heat [J kg^{-1} K^{-1}]	Thermal Conduct [W m^{-1} K^{-1}]
Water	997	8.899 10^{-4}	4181.7	0.6069
DC200(R) 5cSt	915	45.75 10^{-4}	1957.49	0.192
Copper	8933	–	385	401
CVD diamond	3500	–	520	2000

2.3. *Boundary Conditions and Heat Load*

The mass flow rate for the actual circuit cooled by water is 0.2 kg/s (i.e. 720 l/h). The same value is first considered for the silicone oil too, but it is also increased up to 0.3 and 0.6 kg/s. A reference pressure of 0bar is applied at the outlet.

The heat load is provided as consequence of the power absorption of the window. A Gaussian shape is assumed for the mm-wave beam. Thus, the volumetric heat generation is given by the following formula:

$$q'''(r) = \frac{P_{tot}}{2\pi\sigma^2 t} e^{-\frac{r^2}{2\sigma^2}} \tag{1}$$

where: q''' is given in W/m^3, P_{tot} is the total absorbed power (10^3 W, in the present case), t is the thickness of the diamond disc, r is the radius, σ is obtained by the beam radius (in this case is $\sigma = R_{beam}/2$, with $R_{beam} = 20$ mm).

The function $q'''(r)$ provides the power density (in SI units) for the diamond disc. The Gaussian distribution is normalized in order to obtain the total absorbed power given by P_{tot}. Most of the heat is expected to be removed by direct contact between the coolant and the diamond (convection). A partial amount diffuses (conduction) through the copper cuffs, while a very small amount is expected to diffuse to the steel enclosure. As a consequence, such structure has been ignored. No contact resistance between solids has been taken into account (perfect contact).

3. Results

3.1. *Pressure Drop: Comparison Between Water and DC200(R) 5cSt*

The total pressure drop, taking into account the inlet/outlet pipes too, amounts to 0.166 bar for the water and 0.196 bar for the DC200(R). The higher value for the silicone oil is, of course, due to the higher viscosity and the lower density compared to the water (lower density means higher average velocity, when the mass flow rate is the same).

Fig. 3 shows the contour-plots of the pressure distribution for water and DC200(R) at the design-value mass flow rate (0.2 kg/s for the entire assembly). The picture shows how the pressure drop is mainly located at the entrance of the outlet pipe, due to the sudden reduction of the section of the channel. A consequence of this result is that the pressure drop through the entire cooling circuit will practically depend on the diameter of the feed pipes only. Using the Blasius correlation

$$\frac{dp}{dl} = 0.316\mathrm{Re}^{-0.25} \tag{2}$$

a distributed pressure drop of 0.37 bar/m is found assuming DC200(R) 5cSt at 0.2 kg/s through a straight 8 mm diameter pipe. For example, if the feed line was 3 meters long, the pressure drop would be more than 1bar, which is much higher than the 0.2bar drop through the window unit.

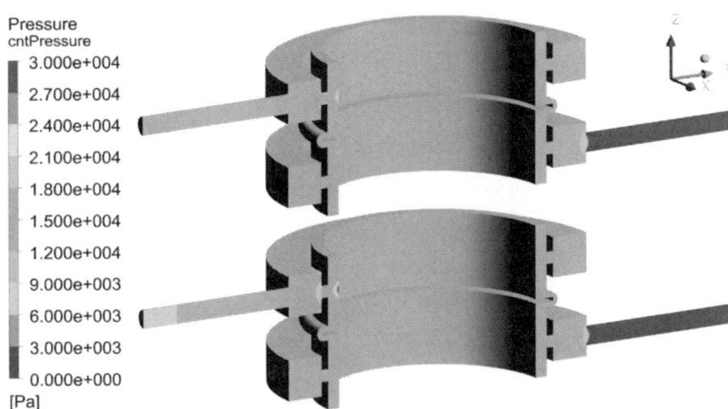

Fig. 3. Contour plots of the pressure drop for water (upper model) and DC200(R) 5cSt (lower model).

When water is considered as coolant, the average pressure acting on the cupper cuffs (with the outlet pressure as reference) is 0.138 bar. This value rises to 0.16 bar if the silicon oil is considered instead. The real value of the outlet pressure has to be considered in order to assess the total pressure on the cuffs and this depends on the setup of the cooling circuit.

Due to the alignment of the inlet pipe with the first small hole of the array, a peak value of the pressure can be noticed on the copper surface. A peak factor can be defined as the ratio between the maximum value on the copper cuff (with respect to the outlet pressure) and the total pressure drop:

$$f_P = \frac{p_{\max} - p_{outlet}}{p_{inlet} - p_{outlet}} \tag{3}$$

thus the maximum pressure on the copper cuff is given by:

$$p_{\max} = p_{inlet} + (f_p - 1)\Delta p \tag{4}$$

where $p_{\max}$ is the peak pressure and Δp is the total pressure drop. The factor f_p amounts to 1.35 for water cooling, while it is 1.48 for the silicon oil. For example: if an inlet pressure of 2bar is assumed for the silicon oil, a maximum pressure of 2.1bar is obtained by Eq. (4), which is only slightly higher than the maximum allowed value and, in any case, it is located on a very small area.

3.2. *Pressure Drop: DC200(R) at Different Flow Rates*

Fig. 4 shows the behavior of the total pressure drop at different mass flow rates of the silicone oil. The values can be well fitted by a parabola passing through the origin and symmetric with respect to the vertical axis, that is:

$$\Delta p = aG^2 \tag{5}$$

where Δp (bar) is the pressure drop, G (kg/s) is the total mass flow rate and $a = 4.42$. This result is a natural consequence of the resistance due to the high reduction of the cross section at the outlet pipe, where the concentrated pressure drop is proportional to the square of the mean velocity (if the regime is turbulent).

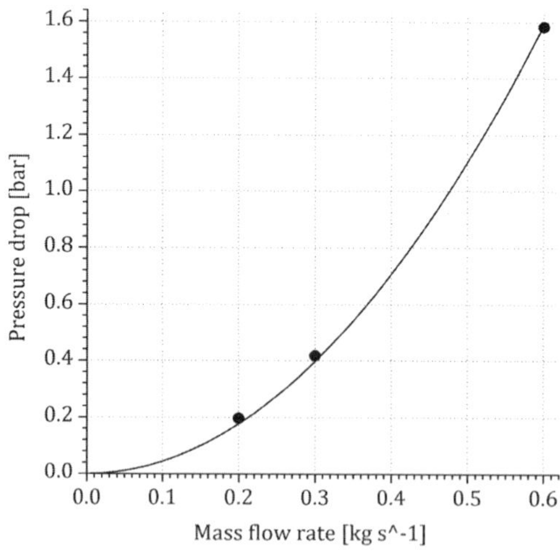

Fig. 4. Total pressure drop at different mass flow rates of the DC200(R) 5cSt silicone oil: • simulations results; —— $y = 4.42x^2$.

3.3. *Temperature: Comparison Between Water and DC200(R) 5cSt*

The temperature profiles for water and silicone oil, both with a mass flow rate of 0.2kg/s, along the diameter of the diamond window are shown in Fig. 5. The convective heat transfer in the case of silicone oil is disadvantaged by the larger viscosity (which causes lower levels of turbulence) and the smaller values of the thermal properties. As a consequence, the window reaches higher temperatures when the DC200(R) is considered as coolant (35 K more, compared to the case of water).

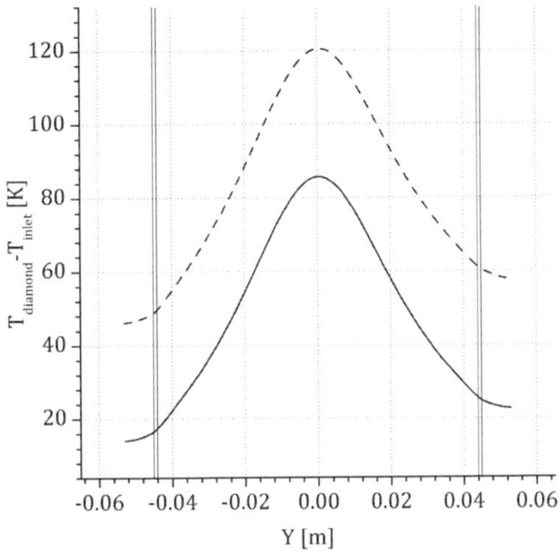

Fig. 5. Temperature profile along the diameter of the diamond disc. Comparison between water and DC200(R) 5cSt. —— water; - - - DC200(R) 5cSt. The vertical lines represent the copper cuffs.

The temperature profiles are not perfectly symmetric with respect to the central axis of the CVD disc. This result depends on the high turbulent kinetic energy (thus better heat exchange) of the fluid right after the inlet pipe. Previous simulations (which have been run for benchmarking purposes and are not reported in the present work) with an inlet pipe of 12 mm diameter have shown a much better symmetry in the temperature profiles.

An interesting result is that the heat is not mostly removed by direct cooling of the disc (as expected at the end of §2); on the contrary, a significant fraction of the generated power diffuses through the copper. As a consequence the heat removal is almost equally distributed among the interfaces coolant–CVD (52%) and coolant–copper (48%).

At the corner formed by the copper cuffs and the window rim a degraded heat exchange occurs. Thereby a maximum wall temperature of 85°C is found when silicon oil at 25°C (inlet temperature) is assumed as coolant. Since the flash point for the DC200(R) 5cSt is about 100°C, this result may arise some concern about safety issues. Usually the DC200(R) is chemically very stable and, according to the Material Safety Data Sheet (MSDS) [4], its NFPA profile

(National Fire and Protection Agency) reports a low flammability value, that is 1 in a scale ranging from 0 to 4. Furthermore, still according to the MSDS, it has no explosive behaviour and in case any flame was ignited, it could be extinguished by different means (CO_2, dry chemical or water spray).

3.4. *Temperature: DC200(R) at Different Flow Rates*

Fig. 6 shows the temperature profiles along the diamond disc cooled by the silicone oil at three different mass flow rates. The temperature profiles keep the same shape: with respect to the case at 0.2 kg/s, the temperature difference is 9.9 K for the case at 0.3 kg/s and 21.7 K for the case at 0.6 kg/s.

The reduction of the temperature is not significant and thus an increment of the mass flow rate, with the purpose of a better cooling, is not justified: a high increase of the mass flow rate would lead to a very high pressure drop, as discussed in §0.

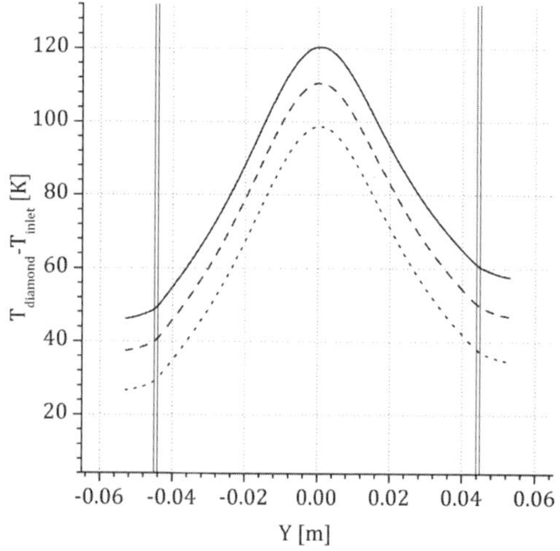

Fig. 6. Temperature profile along the diameter of the diamond disc cooled by the DC200(R) 5cSt silicone oil at three different mass flow rates: —— 0.2 kg/s; - - - 0.3 kg/s; ········· 0.6 kg/s. The vertical lines represent the copper cuffs.

4. Conclusion

The Dow Corning 200(R) 5cSt is a low viscosity silicone oil characterized by very good chemical properties. For such reason, it could be a good alternative to water for all those application where corrosion problems are a critical issue. This is the case of the high power window unit of the Wendelstein 7-X gyrotron.

A comparison between the performances of both water and DC200(R) 5cSt has been made by CFD analysis. As a general result, the silicone oil provides higher pressure drop in the cooling circuit and higher temperatures on the CVD diamond disc with respect to the case of water. However, its performance are not so degraded to compromise the overall functions of the diamond unit.

The pressure drop is about 20% higher when the DC200(R) 5cSt is used in place of water, but it is still 0.2 bar only, that is a small value. The pressure difference between inlet and outlet is mainly due to the small diameter of the connection pipes (8 mm) and quite high values can be obtained by taking into account of the complete feed line. This could be mitigated by considering a larger diameter for such pipes.

The temperature distribution on the diamond window is 35 K higher when water is replaced by the silicone oil. Of course, the heat exchange is degraded by the different material properties of the silicone oil (lower heat capacity and conductivity, higher viscosity). On the other hand, the CVD diamond is a material that can withstand very high temperatures, thus there is no real concern about these higher temperatures. The heat exchange could be improved by increasing the mass flow rate of the coolant, but this would lead to very high pressure losses as already discussed. Increasing the diameter of the piping would also deliver a better symmetry in the temperature distribution.

If an inlet temperature of 298.15 K (25°C) is assumed for the silicone oil, the room between the highest wall temperature and the flash point will be only 15 K. Thus, some attention should be addressed to eventual safety requirements. Because of the good chemical stability of this fluid, the possibility of a flame ignition is a non-critical issue (in any case this oil has no explosive behaviour), but it should not be omitted.

In general, the silicon oil Dow Corning 200(R) 5cSt has proven to be a good alternative to water for the cooling of the Wendelstein 7-X high power gyrotron output window. Its chemical properties can effectively prevent corrosion problems and, despite its different thermal and mechanical properties, it is able to remove the heat keeping the temperature of the CVD diamond disc at an acceptable level. An important increment of the pressure drop is expected along the complete cooling circuit, thus a larger diameter for the feed pipes could be a key requirement.

Acknowledgment

This work, supported by the European Communities, was carried out within the framework of the European Fusion Training Scheme.

References

1. Gunter Neffe, Gunter Dammertz, Manfred Thumm, *Fusion Eng. and Design* **56-57**, 627-632 (2001), ISSN 0920-3796, DOI: 10.1016/S0920-3796(01)00265-4.
2. O. Sandberg and B. Sundqvist; *J. Appl. Phys.* **53**, 8751-8755 (1982).
3. D. C. Harris, Infrared window and dome materials, SPIE – The International Society for Optical Engineering, Washington, 1992, ISBN 0-8194-0998-7.
4. Safety Management Data Sheet for the DOW CORNING 200(R) Fluid 5cSt, Dow Corning Corporation. Available at http://www1.dowcorning.com/DataFiles/090007b281062719.pdf